A Roadmap to the Successful Development and Commercialization of Microbial Pest Control Products for Control of Arthropods

Progress in Biological Control

Volume 10

Published:

Volume 1
H.M.T. Hokkanen and A.E. Hajek (eds.):
Environmental Impacts of Microbial Insecticides – Need and Methods for Risk Assessment. 2004
ISBN 978-1-4020-0813-9

Volume 2
J. Eilenberg and H.M.T. Hokkanen (eds.):
An Ecological and Societal Approach to Biological Control. 2007
ISBN 978-1-4020-4320-8

Volume 3
J. Brodeur and G. Boivin (eds.):
Trophic and Guild Interactions in Biological Control. 2006
ISBN 978-1-4020-4766-4

Volume 4
J. Gould, K. Hoelmer and J. Goolsby (eds.):
Classical Biological Control of *Bemisia tabaci* in the United States. 2008
ISBN 978-1-4020-6739-6

Volume 5
J. Romeis, A.M. Shelton and G. Kennedy (eds.):
Integration of Insect-Resistant Genetically Modified Crops within IPM Programs. 2008
HB ISBN 978-1-4020-8372-3; PB ISBN 978-1-4020-8459-1

Volume 6
A.E. Hajek, T.R. Glare and M. O'Callaghan (eds.):
Use of Microbes for Control and Eradication of Invasive Arthropods. 2008
ISBN: 978-1-4020-8559-8

Volume 7
H.M.T. Hokkanen (ed.):
Relationships of Natural Enemies and Non-Prey Foods. 2008
ISBN: 978-1-4020-9234-3

Volume 8
S.S. Gnanamanickam:
Biological Control of Rice Diseases
ISBN: 978-90-481-2464-0

Volume 9
F.L. Cônsoli, J.R.P. Parra and R.A. Zucchi (eds.):
Egg Parasitoids in Agroecosystems with Emphasis on *Trichogramma*
ISBN: 978-1-4020-9109-4

For further volumes:
http://www.springer.com/series/6417

Willem J. Ravensberg

A Roadmap to the Successful Development and Commercialization of Microbial Pest Control Products for Control of Arthropods

Springer

Dr. Willem J. Ravensberg
Koppert Biological Systems
R & D Manager Microbials Department
Populierzoom 1
2719 HL Zoetermeer
The Netherlands
wjravensberg@hetnet.nl

ISBN 978-94-007-0436-7 e-ISBN 978-94-007-0437-4
DOI 10.1007/978-94-007-0437-4
Springer Dordrecht Heidelberg London New York

Library of Congress Control Number: 2011920996

Cover illustration: phialides and conidia formation in heads of *Lecanicillium muscarium* (formerly *Verticillium lecanii*) (photograph: Koppert BV; design: Martin Ravensberg)

Printed on acid-free paper

Springer is part of Springer Science+Business Media (www.springer.com)

Progress in Biological Control

Series Preface

Biological control of pests, weeds, and plant and animal diseases utilising their natural antagonists is a well-established and rapidly evolving field of science. Despite its stunning successes world-wide and a steadily growing number of applications, biological control has remained grossly underexploited. Its untapped potential, however, represents the best hope to providing lasting, environmentally sound, and socially acceptable pest management. Such techniques are urgently needed for the control of an increasing number of problem pests affecting agriculture and forestry, and to suppress invasive organisms which threaten natural habitats and global biodiversity.

Based on the positive features of biological control, such as its target specificity and the lack of negative impacts on humans, it is the prime candidate in the search for reducing dependency on chemical pesticides. Replacement of chemical control by biological control – even partially as in many IPM programs – has important positive but so far neglected socio-economic, humanitarian, environmental and ethical implications. Change from chemical to biological control substantially contributes to the conservation of natural resources, and results in a considerable reduction of environmental pollution. It eliminates human exposure to toxic pesticides, improves sustainability of production systems, and enhances biodiversity. Public demand for finding solutions based on biological control is the main driving force in the increasing utilisation of natural enemies for controlling noxious organisms.

This book series is intended to accelerate these developments through exploring the progress made within the various aspects of biological control, and via documenting these advances to the benefit of fellow scientists, students, public officials, policymakers, and the public at large. Each of the books in this series is expected to provide a comprehensive, authoritative synthesis of the topic, likely to stand the test of time.

Heikki M.T. Hokkanen, Series Editor

Foreword

When working full-time for a commercial producer of biological control agents, one seldom has time to reflect on past successes and failures, to carefully redesign project approaches and to speculate about future methodology. Still, this would be particularly relevant for the field of microbial pest control, because in the past, many projects provided interesting candidate microbial control agents, but very few made it to the market. Therefore, I was very happy and pleasantly surprised that Willem Ravensberg, the author of this book, reacted positively when I proposed to him to write this book about his rich and long-term experiences in the field of microbial biological control.

He started with an intensive literature search and quickly realized that very few practical, overall research approaches existed for the selection, mass production, quality control, and development and marketing of microbial pest control agents. One aim of this book is to present such an overall approach. Another aim is to clearly summarize the developments, possibilities and limitations with regard to each group of microbials: nematodes, viruses, bacteria and fungi. Next, the complicated and frustrating area of registration of microbial control agents is analyzed. Ravensberg then proceeds with describing the difficulties one encounters when implementing a microbial control agent into an existing Integrated Pest Management programme. In the final chapters, he discusses critical factors in the successful commercialization of microbial pest control products and provides a roadmap to success. In addition, the book provides a very thorough overview of the literature related to microbial pest control. Thus, I expect that this book will be a landmark publication for the years to come!

With the experiences summarized in this book and the road map for identification of new microbial pest control products, development and marketing of products should be realized quicker than before. My hope is that simplification and harmonization of regulation will give another boost to the use of microbial agents. The book appears at a critical moment in the field of crop protection. Many countries strive to reduce the application of conventional pesticides, and in Europe, for example, 750 of the current 1000 chemical active ingredients used in pest control today will be withdrawn from the market. Microbial control agents are important, environmentally safe, and sustainable alternatives!

Wageningen, The Netherlands Joop C. van Lenteren

Preface

Microbial pesticides have been developed for a 100 years, but many of these biological crop protection products have not been successful in the market. This is illustrated in Chapter 1 by the history of microbial pest control products and the biopesticide companies. Many factors influence the adoption of biopesticides in the marketplace. At present, we see changes in regulations, and in political, cultural and social perceptions that all create an increased demand for sustainable crop protection agents. Also, research and technological discoveries create new possibilities for development of better products. Globalization opens new markets and offers potential usages. These macro-environmental trends support the assumption that the demand for alternative crop protection products will grow rapidly to replace conventional chemical pesticides.

It is at this time that I recognized the need for a model that would facilitate the development and commercialization of biopesticides based on entomopathogenic bacteria, fungi, viruses, and nematodes. The development and commercialization of a microbial pest control product is an extremely complex process. The total project from idea to market entry can take many years, and costs may amount to a few million Euros. Therefore, it is necessary to have from the outset an extensive overview on how such a project will be developed, and which critical issues will be encountered. A complete and structured project plan is required which covers all phases, all decisions, and the consequences thereof. Such a complete project plan has not yet been described in the literature. This book aims to produce a complete roadmap for the development and commercialization of an entomopathogenic microbial pest control product, with an emphasis on commercial and economic issues.

The initial step in such a project is finding a microbial pest control agent which has the potential to control the pest (Chapter 2). This study identifies three decisive selection criteria for a commercial microbial insecticide. The selection process should deliver determinative information on which one or at the most two to three strains are chosen for further development.

The next phase is the investigation of the feasibility of economic mass production of the selected strain(s) and the development of a stable product (Chapter 3). Two phases are distinguished, the development of the production process, including medium development and downstream processing, and the development of the product, including formulation, packaging and field testing. The critical technical and economic factors are identified and evaluated for the four types of pathogens. As the price of a product is an essential element, a cost price model for biopesticides is presented. The model provides a perspective on the makeup of the end-user's price.

Quality control (Chapter 4) provides feedback on the production and formulation processes, and on the final product. Quality control is also required for registration. The continuous process of improvements will ultimately decrease costs and improve performance of the production system and the product. I show that both the biocontrol industry and its customers benefit from proper quality control.

Registration is perceived as the main hurdle to the development of a biopesticide. In Chapter 5 regulations for microorganisms are reviewed, with an emphasis on Europe. Microorganisms, except nematodes, need to be registered as plant protection products for crop protection. The procedures in the EU are presented and difficulties discussed. Registration is expensive and takes many years. I present registration cost estimates for each type of entomopathogenic product. I also provide recommendations for improvements for data requirements and regulatory procedures. Various import and export regulations are discussed, including the possible impact of the Convention of Biological Diversity through Access and Benefit Sharing regulations. The patentability of an entomopathogen is discussed as well as costs and other considerations whether to apply for a patent for a biopesticide.

The implementation strategy of the product in an IPM programme is a basic element of the use of any microbial pest control product (Chapter 6). Three phases are distinguished: the optimal application strategy of the product, the incorporation of the microbial pest control product in an IPM system, and a carefully designed adoption strategy. Determinative parameters for each phase, and for each type of product are identified. Many companies underestimated or even neglected this part of product development. In my opinion, these phases are paramount for good market introduction. Therefore, I report the most relevant requirements for successful use of a microbial pest control product.

Commercialization is the final and most difficult step in the development and the market introduction of a microbial pest control product. The factors that determine success or failure are identified for a company as well as for a product, and recommendations are provided (Chapter 7) that will facilitate success. Companies which contemplate the development and commercialization of a biopesticide need realistic data on five key aspects to make their decision: market demand, market size, profit margin, time to market, and time to volume. The biggest mistake companies still make today is a misjudgement of the potential market size and the expected market adoption rate. I propose the use of a stage-gate process with objective, quantifiable, and transparent tools in decision-making. The business model that performs best at present is identified. Total developmental costs and time to market are significant factors of a company's success, estimates of costs and time are presented. Furthermore, I have identified five determinants for successful commercialization.

A new product development project is extensive and difficult to oversee. In Chapter 8, I make an analysis of the various phases and highlight the most important topics in the development and commercialization of a microbial pest control product. This book demonstrates that the development of a microbial pest control product requires a structured project plan. Finally, a future perspective on the biopesticide market is presented with limiting and promotional factors and trends. The significant drivers for success are food safety concern, changes in the regulatory climate, biodiversity and environmental issues, new research and technology, and the occurrence of new invasive pests.

Although this book focuses on biocontrol products for control of arthropods, I believe that the roadmap can also be used for development of disease and weed control products, based on either microorganisms, natural substances, plant extracts, or even semio-chemicals. I hope that this proposed roadmap will assist researchers and developers of biopesticides in accomplishing their goals in a cost- and time-effective way, and that this will result in successful and sustainable products and expanding biocontrol companies.

Zoetermeer, Netherlands Willem J. Ravensberg
October 2010

Acknowledgements

The writing of this book was initiated as a PhD dissertation and it was published as a PhD thesis of the Wageningen University, the Netherlands, in September 2010. Therefore, and first of all, I would like to thank my promotor Prof. Dr. Joop van Lenteren who provided me the opportunity of writing this thesis for my doctorate. Joop, thank you for inviting me and for inspiring me to start this project, and for your guidance over the years. When you asked me at the end 2005 whether I was interested to conduct a PhD study on the development of microbial pesticides, I did not need much time contemplating this proposal. I started enthusiastically, knowing, however, that it would be quite a challenge to combine this with a busy job at Koppert Biological Systems. But, Joop, you motivated me and kept me on track during these years. I have appreciated your valuable comments and suggestions. I want to thank you for the pleasant collaboration and for your rapid responses when we communicated by email. We have known each other since my student time in Leiden, and during my career at Koppert. It was especially pleasant to have a promotor which with I have collaborated for so long , and at the same time being friends. The result is the publication of this thesis.

Secondly, I like to thank my partner Carol Waddington who motivated and supported me during all these weekends of writing and other spare hours and holidays. With your background as an entomologist and your experience and passion for biological control, your comments and suggestions have been very valuable to me. Carol, I am very grateful too for your help in improving and editing my English.

I also want to thank the management of Koppert Biological Systems for their support and for allowing me to write and discuss many subjects which are closely linked to my daily activities for the company. Intellectual property is a major asset of the company and they gave me their confidence to deal with the company's IP in an appropriate way. They also encouraged me to conduct and accomplish this study. I particularly thank Paul Koppert for his support over this period.

Many colleagues have assisted me with small and large things, I express my gratitude to all of them. A special word of thanks goes to Heleen van der Steen and Monique den Dekker, my consecutive secretaries in this period, for helping me with searching literature, books and articles, and for their assistance. Members of my team in the Department R & D Microbials have been very helpful, and I specifically thank Marlies Dissevelt, Frans Weber, and Katja Hora for their assistance and suggestions. Further, I would like to acknowledge Karel Bolckmans, Roland de Bruyne, Peter Couwels, Markus Huurman, Mark Jongen, Johannette Klapwijk, Peter Maes, Harald Mikkelsen, Ernst ter Meulen, and Jan Rodenrijs for their help and suggestions.

I also thank my colleagues from the biopesticide industry from whom I sometimes wanted to have delicate information. I acknowledge useful suggestions and information from Martin Andermatt and Daniel Zingg (Andermatt BioControl), Dirk Avé (Valent BioScience), Don Edgecomb (Agraquest), Ralf-Udo Ehlers (Enema), Steve Lisansky (CPL), Olav Messerschmidt (OLC), Bernard Blum and Ulf Heilig (IBMA), Sergio Franceschini (Intrachem Production), Jan van der Blom (Coexphal), Joop de Hoog (Cebeco), and Eric Kiers (BASF). A word of thank also for academic colleagues from various universities and institutes, including regulatory authorities who provided me with useful information: Just Vlak (University Wageningen), Jürg Huber (Julius Kühn Institute), Tariq Butt (Swansea University), and Jeroen Meeussen (Ctgb).

I also thank Springer for providing me the opportunity to publish my thesis in a slightly different form as a commercial edition making the information available for a broader audience.

Contents

List of Acronyms and Abbreviations

ABIM	Annual Biocontrol Industry Meeting
ABS	Access and Benefit Sharing
ATCC	American Type Culture Collection
BCA	Biological Control Agent
BIPESCO	Biocontrol of important soil dwelling pests by improving the efficacy of insect pathogenic fungi (EU research project)
BPIA	Biopesticide Industry Alliance (USA)
BPPD	Biopesticides and Pollution Prevention Division (of the US-EPA)
BPSG	BioPesticides Steering Group (OECD)
BVL	Bundesamt für Verbraucherschutz und Lebensmittelsicherheit (The Federal office for Consumer Protection and Food Supply), Germany
CABI	Commonwealth Agricultural Bureau International
CBD	Convention on Biological Diversity
CBS	Centraal Bureau voor Schimmelcultures (Fungal Biodiversity Centre)
CFU	Colony Forming Unit
CIPAC	Collaborative International Pesticides Analytical Council
COST	Cooperation in Science and Technology (EU projects)
CRD	Chemicals Regulation Directorate (United Kingdom)
CRO	Contract Research Organization
CTB	College voor de Toelating van Bestrijdingsmiddelen, the Netherlands
CTGB	College voor de toelating van gewasbeschermingsmiddelen en biociden (Board for the Authorisation of Plant Protection Products and Biocides), the Netherlands
DAR	Draft Assessment Report
DG ENVI	Directorate-General Environment (EU)
DG SANCO	Directorate-General for Health and Consumers (EU)
DJ	Dauer juvenile (of entomopathogenic nematode)
DSMZ	Deutsche Sammlung von Mikroorganismen und Zellkulturen (German Collection of Microorganisms and Cell Cultures)

EC	European Commission
ECS	European Committee for Standardization
EFSA	European Food Safety Authority
EPA	Environmental Protection Agency (USA)
EPC	European Patent Convention
EPN	Entomopathogenic Nematode
EPO	European Patent Office
EPPO	European Plant Protection Organisation
ERA	Environmental Risk Assessment
EU	European Union
EUROSTAT	Statistical Office of the European Communities
FAO	Food and Agriculture Organization (United Nations)
GEP	Good Experimental Practices
GLP	Good Laboratory Practices
GMO	Genetically Modified Organisms
GMP	Good Manufacturing Practices
GV	Granuloviruses
HACCP	Hazard Analysis and Critical Control Points
IBCA	Invertebrate Biological Control Agent
IBMA	International Biocontrol Manufacturers Association
IGR	Insect Growth Regulator
IOBC	International Organization for Biological and Integrated Control of Noxious Animals and Plants
IOBC/EPS	IOBC East Palaearctic Section
IOBC/WPRS	IOBC West Palaearctic Regional Section
IP(R)	Intellectual Property (Rights)
IPM	Integrated Pest Management
ISO	International Organization for Standardization
ISPM	International Standards for Phytosanitary Measures (FAO)
LSF	Liquid State Fermentation
MPCA	Microbial Pest Control Agent
MPCP	Microbial Pest Control Product
MSDS	Material and Safety Data Sheet
MRL	Maximum Residue Level
MS	Member State of the European Union
NPV	Nucleopolyhedroviruses
OB	Occlusion Bodies
OECD	Organisation for Economic Co-operation and Development
PCT	Patent Cooperation Treaty
PIB	Polyhedral Inclusion Bodies
PMRA	Pest Management Regulatory Agency (Canada)
PSD	Pesticide Safety Directorate (United Kingdom)
PPP	Plant Protection Product
QC	Quality Control

RAFBCA	Risk Assessment of Fungal BioControl Agents (EU research project)
REBECA	Regulation of Biological Control Agents (EU research project)
RMS	Rapporteur Member State
SIP	Society for Invertebrate Pathology
SSF	Solid State Fermentation
TGAI	Technical Grade Active Ingredient
TQC	Total Quality Control
USDA	United States Department of Agriculture
VWA	Voedsel en Waren Autoriteit (in Dutch) (Food and Consumer Product Safety Authority)
WHO	World Health Organization (United Nations)

Chapter 1
General Introduction and Outline

Abstract Microbial pesticides have been developed for a 100 years, but many of these crop protection products have not been successful in the market. The history of microbial pest control products is summarized, the companies involved are reviewed, and the reasons for success and failure are briefly described. The need for a model that facilitates the development and commercialization of these products is recognized. The entomopathogenic bacteria, fungi, viruses and nematodes utilized as biocontrol agents are briefly introduced. For the first time, a rational and structured roadmap will be designed that will increase the chances of achieving success with microbial pest control products. Finally, this chapter presents an outline of this book. Terminology and definitions used are provided in the Appendix for products, methods of control, and control strategies.

1 Introduction

1.1 Microbial Pest Control Products

Numerous microbial pest control products for control of arthropod pests have been developed in the past and are developed at present. In theory, they offer one of the most sustainable and ecologically acceptable means of crop protection for modern agriculture. Generally, they are intended to replace synthetic chemical pesticides as our society requires sustainable pest control solutions that are safe to humans and to the environment. Research on new biocontrol agents, frequently initiated by academic scientists, promises elegant solutions. Companies translate these research results into the development of a commercial product. Many of these new products, however, do not become successful in the market, and companies fail to achieve a profitable business. Annual sales of microbial pesticides are reported to be $750 million globally, amounting only to 2.5% of the chemical market (Evans 2008). An imperative question is whether it is possible to identify the obstacles and constraints to the development of a reliable, efficacious and commercially-viable product? Can the currently frustrating situation be improved? I believe that successful products can be developed when a rational and structured pathway is followed for the entire

W.J. Ravensberg, *A Roadmap to the Successful Development and Commercialization of Microbial Pest Control Products for Control of Arthropods*, Progress in Biological Control 10, DOI 10.1007/978-94-007-0437-4_1, © Springer Science+Business Media B.V. 2011

process of development and commercialization of a microbial pest control product. A model describing such a pathway has not been presented to date. In this study I will critically evaluate the situation at present from a commercial perspective, and propose a systematic model that will guide future product developers to a successful microbial product for control of arthropod pests. For terms and definitions used, see Appendix at the end of this chapter. For acronyms and abbreviations used, see List of Acronyms and Abbreviations.

1.2 The Early History of Microbial Control

The idea of using entomopathogenic microorganisms to control insect pests is about 150 years old. The discovery of disease-causing microorganisms in the silkworm industry was the first step. Spreading these microorganisms in the field to reduce a pest insect was the second step towards the concept of microbial control. Steinhaus (1956) described the early observations on diseased insects and the rise of the idea of the use of microbial control in the nineteenth century in a fascinating paper. Famous men as Agostini Bassi in Italy, Louis Pasteur in France, J.L. Leconte and H.A. Hagen in the USA were the first to suggest the use of diseases of insects to kill harmful insects in crops (Steinhaus 1956). In his paper in 1879, Hagen suggested mass production of the disease-causing organism and spraying those over the infested plants and to wait for the epizootic effect to reduce the pest. He suggested a greenhouse trial against worms and the potato bug, and considered the dose and the application method. Furthermore, he stated that ". . . . the remedy is very cheap, easy to be prepared, not obnoxious to man or domestic animals, and if successful a benefit to mankind" (Steinhaus 1956). This statement can probably be seen as the first conceptual idea of the development of a microbial pest control product including biological, technical, economical and safety-related considerations. All these aspects then considered by Hagen are still relevant today for the development of a biopesticide.

Simultaneously, Metchnikoff envisaged the use of a fungus, *Metarhizium anisopliae*, to control insects in Russia's agriculture. He succeeded in growing the fungus on a sterilized, artificial medium. Large-scale production and field applications were conducted by Krassiltschik for control of various insect pests in the 1880s (Steinhaus 1956). Similar developments occurred in the 1880–1890s with *Beauveria bassiana* in the USA (Lord 2005).

1.3 The First Microbial Pest Control Products

In the first decade of the 1900s bacteria were discovered as insect pathogens, the first being *Bacillus thuringiensis* (Bt) in 1911, shortly followed by *B. popilliae* (Lord 2005). Both pathogens were further investigated, production was started, and products developed and used. The first commercial biopesticide was Sporeine, based on *B. thuringiensis* and available in 1938 in France. The first attempts to use entomopathogenic nematodes were conducted with *Steinernema glaseri* in the late 1930s in the USA. The programme stopped because of World War II and results of field applications were not assessed. The first fungal product was developed in the former USSR in 1965. This product, Boverin, was based on *B. bassiana*, and used to control the Colorado potato beetle and the codling moth (De Faria and Wraight 2007). Although baculoviruses were found to cause epizootics in the 1940s, a commercial product was only developed in the 1970s (Ignoffo 1973). The first product was VironH for control of *Helicoverpa zea*. These events are highlighted by Lord (2005) and earlier authors referred to by Lord.

1.4 The Rise and Struggle of the Biopesticide Industry

Serious industrial developments with biopesticides started in the 1960–1970s with products based on Bt. Large agrichemical companies such as Sandoz and Solvay started to produce products and the first commercial biopesticides were registered and launched on the market in the USA and in Europe in the early 1960s.

The research on insect pathogens has grown extensively the last four decades and the literature on insect pathogens and on their potential as microbial control agents is vast. During this period, many companies have employed activities with biopesticides and numerous products have been developed, registered, and introduced on the market. The use of biopesticides is constantly increasing. However, their overall use is merely a few percent of the total worldwide use of plant protection products. Many products have not been successful, and numerous companies have failed. According to Lisansky (CPL 2006) over four hundred companies have been active at different times with commercialization of microbial pesticides, and the majority of these companies has left the field of biopesticides. Gelernter (2005) provided an overview of these commercial developments within the biopesticide industry from 1950 to 2005. The challenges that entrepreneurs face when they try to develop and commercialize microbial pest control products have been outlined by many authors working in the biopesticide industry (Jaronski 1986; Cross and Polonenko 1996; Marrone 1999; Van der Pas et al. 2000; Warrior 2000; Benuzzi 2004; Georgis 2002; Krause et al. 2006). The reasons for failure have been reported many times and the dominant issues are: variable quality and efficacy of biocontrol products, their costs compared to their efficacy, competition with chemicals, registration, overestimation of the market size, underestimation of the assumption that market adoption will be easy, underestimation of the total required budget and the time to market, and the often less than optimal collaboration between academic scientists and the product developers in the industry.

1.5 Hurdles and Constraints

The failure of many of these initiatives with biopesticides has been a serious problem. Vast sums of money have been spent and were largely wasted. This still occurs today. As a result, the development of biological and sustainable plant protection products is hardly accomplished. The demand for these kinds of products is rising in our societies, but the potential that microorganisms offer to control pests is not met. Lessons should be learned from this history with the commercialization of biopesticides. In the scientific literature, many authors report research topics related to the development and commercialization of a microbial control agent. Usually, this refers to biological and technical subjects, and sometimes to regulatory aspects. Few authors addressed market and commercial aspects, and even fewer focussed on economic considerations. There is no extensive treatment of the entire developmental process in the literature. Several authors have observed too that, despite abundant research on microbial pest control agents, commercialization of products has only been accomplished in a limited number of cases (Cross and Polonenko 1996; Fravel 1999; Stewart 2001; Hallett 2005; CPL 2007; Droby et al. 2009).

A statement of Dent (1997) emphasizes the need for a systematic approach to efficiently develop new biological products: "Microbial insecticide research is usually funded piecemeal, largely by the public sector, and rarely involves multidisciplinary teams that develop a microbial insecticide from start to finish. The general knowledge base in microbial insecticides is built up in a haphazard way, through uncoordinated efforts of many scientists all pursuing their own individual research objectives and interests. This contrasts markedly with the more focussed factory-like screening and development process which characterizes agrochemical R & D that produces new chemical insecticides." Later, other researchers have expressed similar self-criticism with regard to their attitude towards research on biocontrol agents and are calling upon colleagues to maximize efforts to attaining more commercial successes (Fravel 1999; Stewart 2001). Scientists again recognized the lack of direction in their research approaches when it relates to biopesticide development. Ash (2009) stated that "researchers should also be encouraged to abandon referring to their "pet pathogen" as a bioherbicide candidate. Long after they have failed to demonstrate economic and biological rationale for the production of a bioherbicide the researcher continues to research and publish on the system they have studied for so long". Not only in bioherbicides is the phenomena of "pet pathogens" noticeable, and resources could be spent more efficiently when the development of a biopesticide is the goal.

At the same time, companies contemplating the development of a microbial pest control product made and still make serious mistakes in their assumptions on their knowledge and skills, the market potential and many other aspects of commercialization of these products (Gelernter 2005; CPL 2006). Many potential products remain on the shelf of the academic scientists because of the enormous challenge of developing it into a successful product. Still today, I see research conducted in the way Dent (1997) described more then 10 years ago. On the other hand, I also see more and more public and private research collaborating with the goal to develop microbial products from the onset of a project. Small biocontrol companies need to adopt appropriate aspects of the approach of large agrochemical R & D companies to enhance their product development. Both academic scientists and biopesticide developers in industry need to collaborate earlier in a project and should be more focussed on the entire process of product development and commercialization. With a well-coordinated research between academics and industry, and a clear focus on the final product and market, more biopesticides are likely to reach the market and become successful.

Registration is often cited as the main hurdle in the commercialization of microbial pest control products. It is a long and expensive procedure. Moreover, many dossier requirements do not address the specific nature of these products appropriately which makes registration even more difficult. Regulatory innovation and harmonization are required at EU level and at national level to remedy this situation (Chandler et al. 2008). Concerning registration worldwide, the OECD's Pesticide Programme strives to facilitate and improve registration procedures of crop protection products in member countries by harmonization (Sigman 2005). The OECD Biopesticide Steering Group is working on several issues related to data requirements for microbials (Meeussen 2007).

1.6 The Need for a Roadmap for the Development of a Microbial Pest Control Product

The development of a biopesticide requires many steps from concept to product, and a well-targeted project strategy is necessary to reach success. This has lead to the question: "can a pathway be developed that leads to a successful biopesticide". This question has been posed in various wordings by many scientists, not only in the field of bioinsecticides (Jaronski 1986; Dent 1997; Dent et al. 1999; Lidert 2001; Jackson et al. 1992; Bateman 2004; Benuzzi 2004), but also the field of biofungicides (Campbell 1986; Woodhead et al. 1990; Mathre et al. 1999), bioherbicides (Cross and Polonenko 1996; Leggett and Gleddie 1995; Stewart 2001; Hallett 2005; Ash 2009), and post-harvest pest and disease control products (Droby et al. 2009). It is obvious that there is a need for a general model on the development and a successful commercialization of a microbial pesticide.

Numerous authors have briefly treated one or more steps in the product development; few have addressed all steps in a systematic and structured way. Several authors have briefly described models and provided flow diagrams with the consecutive steps of product development (Lisansky 1985; Törmälä 1995; Hofstein and Chapple 1999; Marrone 1999; Butt et al. 2001; Montesinos 2003; Jijakli 2003; Fravel 2005; Kiewnick 2007). Most of these attempts focused on biological features of the candidate biocontrol agents, some included biological and technical aspects of mass production and formulation, and others reviewed aspects of regulatory importance. Few addressed economic, market-related and commercial elements, but these are key considerations in each step of the product development. Furthermore, the developmental process up to regulatory approval to sell is just half the story, only hereafter the product and company are really tested to perform well in the market.

There is one well-documented example originating from the academic world. It concerns the Lubilosa project where a mycoinsecticide was developed for control of locusts and grasshoppers by a large multi-disciplinary team of researchers. From the outset, the development of a mycoinsecticide was foreseen and all required steps were identified and executed in a systematic way (Lomer 1999; Dent 1998). Also, continuous feedback with regard to economical, sociological and practical application aspects was provided between the teams (Bateman 1997; De Groote 1997). The flow scheme (Fig. 1.1) used in the Lubilosa project (Dent 1998) is similar to the one developed earlier by Lisansky (1985) and based on his industrial experience. The Lubilosa flow scheme is the most complete one that I have found in the literature. Industry, however, has not described their way of working with regard to biopesticide development. Understandably, proprietary considerations have prevented an elaborate process description, and secondly, writing publications is low on the priority list within industry.

Clearly, a model that describes and critically evaluates all steps of the developmental process and includes biological, technical, regulatory and economic factors is lacking. Such a general model would facilitate this process and provide guidance for product developers. This information may also increase the chance that new product ideas make it to the market in a successful way. In order to be successful with a product in the market, the economics of the whole process and the final product must result in a profit for the company.

2 Insect Pathogens as Microbial Control Agents

Four groups of microbial pathogens are typically found in insects: bacteria, fungi, viruses, and protozoa. Their biological properties determine their utility as agents for the control of insects (Federici 1999). In addition to these microorganisms, entomopathogenic nematodes will be treated because many of the same principles apply to the use of these organisms as biocontrol agents in the form of a microbial pest control product. Genetically modified organisms (GMOs) and those control products based on GMOs are not reviewed and discussed, although there are GMO-based microbial pest control products (MPCPs) developed and available. A small number of MPCPs with genetically modified *Bacillus thuringiensis* strains are approved in the USA. Genetically modified plants incorporated with Bt genes are also available on the market. Sometimes these are also categorized as biopesticides. GMO-based biopesticides are at present not allowed in Europe, and even when allowed, registration would be extremely expensive and complicated. These kinds of biopesticides are not foreseen for the European market for the near future.

2.1 Bacteria

Most bacteria that provide potential as a bioinsecticide belong to the spore-forming bacteria of the genus *Bacillus* (Bacillaceae). The major species used in biocontrol is *B. thuringiensis*. Other species are *B. sphaericus* and *Paenibacillus popilliae*, and

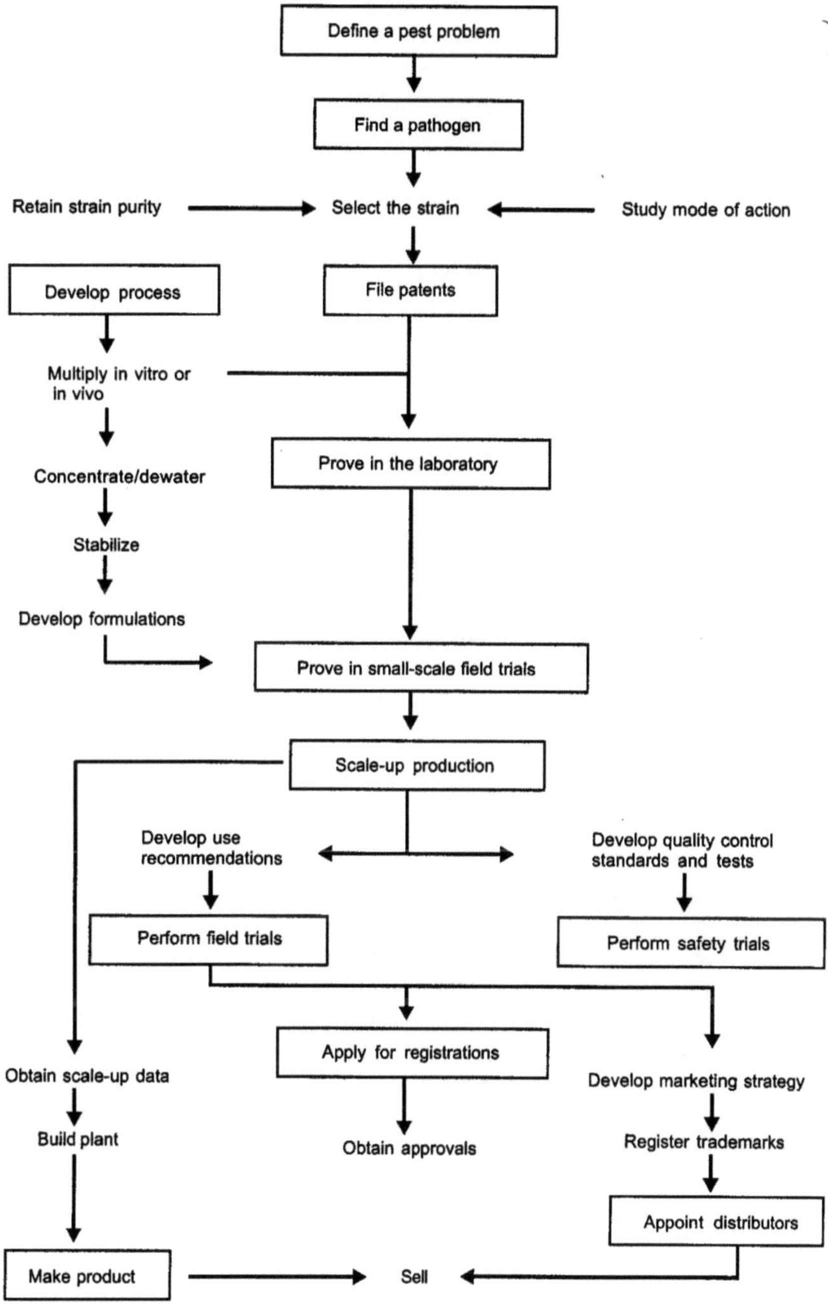

Fig. 1.1 Scheme of development of a microbial pest control product used in the Lubilosa programme (Dent 1998). © CAB International 1998. All rights reserved

one species from the genus *Serratia*. A brief overview of bacteria, their biology, mode of action, and ecology in relation to their potential as biocontrol agents is provided by Garczynski and Siegel (2007). The number of bacterial species used in biocontrol to date is limited. The number of isolates is numerous, whereby any isolate may present unique features that offer potential for a new product. Hundreds of products have been developed with *B. thuringiensis*, whereas only a handful based on other bacteria (Garczynski and Siegel 2007; Copping 2009). Bacterial insecticides have been primarily developed for control of lepidopteran and dipteran pest. A few products have been developed for control of beetle larvae.

2.2 Fungi

Two groups of fungi are generally found to cause diseases in insects. These pathogens belong to the orders Entomophthorales and Hypocreales (formerly called Hyphomycetes). The higher-level phylogenetic classification of fungi, including entomopathogenic fungi, has been modified considerably, for more information I refer to Hibbett et al. (2007) and Humber (2007, 2008). Several other entomopathogenic fungi from other taxonomic groups are known. Over 700 species of fungi are known to infest insects (Wraight et al. 2007). For more information on the entomopathogenic fungi, their biology, taxonomy and ecology with regard to biocontrol I refer to Chandler et al. (2000), Shah and Pell (2003), and Wraight et al. (2007).

Entomophthoralean fungi are important in nature as insect pathogens and they play an significant role in suppression of insects, often through epizootics. They are primarily found on aphids, caterpillars and flies. More information on the biology and occurrence of these fungi can be obtained from Pell et al. (2001). These fungi are not used as microbial insecticides to date because of the difficulty of mass production. They are therefore not further discussed.

Hypocrealean fungi are interesting fungi for use as biocontrol agents. An overview of this group of entomopathogenic fungi and their biology and ecology in relation to their use as biocontrol agents is presented by Inglis et al. (2001). They are found on members of many insect orders, and on other arthropods such as mites and ticks. Numerous mycoinsecticides have been developed based on these fungi, particularly on the species *B. bassiana* and *M. anisopliae* (Copping 2009). This group offers great potential for commercial use.

2.3 Viruses

Viruses are obligate intracellular parasites, and numerous viruses (around 1,000) are specific to insects. Three principal groups if insect viruses have been found in the families of *Baculoviridae*, *Entomopoxvirinae* and *Reoviridae*. These families are characterized by the formation of occlusion bodies, in which the virus particles (virions) are occluded in a protective proteinaceous matrix. This characteristic provides these viruses the possibility to be used as a biocontrol agent. A brief overview of viruses, their biology and ecology in relation to their potential as biocontrol agents is provided by Cory and Evans (2007). Baculoviruses are the most often used and most successful biocontrol agents from this group of pathogens. The family *Baculoviridae* is divided in four genera: *Alphabaculovirus, Betabaculovirus, Deltabaculovirus,* and *Gammabaculovirus* (Jehle et al. 2006; www.ictvonline.org/virus TaxInfo.asp). There are two types of rod-shaped viruses, known as nucleopolyhedroviruses (NPVs) and granuloviruses (GVs). NPVs (over 500 isolates) are infectious to Lepidoptera, Hymenoptera, and Diptera. GVs (over 100 isolates) are only known from Lepidoptera (Federici 1999; Cory and Evans 2007). About fifteen products have been developed based on baculoviruses (Copping 2009). Only the group of baculoviruses will be dealt with.

2.4 Protozoa and Microsporidia

Protozoa consists of a large and diverse group of unicellular microorganisms. Some protozoans are parasitic on insects and cause chronic diseases. Epizootics cause a decline in insect populations in nature and this offers potential to use them as biocontrol agents, particularly members of the Microsporidia. They are, however, no longer grouped with Protozoa. Hibbett et al. (2007) reclassified Microsporidia as a phylum within the kingdom of Fungi. Thirteen thousand Microsporidia have been described. A brief overview of entomopathogenic microsporidia, their biology and ecology in relation to their potential as microbial insecticides is provided by Solter and Becnel (2007). Only one microsporidium has been developed for use as an insecticide: *Nosema locustae*, for control of grasshoppers. A few products based on this microorganism have been registered in the USA. Another species, *Vairimorpha necatrix*, has been developed for control of Lepidopteran pests, but was never commercialized (Solter and Becnel 2007). The potential for the Protozoa and Microsporidia as microbial pest control products is limited due to the difficulty of the mass production of these obligate parasites, and the slow effect on the pest population. Therefore, these two group of pathogens are not further covered in this book.

2.5 Nematodes

More than 30 nematode families (Nematoda, roundworms) are associated with insects. Only members of seven families are considered to have potential as biocontrol agents. Species of the Heterorhabditidae and Steinernematidae have received most attention in research and in application for control of insect pests (Stock and Hunt 2006). A member of the family Rhabditidae, *Phasmarhabditis hermaphrodita*, is parasitic to slugs and has been developed as a biological molluscide. There are about ten species of the Heterorhabditidae known and over thirty species of the Steinernematidae (Stock and Hunt 2006), several *Heterorhabditis* and *Steinernema* species have been developed as biological insecticides. They contain symbiotic bacteria, respectively species of the genera *Photorhabdus* and *Xenorhabdus*, which cause rapid death in infected insects. More information on these nematodes and their symbiotic bacteria can be found in Gaugler (2002) and Grewal et al. (2005). The possibility of mass production in an artificial medium gives these species utility as biological insecticides. These entomopathogenic nematodes are the subject of the development of microbial pest control products covered in this book.

3 Product Development Process

The development and commercialization of a biopesticide is an expensive and time-consuming process. Biocontrol product developers are dealing with a complicated project in which a living organism is the key element. Such a project entails many phases where each phase is a research project by itself, and successive phases are strongly interconnected. The new product development project often involves the development of new technology, and presents a highly complex matrix of biological, technical, regulatory and commercial challenges. A continuous feedback within the multi-disciplinary project team, with persons with biological skills, technical skills, and market knowledge, is required in order to optimize time and resources. Along the way, numerous decisions have to be taken on hundreds of subjects. The entire process will take many years and costs many millions of Euros. In a business environment, the ultimate objective is to sell the product in such a way that sufficient revenues are generated to achieve return on investment, and to make a profit that allows the company to develop more products, to accomplish a strong position in the marketplace, and to become a sustainable and profitable enterprise. Ultimately, the financial aspects of the product development are determinative for a company. This requires an accurate development and business plan that addresses all steps in a systematic way and that focuses on a cost-effective product that brings the grower an effective solution and meets his expectations. The path to the market is full of obstacles, and a potential product can fail at any point along the way. The time period from idea-to-launch should be as short as possible, and the required budget as low as possible. These are crucial business elements. The time from market introduction to sales volume is the other relevant phase where return on investment should be achieved. The biocontrol industry is small and young, often new players "invent the wheel again", and time and money are lost. Therefore, there is a need for a general pathway that describes all phases including the critical steps and decisions that guides an enterprise through the development of new products in an optimal way.

The principal aim of this book is to examine and evaluate the current situation, to understand why projects and products fail, to understand the needs of the market, and to examine what knowledge, techniques and resources are required to improve the development of a microbial pest control product. With this understanding I will develop a complete and structured roadmap for the successful development and commercialization of a microbial pest control product for the control of insect and mite pests. All subjects will be covered, from the identification and description of the pest problem for which a solution is needed, to successfully selling the product. The emphasis will be on the entire process of development and commercialization from a business point of view. Economic considerations will be highlighted, next to biological, technical and regulatory aspects. Special emphasis is given on criteria for decisions that have to be made continuously in such a complex process. A systematic approach is provided in which each phase is elaborately discussed, and for each step relevant considerations and decision criteria will be presented along with useful recommendations.

This pathway will be divided in the following processes: selection of the microbial agent, production and product development, quality control, registration, implementation, and commercialization. Other topics treated are referring to these processes:

- identification of selection criteria for a microbial pest control agent;
- identification of critical parameters in production and product development;
- establishment of a reliable quality control procedure;
- recommendations for the generation of a registration dossier, and a review of other relevant regulations;
- identification of the most important issues with regard to successful implementation of a new product in an Integrated Pest Management (IPM) system;
- identification of critical success and failure factors in the commercialization of a microbial pest control product, and identification of the most successful business model;
- and a synthesis of the processes that will facilitate the product development and commercialization.

Each process will be considered in detail in the consecutive chapters. Both product-related and company-related factors will be treated. Information available in the literature, from academia as well as from industrial scientists, will be reviewed and analyzed. Key elements and considerations will be identified and highlighted. Decision criteria will be defined and outlined, alongside with transparent tools that help take the right decisions. Requirements and recommendations will be provided, both for the new product and the company, to increase the chance of a successful outcome. In the last chapter a synthesis will be provided along with flow diagrams that illustrate the various steps and interconnections between phases in a structured way. These diagrams present the roadmap from idea to successful marketing of the product. This proposed systematic roadmap should facilitate the development and commercialization of new microbial pest control products, and provide guidance for entrepreneurs who contemplate to initiate activities in this field of biopesticides. Ultimately, it may increase the chances for success in the market for newly developed microbials. A future outlook for biopesticides will be provided based on the current market and society developments.

I focus on microbial pesticides that are based on entomopathogenic bacteria, fungi, baculoviruses, and nematodes. Protozoa are only briefly mentioned since they are rarely used as a control agent for a biopesticide. Furthermore, I emphasize the situation in Europe, particularly with reference to registration, and products developed for use in protected crops. The systematic approach, however, could also be applied for other microbial products for control of plant diseases, post-harvest diseases, weeds and other pest organisms. This approach is also relevant for biocontrol products developed for other markets. It could even be used in the development of a plant protection product based on natural substances, plant extracts, pheromones, and others. Many developmental and commercial factors are similar for each type of pesticide, including even chemical pesticides.

4 Outline of this Book

Chapter 1 introduces the challenge of the development of microbial pest control products and microbial control. The history of the concept of microbial control is presented and the industrial product developments over time. The need for a roadmap to develop biopesticides is proposed in order to provide successful products in the market. The aim of this book is to present a model that guides product developers towards a successful project. The terminology used is provided. The groups of insect pathogens used in microbial plant protection products are presented: bacteria, fungi, baculoviruses, and entomopathogenic nematodes. Finally, an outline of the chapters in this book is given.

Chapter 2 discusses the exploratory phase of finding a new microbial pest control agent, and the actual screening phase of species and strains of collected entomopathogenic microorganisms. Crucial characteristics of each type of pathogen relevant to their utility as a crop protection product are noted. Three decisive selection criteria will be identified. The importance of various biological parameters related to these criteria is discussed. Testing methods will be critically reviewed and recommendations for standardization of methods and test conditions will be presented. The consecutive steps in the screening process will be listed with emphasis on economic considerations.

Chapter 3 deals with the mass production and product development of a microbial pest control agent. Mass production is reviewed for each category of pathogen and the economic feasibility of mass production systems is identified. The vital issues in downstream processing, medium optimization, equipment, and inoculum stability are examined and recommendations are provided for a cost-effective mass production. Four functions of formulation are identified, formulation requirements are analyzed, and recommendations are provided for each type of pathogen. The economics of mass production and the end-use product are analyzed and a cost price model will be presented. Finally, key factors are determined for successful production and product development, and recommendations are reported to achieve optimal production, formulation and end-use product.

Chapter 4 covers quality control of the production process and the final product, with emphasis on the latter. A manufacturer establishes product specifications and these are checked for every batch. Pivotal aspects of product control are discussed for each type of pathogen. Registration requirements are reviewed. Complete quality control procedures and data for validation must be established, although there are no officially recognized criteria. Practical challenges in quality control procedures are reviewed per type of pathogen. Recommendations for standardization and criteria will be provided. Research needs are identified that may facilitate quality control in the future. The benefits of quality control for the manufacturer and the end-user will be identified.

Chapter 5 reports on regulations related to the use of microbial pest control agents and products. Microorganisms are subject to registration as a "plant protection product" (PPP), whereas nematodes are usually covered by different regulations. Registration requirements are reviewed, particularly for the European Union. For several topics, data requirements are unclear or even lacking; procedures are both long and expensive. Initiatives to ameliorate these impediments are reviewed. Various other regulations, such as laws on biodiversity, importation and release of exotic organisms, and safe handling, apply to the use of microorganisms, and are briefly treated. Intellectual property rights are important for a company. The possibility to obtain a patent for a biopesticide and the value thereof will be discussed. Regulations form a major obstacle in the commercialization of biopesticides, and improvements and recommendations are provided to facilitate the approval processes in the future. The role that the biocontrol industry should have in this political field is outlined.

Chapter 6 presents key factors to cost-effective implementation of a microbial pest control product, namely the application strategy, compatibility, and knowledge transfer to the user. This requires the design of a comprehensive integrated pest management programme in which the microbial pest control product is to be incorporated. The first element is an optimal application strategy per targeted pest and cropping system. The second element is the incorporation of the microbial pest control product in an IPM system. Compatibility with chemical pesticides and with beneficial organisms must be established and critical aspects will be mentioned for each type of pathogen. The risk of resistance will be briefly discussed. Adoption of new products needs knowledge transfer in the entire supply chain to the grower, and pivotal considerations including economic aspects will be reviewed. Requirements and recommendations will be provided on how to accomplish a successful implementation. Some considerations will be presented with regard to the compatibility profile, the advantages and the disadvantages of biopesticides, the cost of the implementation process, and to the potential of combined use of biopesticides.

Chapter 7 reviews the critical factors in the successful commercialization of microbial pest control products. The history of the biopesticide industry will be highlighted. Currently available products will be presented, as well the biopesticide markets in Europe and the Netherlands. The main crops in which biopesticides are used are presented as well as the most successful products. Profitability of products in relation to their market size will be analyzed. Critical success and failure factors for a biopesticide company and for a microbial pest control product are analyzed, and recommendations will be provided on essential factors that need to be considered. Tools will be provided to facilitate decision-making in the commercialization process. Company profiles will be reviewed, and a business model will be identified that currently performs best. An estimate of the developmental time and costs for a microbial pest control product is provided. The distribution and sale strategy is reviewed, and various salient options are discussed. Crucial factors will be identified, and recommendations will be provided that will increase the chances to develop a successful biopesticide and to become a profitable and sustainable biopesticide company.

Chapter 8 provides a roadmap to success, and future prospects for use of microbial control products. The roadmap to successful development and commercialization of a microbial pest control product is amply illustrated in newly designed flow diagrams. Diagrams are presented for the screening phase, the product development phase, and the implementation phase. Relevant input criteria and requirements are provided. Output information leads to consecutive steps and "go/no go"-decisions. A future perspective on the biopesticide market is presented with limiting and promoting factors and trends. The most important drivers are concerns about food safety and the environment, new research and technology, changes in the regulatory climate, and the arrival of exotic pests. The biopesticide industry has reached a level of maturity and critical mass that forms a solid base for further expansion. The future of the market size and growth is estimated, and an outlook of the way forward is presented.

5 Appendix: Terminology and Definitions

5.1 Microbial Pest Control Agents and Products

The term biopesticide is widely used, and definitions vary. Definitions of the term biopesticide may include microbial organisms, pheromones, insect and plant regulators, plant extracts, transgenic plants as well as macroorganisms. Various institutions have defined biopesticides and sub-sets thereof such as microbial pesticides and microorganisms. Regulatory agencies have provided exact definitions of microorganisms, and for plant protection products containing microorganisms as the active ingredient.

The EU has defined "microorganism" as: "a microbiological entity, cellular or non-cellular, capable of replication or of transferring genetic material". The definition applies to, but is not limited to, bacteria, fungi, protozoa, viruses and viroids (EC 1991, 2001). This definition has been slightly modified in the new Regulation on plant protection products (EC 2009): "any microbiological entity, including lower fungi and viruses, cellular or non-cellular, capable of replication or of transferring genetic material". The Directives nor the new Regulation define a microbial plant protection product exactly, but do provide the following description: a microbial plant protection product may contain viable and nonviable microorganisms (including viruses) and formulation substances. It may also contain relevant metabolites/toxins produced during growth, residues from the growth medium, and microbial contaminants. The terms biopesticide and microbial pesticide are not used in the Directives and the Regulation.

The OECD (2003, 2004a, b) uses the following definitions:

- microbial pest control agent (MPCA) (= active substance): a microorganism (bacterium, alga, fungus, protozoan, virus, mycoplasma, rickettsia) and any associated metabolites, to which the effects of pest control are attributed;
- microbial pest control product (MPCP) (= microbial plant protection product): a product containing a MPCA that is registered or labeled with instructions for direct use or application for pest control purposes.

Further, the OECD uses the term "microbial pesticide", however, it is not precisely defined. The word "biopesticide" is not used in their guidance documents; it is only used in the naming of the OECD BioPesticide Steering Group (BPSG).

In the definition of the Environmental Protection Agency (EPA) in the USA, biopesticides include naturally occurring substances that control pests (biochemical pesticides), microorganisms that control pests (microbial pesticides), and pesticidal substances produced by plants containing added genetic material (plant-incorporated protectants) (ww.epa.gov). The definition of a microbial pesticide is as follows: a microbial agent intended for preventing, destroying, repelling, or mitigating any pest, or intended for use as a plant regulator, defoliant, or desiccant, that:

(1) is a eucaryotic microorganism, including, but not limited to, protozoa, algae, and fungi;
(2) is a procaryotic microorganism, including, but not limited to, Eubacteria and Archaebacteria; or
(3) is a parasitically replicating microscopic element, including, but not limited to, viruses.

The definition applies to, but is not limited to, bacteria, fungi, protozoa, viruses and viroids (EPA 2007). The term microbial pest control agents (MPCA) is used in the data requirements, the term microbial pest control product (MPCP) is not used, however.

I will use the term *Microbial Pest Control Agent (MPCA)*. The term MPCA includes microorganisms such as algae, bacteria, fungi, protozoa, yeasts, viruses and viroids, and any associated metabolites that are involved in the mode of action. I also include entomopathogenic nematodes under this definition for reasons explained below and for the sake of convenience.

I will use the term *Microbial Pest Control Product (MPCP)* for plant protection products that contain a MPCA, as I have defined above, as the active ingredient or active substance. It includes all microbial pesticide uses such as control of insects, mites, plant diseases, plant nematodes, and weeds.

I will also use the term *biopesticide* for these products; other synonyms used are *biological pesticide*, *microbial pesticide* or simply a *microbial*. I will use these four terms interchangeably with MPCP. In my opinion, the term biopesticide should strictly be used for products based on living organisms, including viruses and viroids, as also recommended by Eilenberg et al. (2001). Products based on plant extracts, pheromones, etc., should not be called biopesticides, but biochemical pesticides (as the US-EPA does) or natural pesticides. Sometimes they are referred to as "biorationals". When I refer to a broader definition of biopesticides which includes pheromones, insect and plant regulators, plant extracts, and macroorganisms, this will be indicated in the text.

5.2 Biological Control and Microbial Control

I will use the terminology with regard to *biological control* as proposed by Eilenberg et al. (2001). The term "biological control" will be used interchangeably with *biocontrol* as this is widely accepted. "Biological control" is defined by Eilenberg et al. (2001) as *the use of living organisms to suppress the population density or impact of a specific pest organism, making it less abundant or less damaging than it would otherwise be.* The authors emphasized the use of living organisms, and include viruses. They excluded the use of genes and metabolites when used alone. Any associated metabolites, however, to which the effects of pest control are (partly) attributed, such as the Bt toxins, should be inclusive in this definition. In that case, I agree with this definition and will follow it. The term *microbial control* is used to describe the use of living microorganisms as biological control agents, and is considered a sub-set of biological control. The term "microbial control" was first used by Steinhaus (1949) and defined as follows: "that phase of biological control concerned with the employment by man of microorganisms for the control and reduction of the number of animals (or plants) in particular area or a given population." An early definition of the term *biopesticide* could not be traced after an extensive search of the literature. EPA defined the term *biopesticide* in the early 1990s, but, as earlier said, in my opinion unsatisfactory.

I will use the term "microbial control" when microorganisms or entomopathogenic nematodes are used for pest control purposes. Nematodes are not generally defined as microorganisms. They are, however, usually grouped along with bacteria, fungi, viruses, etc. in microbial crop protection, and this probably originates from the research discipline, insect pathology, in which they are studied. This separates them from the entomological research field on natural enemies/beneficial arthropods such as predators and parasitoids. On the other hand, nematodes as well as parasitic and predacious arthropods are referred to as invertebrate biocontrol agents (IBCA) and "macrobials". Regulatory authorities usually regulate nematodes in the same way as "macrobials". In most countries, pesticide regulations for microorganisms do not include nematodes. The positioning of entomopathogenic nematodes is not always logical considering their taxonomy, biology and regulatory position. In the biocontrol industry, nematodes are mostly produced by companies which are also involved in microorganisms. The mass production, product development, and method of application of bacteria, fungi, baculoviruses and nematodes have many similarities. I will position them as microbial pest control agents along with the microorganisms used as biocontrol agents, unless specified differently. Where relevant, I will separate them from true microorganisms as, for instance, in a regulatory context.

5.3 Biocontrol Strategies

Biological control can be divided in four strategies: *Classical biological control, Inoculation biological control, Inundation biological control* and *Conservation biological control.* I follow the definitions as provided by Eilenberg et al. (2001). I will only discuss the use of biocontrol agents according to the strategy of "inundation biological control". Classical biocontrol, inoculation biocontrol and conservation biocontrol with insect pathogens will not be covered. Inundation biological control is defined as *The use of living organisms to control pests when control is achieved exclusively by the released organisms themselves.* Eilenberg et al. (2001) stated that the success solely depends on the released population and not on their progeny, and that this involves storage, formulation and application. The mass release provides immediate effects on the pest, and there is no expectation of long term control. That is precisely my perspective on biological insecticides. The word "released" is debatable in this definition since it suggests release from certain release points from where active dispersion and successive reproduction is expected. This is generally not the case with insect pathogens which are broadly applied (often sprayed), and contradicts the assumption that the released generation will have an immediate effect on the pest population. Inundation biocontrol implies an intervention by man via a certain way of application. In the field of microbial control agents many authors refer to this as "inundative applications" (Federici 1999; Lacey et al. 2001; Inglis et al. 2001). Therefore the word "released" in the definition should be replaced by "applied" to my opinion when we speak about inundative microbial control.

Biopesticides are often used within IPM systems, and can be important pest control tools in such systems. I will use the definition of IPM as provided by van Lenteren (1993): "Integrated Pest Management is a durable, environmentally and economically justifiable system in which damage caused by pests, diseases and weeds is prevented through the use of natural factors which limit the population growth of these organisms, if needed, supplemented with appropriate control measures".

5.4 Taxonomy

Examples of microbial pest control agents will be used to illustrate certain cases. Some of these will refer to the fungus *Verticillium lecanii*. Many isolates formerly classified as *V. lecanii*, however, have recently been re-classified and are considered to represent the new species *Lecanicillium attenuatum, L. lecanii, L. muscarium, L. longisporum* and *L. nodulosum* (Zare and Gams 2001). I will use the new names when it is known that the species or strains in question were verified using the new nomenclature. Not all strains formerly identified as *V. lecanii* have, however, been re-classified by Zare and Gams (2001) or others. I will use the old name *Verticillium lecanii* when it concerns taxa from which it is not clear whether the isolates have been re-identified and re-classified. Furthermore, I will use the latest taxonomic nomenclature where possible. But due to ongoing re-classifications of microorganisms and the problems of re-identification of isolates as described for *V. lecanii*, there may be reference to the literature that no longer refers to the correct taxa, and may therefore not be appropriate. The reader should note that this could also apply to other taxa.

References

Ash, G.J., 2009. The science, art and business of successful bioherbicides. *Biol. Control* 52(3): 230–240.

Bateman, R., 1997. The development of a mycoinsecticide for the control of locusts and grasshoppers. *Outlook Agric.* 26(1): 13–18.

Bateman, R., 2004. Constraints and enabling technologies for mycopesticide development. *Outlooks Pest Manag.* 15(2): 64–69.

Benuzzi, M., 2004. What will be the future for BCAs? The industry's point of view on problems in developing BCAs. *IOBC/WPRS Bull.* 27(8): 429–431.

Butt, T.M., C. Jackson, and N. Magan, 2001. Introduction – fungal biological control agents: progress, problems and potential. In: T.M. Butt, C. Jackson, N. Magan (eds), *Fungi as biocontrol agents: progress, problems and potential.* CAB International, Wallingford. pp. 1–8.

Campbell, R., 1986. The search for biological control agents against plant pathogens: a pragmatic approach. *Biol. Agric. Hortic.* 3: 317–327.

Chandler, D., G. Davidson, J.K. Pell, B.V. Ball, K. Shaw, and K.D. Sunderland, 2000. Fungal biocontrol of Acari. *Biocont. Sci. Technol.* 10: 357–384.

Chandler, D., G. Davidson, W.P. Grant, J. Greaves, and G.M. Tatchell, 2008. Microbial biopesticides for integrated crop management: an assessment of environmental and regulatory sustainability. *Trends Food Sci. Technol.* 19: 275–283.

Copping, L.G., 2009. *The manual of biocontrol agents*, 4th edition. BCPC Publications, Alton.

Cory, J.S. and H.P. Evans, 2007. Viruses. In: L.A. Lacey, H.K. Kaya (eds), *Field manual of techniques in invertebrate pathology*, 2nd edition. Springer, Dordrecht. pp. 149–174.

CPL, 2006. *Biopesticides 2007.* CPL Business Consultants, Wallingford.

CPL, 2007. *Biopesticides 2007, Vol. 5. A how to do it guide to biopesticides.* CPL Business Consultants, Wallingford.

Cross, J.V. and D.R. Polonenko, 1996. An industry perspective on registration and commercialization of biocontrol agents in Canada. *Can. J. Plant Pathol.* 18: 446–454.

De Faria, M.R. and S.P. Wraight, 2007. Mycoinsecticides and mycoararicides: a comprehensive list with worldwide coverage and international classification of formulation types. *Biol. Control* 43: 237–256.

De Groote, H., 1997. Potential for mycopesticide use in Africa: socioeconomic analysis. *LUBILOSA Socioeconomic Working Paper Series* 98(5): 1–27.

Dent, D.R., 1997. Integrated pest management and microbial insecticides. In: H.F. Evans (ed), *Microbial insecticides: novelty or necessity? BCPC symposium proceedings No. 68, Coventry, April 16–18, 1997.* BCPC, Farnham. pp. 127–138.

Dent, D.R., 1998. Removing the barriers to commercialisation: summary of the process and key issues for a mycoinsecticide for locust and grasshopper control. *LUBILOSA brochure*, CAB International, Ascot.

Dent, D., H. Evans, B. Federici, A. Rath, L. Ubsrasith, and K. Jones, 1999. Panel discussion: factors in the success and failure of microbial insecticides. *Integr. Pest Manag. Rev.* 4: 313–316.

Droby, S., M. Wisniewski, D. Macarisin, and C. Wilson, 2009. Twenty years of postharvest biocontrol research: is it time for a new paradigm? *Postharvest Biol. Technol.* 52: 137–145.

EC, 1991. Council Directive 91/414/EEC of 15 July 1991 concerning the placing of plant protection products on the market. *Official Journal of the European Union L 230*, 19.08.1991: 1–32.

EC, 2001. Commission Directive 2001/36/EC of 16 May 2001 amending Council Directive 91/414/EEC concerning the placing of plant protection products on the market (Data requirements microorganisms). *Official Journal of the European Union L 164*, 20.6.2001: 1–38.

EC, 2009. Regulation (EC) No 1107/2009 of the European Parliament and of the Council of 21 October 2009 concerning the placing of plant protection products on the market and repealing Council Directives 79/117/EEC and 91/414/EEC. *Official Journal of the European Union 309*, Vol. 52, 24.11.2009: 1–50.

Eilenberg, J., A. Hajek, and C. Lomer, 2001. Suggestions for unifying the terminology in biological control. *BioControl* 46: 387–400.

EPA, 2007. Pesticides: data requirements for biochemical and microbial pesticides. Federal Register: 40 CFR Part 158, October 26, 2007, Vol. 72, Nr. 207: Section 158.2100.

Evans, J., October 2008. Biopesticides: from cult to mainstream. *Agrow Magazine* 27: 11–14.

Federici, B.A., 1999. A perspective on pathogens as biological control agents for insect pests. In: T.S. Bellows, T.W. Fisher (eds), *Handbook of biological control*. Academic Press, San Diego. pp. 517–548.

Fravel, D., 1999. Hurdles and bottlenecks on the road to biocontrol of plant pathogens. *Australas. Plant Pathol.* 28(1): 153–157.

Fravel, D.R., 2005. Commercialization and implementation of biocontrol. *Annu. Rev. Phytopathol.* 43: 337–359.

Garczynski, S.F. and J.P. Siegel, 2007. Bacteria. In: L.A. Lacey, H.K. Kaya (eds), *Field manual of techniques in invertebrate pathology*, 2nd edition. Springer, Dordrecht. pp. 175–197.

Gaugler, R., 2002. *Entomopathogenic nematology*. CAB International, Wallingford.

Gelernter, W.D., 2005. Biological control products in a changing landscape. *Proceedings of the BCPC Int. Congress – Crop Science & Technology 2005, 31 October–2 November 2005, Glasgow*. pp. 293–300.

Georgis, R., 2002. The biosys experiment: an insider's perspective. In: R. Gaugler (ed), *Entomopathogenic nematology*. CAB International, Wallingford. pp. 357–372.

Grewal, P.S., R.-U. Ehlers, and D.I. Shapiro-Ilan, 2005. *Nematodes as biocontrol agents*. CAB International, Wallingford.

Hallett, S.G., 2005. Where are the bioherbicides? *Weed Sci.* 53: 404–415.

Hibbett, D.S. et al., 2007. A higher-level phylogenetic classification of the fungi. *Mycol. Res.* 111: 509–547.

Hofstein, R. and A. Chapple, 1999. Commercial development of biofungicides. In: F.R. Hall, J.J. Menn (eds), *Biopesticides: use and delivery*. Humana Press, Totowa. pp. 77–102.

Humber, R.A., 2007. Recent phylogenetically based reclassifications of fungal pathogensof invertebrates. Insect Mycologist and Curator, ARSEF. November 2007. http://arsef.fpsnl.cornell.edu/mycology/phyloreclass.pdf. Cited December 20, 2009.

Humber, R.A., 2008. Evolution of entomopathogenicity in fungi. *J. Invertebr. Pathol.* 98: 262–288.

Ignoffo, C.M., 1973. Development of a viral insecticide: concept to commercialization. *Exp. Parasitol.* 33: 380–406.

Inglis, G.D., M.S. Goettel, T.M. Butt, and H. Strasser, 2001. Use of hyphomycetous fungi for managing insect pests. In: T.M. Butt, C. Jackson, N. Magan (eds), *Fungi as biocontrol agents: progress, problems and potential*. CAB International, Wallingford. pp. 23–69.

Jackson, T.A., J.F. Pearson, M. O'Callaghan, H.K. Mahanty, and M.J. Willocks, 1992. Pathogen to product – development of Serratia entomophila (Enterobacteriaceae) as a commercial biological control agent for the New Zealand grass grub (Costelytra zealandica). In: T.A. Jackson, T.R. Glare (eds), *Use of pathogens in scarab pest management*. Intercept, Andover. pp. 191–198.

Jaronski, S.T., 1986. Commercial development of Deuteromycetous fungi of arthropods: a critical appraisal. In: R.A. Samson, J.M. Vlak, R. Peters (eds), *Fundamental and applied aspects of invertebrate pathology*. Foundation of the Fourth International Colloquium of Invertebrate Pathology, Wageningen. pp. 653–656.

Jehle, J.A., G.W. Blissard, B.C. Bonning, J.S. Cory, E.A. Herniou, G.F. Rohrmann, D.A. Theilmann, S.M. Thiem, and J.M. Vlak, 2006. On the classification and nomenclature of baculoviruses: a proposal for revision. *Arch. Virol.* 151: 1257–1266.

Jijakli, M.H., 2003. La lutte biologique en phytopathologie. In: P. Lepoivre (ed), *Phytopathologie: bases moléculaires et biologiques des pathosystèmes et fondements des stratégies de lutte*. De Boeck Université, Brussels. pp. 289–317.

Kiewnick, S., 2007. Practicalities of developing and registering microbial biological control agents. *CAB Rev.: Perspectives in Agriculture, Veterinary Science, Nutrition and Natural Resources* 2(No. 013): 1–11.

Krause, M.S., A.C.R.C. Vanachter, and T.J.J. de Ceuster, 2006. Commercial research and development of disease-suppressive microorganisms. *IOBC/WPRS Bull.* 29(2): 63–66.

Lacey, L.A., R. Frutos, H.K. Kaya, and P. Vail, 2001. Insect pathogens as biological control agents: do they have a future? *Biol. Control* 21: 230–248.

Leggett, M.E. and S.C. Gleddie, 1995. Developing biofertilizers and biocontrol agents that meet farmers' expectations. In: J.H. Andrews, I.C. Tommerup (eds), *Advances in plant pathology*, Vol. 11. Academic Press, San Diego. pp. 59–74.

Lidert, Z., 2001. Biopesticides: is there a path to a commercial success? In: M. Vurro, J. Gressel, T. Butt, G. Harman, R. St. Leger, D. Nuss, A. Pilgeram (eds), *Enhancing biocontrol agents and handling risks*. NATO Science Series, I: Life and Behavioural Sciences, Vol. 339. IOS Press, Amsterdam. pp. 284.

Lisansky, S.G., 1985. Production and commercialization of pathogens. In: N.W. Hussey, N. Scopes (eds), *Biological pest control. The glasshouse experience*. Blanford Press, Poole. pp. 210–218.

Lomer, C., 1999. LUBILOSA Phase 3 Final Report, March 1999. http://www.lubilosa.org.

Lord, J.C., 2005. From Metchnikoff to Monsanto and beyond: the path of microbial control. *J. Invertebr. Pathol.* 89: 19–29.

Marrone, P.G., 1999. Microbial pesticides and natural products as alternatives. *Outlook Agric.* 28(3): 149–154.

Mathre, D.E., R.J. Cook, and N.W. Callan, 1999. From discovery to use. Traversing the world of commercializing biocontrol agents for plant disease control. *Plant Dis.* 83(11): 972–983.

Meeussen, J.J., 2007. OECD Biopesticide Steering Group. Past, presence and future. *Paper presented at the Annual Biocontrol Industry Meeting,* Lucerne, Switzerland, October 22–23, 2007, www.abim-lucerne.ch. Cited November 20, 2008.

Montesinos, E., 2003. Development, registration and commercialization of microbial pesticides for plant protection. *Int. Microbiol.* 6: 245–252.

OECD, 2003. Guidance for registration requirements for microbial pesticides. *Series on Pesticides* No. 18. ENV/JM/MONO(2003)5. OECD, Paris (available at www.oecd.org/ehs/).

OECD, 2004a. OECD guidance for country data review reports on microbial pest control products and their microbial pest control agents (Monograph Guidance for microbials), February 2004. *Series on Pesticides* No. 22. OECD, Paris (available at www.oecd.org/ehs/).

OECD, 2004b. OECD guidance for industry data submissions for microbial pest control products and their microbial pest control agents (Dossier Guidance for microbials), February 2004. *Series on Pesticides* No. 23. OECD, Paris (available at www.oecd.org/ehs/).

Pell, J.K., J. Eilenberg, A.E. Hajek, and D.C. Steinkraus, 2001. Biology, ecology and pest management potential of Entomophthorales. In: T.M. Butt, C. Jackson, N. Magan (eds), *Fungi as biocontrol agents: progress, problems and potential*. CAB International, Wallingford. pp. 71–153.

Shah, P.A. and J.K. Pell, 2003. Entomopathogenic fungi as biological control agents. *Appl. Microbiol. Biotechnol.* 61: 413–423.

Sigman, R., 2005. OECD's vision for the regulation of crop protection products. *Proceedings of the BCPC International Congress – Crop Science & Technology 2005, 31 October–2 November 2005, Glasgow*. pp. 329–336.

Solter, L.F. and J.J. Becnel, 2007. Entomopathogenic microsporidia. In: L.A. Lacey, H.K. Kaya (eds), *Field manual of techniques in invertebrate pathology*, 2nd edition. Springer, Dordrecht. pp. 199–221.

Steinhaus, E.A., 1949. *Principles of insect pathology*. McGraw Hill Co, New York.

Steinhaus, E.A., 1956. Microbial control – the emergence of an idea. *Hilgardia* 26(2): 107–160.

Stewart, A., 2001. Commercial biocontrol – reality or fantasy? *Australasian Plant Pathol.* 30: 127–131.

Stock, S.P. and D.J. Hunt, 2006. Morphology and systematics in nematodes used in biocontrol. In: P.S. Grewal, R.-U. Ehlers, D.I. Shapiro-Ilan (eds), *Nematodes as biocontrol agents*. CAB International, Wallingford. pp. 3–43.

Törmälä, T., 1995. Economics of biocontrol agents: an industrial view. In: H.M.T. Hokkanen, J.M. Lynch (eds), *Biological control: benefits and risks*. Cambridge University Press, Cambridge. pp. 277–282.

Van der Pas, R.K., C.J. Waddington, and W.J. Ravensberg, 2000. Commercialisation of a microbial pesticide: "challenges and constraints". *Bull. IOBC/WPRS* 23(2): 15–18.

van Lenteren, J.C., 1993. Integrated pest management: the inescapable future. In: J.C. Zadoks (ed), *Modern crop protection: developments and perspectives*. Wageningen Pers, The Netherlands. pp. 217–225.

Warrior, P., 2000. Living systems as natural crop-protection agents. *Pest Manag. Sci.* 56: 681–687.

Woodhead, S.H., A.L. O'Leary, D.J. O'Leary, and S.C. Rabatin, 1990. Discovery, development, and registration of a biocontrol agent from an industrial perspective. *Can. J. Plant Pathol.* 12: 328–331.

Wraight, S.P., G.D. Inglis, and M.S. Goettel, 2007. Fungi. In: L.A. Lacey, H.K. Kaya (eds), *Field manual of techniques in invertebrate pathology*, 2nd edition. Springer, Dordrecht. pp. 223–248.

Zare, R. and W. Gams, 2001. A revision of *Verticillium* section *Prostata*. IV. The genera *Lecanicillium* and *Simplicillium* gen.nov. *Nova Hedwigia* 73(1–2): 1–50.

Chapter 2
Selection of a Microbial Pest Control Agent

Abstract From a comparison between academic research and industrial research with regard to the finding and the development of a microbial pest control agent, it is concluded that there is a need for well-defined selection criteria and a complete process description for the entire development of a microbial pest control product. This chapter discusses the exploratory phase of finding a new microbial pest control agent, and the actual screening phase of species and strains of collected insect pathogens, potentially bacteria, fungi, viruses, protozoa, and entomopathogenic nematodes. Three decisive selection criteria are identified for a commercial microbial insecticide: mortality, production efficiency, and safety. Relevant factors in the screening process related to mortality are dose rate, mode of action, speed of kill, host range, sensitivity to abiotic factors, and persistence. Second, the possibility of mass production with high yields is a critical criterion. Third, safety to humans and to the environment is a selection criterion in relation to registration requirements and expenses. Crucial characteristics of each type of pathogen relevant to their utility as crop protection product are noted. Testing methods will be critically reviewed and recommendations for standardization of methods and test conditions will be presented. The consecutive steps in the screening process will be listed with emphasis on economic considerations. The best strain for further development is the one that provides a well-balanced compromise between efficacy and economics.

1 Introduction

The search for and the selection of a new agent is the first step in the developmental process of a microbial pest control product (MPCP). It requires a screening procedure with relevant evaluation criteria for a commercial product. This selection step should address appropriate biological, technical and economic aspects that play a key role in the entire product development, and in the marketing of the product. The criteria differ per type of organism and depend on the ultimate use of the product. Although each product demands its specific criteria, several general criteria can be identified. The actual screening process is preceded by an exploratory phase in

W.J. Ravensberg, *A Roadmap to the Successful Development and Commercialization of Microbial Pest Control Products for Control of Arthropods*, Progress in Biological Control 10, DOI 10.1007/978-94-007-0437-4_2, © Springer Science+Business Media B.V. 2011

which the problem that needs to be solved is well-described as well as the foreseen solution. During this phase the research direction must be established and applied to the entire process.

Approaches used to find and select a microbial pest control agent by academic and industrial scientists are compared. Apparently, well-defined selection criteria are not provided in the current literature. Therefore, these will be identified in this chapter. Biological characteristics of the four types of insect pathogens are reviewed in relation to screening methods as well as to critical factors in the selection of isolates. The most important selection criteria and biological parameters related to these criteria will be described, and arguments will be provided as to why these criteria are determinative for a commercial product. In the conclusion, I will discuss the results of the selection procedure in relation to the decision whether to continue the development of a new product based on (one of) the selected isolate(s). Further, I will provide recommendations for the use of bio-assays in the screening process and the other steps in the selection procedure.

1.1 Different Approaches in Finding a New Pest Control Agent

Predators, parasitoids and insect pathogens are used as biocontrol agents in biological control of pests. In the finding of a biocontrol agent and the development of it into a biocontrol product, the first two require a similar approach. Pathogens, however, demand different evaluation criteria. On the other hand, many aspects are similar for arthropods and insect pathogens, particularly commercial considerations.

For arthropods, Van Lenteren and Manzaroli (1999) described the various steps and the strategy in the development of a biological control programme for a greenhouse crop with a specific natural enemy. In biocontrol programmes, natural enemies were often found by trial and error, chiefly by empirical procedures. Van Lenteren and Manzaroli described pre-introductory evaluation criteria and tried to come to a more efficient and less time-consuming procedure, reducing the number of potential candidates for field testing to a few. They presented a flow diagram with the most important criteria in an evaluation programme for natural enemies for inundative releases.

Similar to the field of biocontrol with arthropods, an evaluation approach for insect pathogens as biocontrol agents has only been briefly presented in the literature by several authors. But few authors described the entire process of finding a new microbial control agent and its development into a successful commercial product. Burges (1981a) described three approaches to the discovery and development of a microbial pest control agent (MPCA):

(1) a serendipity approach, where a promising agent is noticed in a particular pest and further developed against that pest. As an example he referred to *Bacillus popilliae* against the Japanese beetle, *Popillia japonica*;

(2) an agent-oriented approach, where a known agent is developed for as many as possible pests. The example here is *Bacillus thuringiensis* (Bt) for control of various lepidopteran pests;
(3) a goal-oriented approach, where a MPCA is sought for control of a certain pest. An example here is a virus against the palm rhinoceros beetle *Oryctes rhinoceros*.

According to Burges the third approach has been the least successful, but one that could be further developed for integrated pest control programmes which have gaps where particular pests are difficult to control when broad-spectrum chemicals are no longer used. Burges (1981a) clearly recognized all the prerequisites for the development of a microbial pest control agent. He realized that "market size and competitiveness are vital" and that "important other commercial considerations include wide host spectrum, long shelf-life and production technology, and safety and registration costs". In the early days of microbial biocontrol this could be described as the applied scientific approach.

In the commercial development of biocontrol agents two approaches can be distinguished:

(1) a problem-solving approach: I would argue that the problem solving approach is the most common and most successful one in commercial biocontrol in the last 25 years, certainly in protected crops. A certain pest cannot be controlled well enough in the existing integrated pest management (IPM) programme and a new solution has to be found in order to safeguard the system or to help control the pest. Any type of biological control agent (BCA) is then, in theory, considered and can be searched for and developed following appropriate selection criteria. This problem-solving approach is seen in IPM programmes in greenhouse crops where natural enemies are already being used and where a new predator or parasitoid is sought when a new pest occurs;
(2) an opportunity approach: an opportunity presents itself when a novel BCA is found fortuitously, and someone recognizes based on experience and general insight, that a new product could be developed for which a market may exist. This could be a new product in a new market or a product for an existing market, which requires penetration among existing products, often chemicals. This approach is not often seen in commercial biocontrol of pests, and certainly not with beneficial arthropods.

One could say that the chemical pesticide industry works according to this opportunity principle. Companies try to make as many new chemical molecules as possible and then they screen them for biological activity against a number of pests, diseases and weeds. Within the agrochemical industry this was often called "the spray and pray" approach (Evans 2006). By means of automated high through-put screening systems, chemical companies can nowadays test an enormous amount of new molecules. The newest systems can even design, create and test more than 150,000 compounds per day by a biochemical screening method, resulting in testing more than 500,000 molecules per year in vivo for biological activity (Stenzel 2004). If a good activity against certain targets is found, the development of a molecule is continued. A broad spectrum of activity increases the chances of a successful product that needs to compete with existing pesticides. Target markets are the worldwide agricultural crops.

The biotechnology company Agraquest, USA, employed this approach and they have been screening over 20,000 microbial organisms in a similar way (see www.agraquest.com). It is the first company active in the field of microorganisms that duplicates this screening approach. They collect, produce and test as many isolates as they can find and also identify the compounds (metabolites) produced by the microorganisms. Further field testing, production and formulation leads to the final product for which markets have to be developed (Marrone 1999). This approach is also used by Marrone Bio Innovations Inc. (www.marronebioinnovations.com). This screening method has clearly been successful with the large chemical pesticide industry, but whether this will succeed with microbial products still remains to be seen.

1.2 Academic Research on the Selection and Development of a Microbial Pest Control Agent

Entomopathogens have been studied for a long time in academic research and numerous authors mention and investigate aspects of the development of a MPCA. The literature on entomopathogens as biocontrol agents is vast and many handbooks and papers have been written on bacteria (Charles et al., 2000; Entwistle et al., 1993), on fungi (Burge 1988; Butt et al., 2001; Butt 2002; Ferron et al., 1991; Hall and Menn 1999; Lacey et al., 1996; Lisansky and Hall 1983; McCoy et al., 1988; Vurro et al., 2001), on baculoviruses (Miller 1997; Moscardi 1999; Shuler et al., 1995), and on entomopathogenic nematodes (Gaugler and Kaya 1990; Gaugler 2002; Grewal et al., 2005a). In most of these publications, however, key selection criteria are lacking as well as a description of the entire process of the development of a MPCA.

For example, in early work in this area, Ferron (1978) presented a detailed overview of the development of biological control of insect pests by entomogenous fungi, and the situation up to the mid 1970s. He described the features of a fungus that has a potential as a biocontrol agent, such as mode of action, development of mycosis and saprophytic development of the fungus, the effects of abiotic factors, the quantity of spores needed to kill the host, etc. He mentioned also the importance of strain selection, the capacity of mass production of spores and the development of a final preparation, and the safety to vertebrates. There is, however, no attempt to order the traits in importance or to define a systematic approach for the entire process. In contrast, Hall (1981) gave an example in which the development of a MPCA, *Verticillium lecanii*, was clearly foreseen. All relevant aspects for the development of a product were investigated (see Box 2.1).

Box 2.1 The Development of Verticillium lecanii as a Biopesticide

The entomopathogenic fungus *Verticillium lecanii*, of which isolates are recently reclassified and included in the new species *Lecanicillium muscarium, L. longisporum, L. lecanii* and other species (Zare and Gams 2001), is frequently found on many kinds of insects, mainly on Homoptera, but also on mites and nematodes, and as a soil saprophyte, and even as a parasite on powdery mildews and rusts. It was noticed by growers in vegetable crops on aphids and whitefly nymphs already in the early 1920s in northern Europe, mainly in the UK. Hussey found it in 1956 in a commercial cucumber greenhouse on whiteflies in the UK. The research group at GCRI (Glasshouse Crops Research Institute at Littlehampton, UK) was the first to recognize that this fungus could be used to control aphids and whitefly in greenhouse crops and Hall studied the fungus in detail (Hall 1981). Resistance to chemical insecticides in *Myzus persicae*, a dominant pest in chrysanthemums, was the incentive for this project. During this study the apparent biological life cycle parameters were assessed as well as the abiotic factors influencing the fungus.

From the context it is clear that the development into a commercial preparation was in the back of the mind of the researcher. Screening of strains was carried out on a limited scale, but the emphasis was clearly on biological growth parameters and host virulence in the laboratory. Hall also studied storage of conidiospores and of blastospores, environmental factors influencing field performance, spore dispersal, and spread of infection. Greenhouse trials were performed in chrysanthemum on several aphid species with "formulated" spore suspensions. Spore concentrations were tested, as were application time and frequency. Trials were also carried out in cucumber against the cotton aphid *Aphis gossypii* and against the greenhouse whitefly *Trialeurodes vaporariorum*. Production was studied and critical aspects were considered such as culture purity, production of conidiospores versus blastospores and the costs of the chosen methods. Commercial mass production was not yet possible and Hall considered the possibility of culturing spores for immediate use by the growers. Quality control was also evaluated. Spore viability and potency were assessed in a bio-assay on the host insect compared to a chemical standard. Compatibility with natural enemies used in the greenhouses was tested, as was compatibility with insecticides and fungicides. Pollinators were considered, but not tested. The toxicological profile of the fungus was checked in literature and toxicological animal tests were performed to obtain an indication about the safety of the product to animals and to humans.

The above example shows that in Hall's research many of the aspects of developing a fungus into a commercial biopesticide were studied (Hall 1981). When we consider that these were the early days of developing a fungal microbial insecticide, this was a solid approach.

The route to commercialization started when Tate & Lyle, a UK based food ingredient and sugar manufacturer, contacted GCRI in 1978. Tate & Lyle tried to diversify their business and one of their new interests was the area of biopesticides. The company was building a new fermentation plant to produce alginate and it was looking for microbial products that could be produced by liquid fermentation. They contacted research institutes and the one on the doorstep happened to be working on a promising fungus. The ensuing cooperation resulted in the development of Vertalec and Mycotal (S. Lisansky, personal communication).

In the extensive overview presented in the book of Burges (1981b) on the progress of "Microbial Control of Pests and Plant Diseases, 1970–1980", on all the types of MPCAs neither Burges nor the contributing authors described any selection criteria or methodical developmental process for BCAs. In later work, like in the overview "Microbial Insecticides: Novelty or Necessity?" (Evans 1997) still no reference was given to a developmental process for a MPCA, despite the 15 years between the book of Burges (1981b) and this book. The same holds for the review of Lacey and Goettel (1996) on the developments in microbial control products of insect pests and their prospects for the twenty-first century.

In the study of Fransen (1987) on the development of *aleyrodis* as a microbial biocontrol agent for whiteflies aspects of the development of a commercial microbial agent are systematically considered (see Box 2.2). She listed fungal characteristics, host characteristics and environmental influences. She investigated primarily the biological parameters of the fungus and its interaction with the greenhouse whitefly and its parasitoid *Encarsia formosa*, however. Some attention was indeed paid to the production of the fungus and the formulation. Storage, safety and integration with chemicals were itemized but not studied. Results on the studied evaluation criteria were summarized with the purpose of making a decision on the potential of *aleyrodis* as a commercial biopesticide.

Box 2.2 Why Aschersonia aleyrodis Was Not Developed as a Biopesticide

In the early 1990s, Koppert tried to develop *aleyrodis* into a mycoinsecticide for control of the greenhouse whitefly. This fungus was expected to perform better at lower relative humidity (RH) conditions, as compared to the high RH requirements of *L. muscarium*. Moreover, whitefly larvae infected with *aleyrodis* turn orange and the effect is therefore clearly visible to growers, in contrast to larvae killed by *L. muscarium*. Visible effects of a treatment were considered to be a convincing feature for the marketing of the product.

We started to produce and conduct field trials with one strain. Spore production is initiated by light, and it was complicated, thus expensive, to build bioreactors in which light could be installed. Secondly, strain attenuation was a problem and the strain easily lost its ability to sporulate. In field trials in tomato and cucumber, mortality by *aleyrodis* was similar (at 60 and 75% RH) to *L. muscarium* (Mycotal) and even lower at 80% RH (van der Pas et al., 1996). These two aspects led to the decision not to continue with the product development, and not to invest money and time for registration. The underlying decisive factors were the difficulty in production and the moderate efficacy.

In retrospect, one could argue that we did not select the most effective strain and we did not pay enough attention to the attenuation of the strain. This example taught us to continuously re-evaluate technical and economic aspects and that decisions have to be taken after every step in the developmental process in order to prevent losses in investment further in that process. Of course, such a decision should be taken the earlier the better, but also involves a risk of discarding a potential product too early.

The study of Fransen, however, concerned only one strain and selection of a virulent strain was not part of it. Meekes (2001) followed the approach of Fransen and specifically focused on the identification of a virulent strain of *Aschersonia* spp. for the control of the greenhouse whitefly *T. vaporariorum* and the tobacco whitefly *Bemisia tabaci.* Two significant aspects were investigated in more detail: the production of spores and virulence to two dominant pest species and based on the results many strains were eliminated. Many aspects of the development of a microbial control agent were considered such as dispersal, persistence, production and formulation, safety and compatibility with other control methods as well as the influence of environmental factors. Finally, a table is presented in which the results are listed and on which a further decision could be based. This study exemplifies how screening of a potential strain for the development of a microbial pesticide should be approached, as it emphasized decisive factors such as efficacy and production efficiency.

1.3 The Development of a Microbial Pest Control Agent from Finding to Product Launch by Industry

In the research to develop an MPCA we often see in academic studies a focus on particular biological aspects of the MPCA. In order to arrive at the launch of a successful product all steps in the developmental process must be conducted. A commercial producer has to carry out all these steps. In the literature, however, we rarely find a description of a complete and systematic approach. In a few projects the entire process is described. An early example from the biocontrol industry is given by Lisansky (1985) based on the experience with *L. muscarium* and *L. longisporum*. He presented a diagram that covers all the steps, from defining the pest problem to selling the product and after-sales service, and he briefly described the considerations per step. Screening of strains is not mentioned, however, nor did he give critical aspects that lead to decision-making in the process.

Bayer initiated the development of a biopesticide around 1990 and developed Bio-1020; a mycoinsecticide based on a strain of *Metarhizium anisopliae* for control of the black vine weevil. A brief development scheme, mainly focused on production, was given on the developmental steps for a microbial product (Andersch 1992). No details were given on what the determinative considerations were, while one expects the largest agrochemical company in the world to have clear decision criteria for such a developmental process.

In the field of entomopathogenic nematodes (EPNs), Biosys, an American EPN producing company, recognized three stages in research in the product development pathway: basic research, applied research and commercialization (Georgis 2002). Remarkably, strain selection is not mentioned. It does not reveal importance or order of research steps, nor criteria that could decide whether to continue with a product.

A good example originating from the academic world is the Lubilosa project (see Chapter 1). Here, the problem-solving approach was clearly taken. The wide use of chemical insecticides had so many negative aspects that public concern demanded an alternative solution, and a number of donor countries initiated the Lubilosa programme. From the beginning, the development of a mycoinsecticide was foreseen and all necessary steps were identified and executed (Lomer 1999; Dent 1998). The scheme (Chapter 1, Fig. 1.1) used in the Lubilosa project (Dent 1998) illustrates all the steps in the project.

Most scientists and biocontrol companies, however, seem to develop their products without a well-described plan. As observed in Chapter 1, there is a need for a general procedure, for a systematic and structured roadmap that guides biocontrol developers through the complex process of developing a new biocontrol agent. A model for the product development of a MPCA based on 25 years of experience in the biocontrol industry will be presented in this study.

2 Critical Steps and Factors in the Selection Process of a Microbial Pest Control Agent

In the development of a BCA into a commercial product, project management is an indispensible tool in order to plan, budget and conduct every step in the right sequence, so that no time and money will be wasted in this complex process. Project management includes a plan with goals and objectives, specifying tasks and timelines and how to achieve all the goals. It is a team effort, and proper management to stick to the "critical path" is essential. The sequence of the relevant steps in a selection programme of a MPCA is presented. The steps are the description of the problem, scouting for a solution, the collection of the MPCA, and the selection of the best species and strain based on well-defined selection criteria. These will be discussed in detail. Finally, the decisive steps and factors in the selection of a MPCA are defined.

2.1 The Description of the Problem

A new product will be developed to solve a problem or to grasp an opportunity. To justify the input in terms of money and time, it needs to be clear from the outset why the project will be started and what the ultimate goal is. In this phase, the deliverable is the project letter that includes the project justification, the project definition, the project plan including the project team, a rough estimate of the budget (resources and costs) and the time frame. When possible, the new product should be described in general terms, so it is clear in which direction the project is heading. It should also address the value of the project for the company, for the strategic and the operational goals, and mission. The reason for the project letter is to clarify to all stakeholders what the investment will be and why this investment is made. Commitment from all the decision-makers (board, share-holders) is crucial, especially for later stages when it may need more time or money. Of course this depends on the size of the project and is a case by case situation.

In this exploratory phase, it is essential to make a detailed characterization of the problem or opportunity. What exactly is the problem and what is the scale of the problem? For pest control in greenhouse crops we need answers to the following questions: which arthropod pest is it; in which crops does it occur; in what country/region; at what time of year; is the problem expected to increase or decrease in the near future; what are the short term solutions, either available or being developed by other biocontrol companies or the chemical industry, or even by the seed-producing industry (are there new tolerant or resistant cultivars); what are the acceptable costs for a solution for the grower; will it become an affordable product. Finally, an impression of the potential market should be developed. If possible the expected solutions should be described briefly so it is understood where the project is going. The project plan should also include evaluation and decision moments, i.e. points of "go/no go". All these aspects will make the overall picture clearer on which a decision can be based and commitment is assured.

2.2 Scouting and Finding a Potential Solution

Where can we find information on biocontrol agents that can possibly control the pest? Or where can we find information on other ways, methods or products to control the problem. There are different areas to look at and these are usually checked in the following order:

(1) literature: screen the literature, including old literature, for possible answers. Nowadays that also includes searching via the internet. The first step is to make a thorough literature search of the natural enemies of the target pest and on related pests, followed by a literature search of the natural enemies which are found in nature on the target species (especially for MPCAs with a broad host range). Maintain a library with all the information found and a knowledge management system to store knowledge and experiences from people in different areas;
(2) academic research: is or was there any research related to the topic conducted at a university or at a research institute, anywhere in the world? What kind of research has been done on the pest or even on related (pest) species in the past or recently? Is somebody already working on a solution? Government funded and organized classical biocontrol projects are also sources of beneficial organisms for augmentation biocontrol (e.g. the 5 year *Bemisia* project from USDA);
(3) industrial research: is any company working on the problem or does a company have a product on the market, including a chemical or biological pesticide, resistant cultivars, supporting measures, monitoring systems, pheromones, physical means (screens, disinfection methods, etc.) that could contribute to a solution? In the biocontrol industry only a few companies really develop new products, so it is relatively easy to know what is going on. Although patents in the world of biocontrol are scarce (since living organisms per se cannot be patented) a patent search could be considered also;
(4) network: not everything is (yet) published, so it is useful to create and to have a network of contacts and to know who is working on what, in research, in industry, anywhere in the world; informing your network about the problem and that you are seeking a solution is useful too. This can be done quickly and widely now through internet, e.g. via list servers such as "Goodbugs" and "Thripnet" or through direct contact. Often research that did not lead to a solution has not been published. It can be useful to know what natural enemies or entomopathogens have been studied and which ones were discarded as a potential candidate and why this was decided;
(5) nature: what is around in nature? And where to search for it? What kind of solution is conceivable: a microbial or macrobial agent, or other measures (pheromones, natural products) or combinations thereof? Collections could be done in the area of origin of the crop, in the area of origin of the pest or in areas where the pests already occurs (e.g. as in the 5-year *Bemisia* project by Kirk and Lacey (1996)). An efficient way has been to collect in (unsprayed) crops in different locations and during different times of the year. In this way many natural enemies, especially arthropods, have been found (*Dacnusa sibirica, Opius pallipes, Diglyphus isaea, Macrolophus caliginosus, Orius laevigatus, Amblyseius calfornicus, Chrysonotomyia formosa*, but also *Verticillium lecanii*) (Enkegaard and Brødsgaard 2006). Another method is to put out trapping plants with the

target pest, outside in different biotopes or to check the same plant species or related plant species in nature, or even to look on wild plants which are a host of the pest.

2.3 *Collection of a Microbial Pest Control Agent*

When specimens of microbial agents are being collected for further research the following aspects need to be taken into consideration: identification, the quantity of genetic variation, and purity. In the case of microorganisms care has to be taken to make a pure culture from the collected material, while secondary microorganisms may also often colonize the dead insect. It is crucial to have the primary pathogen in hand. This needs to be tested then by satisfying Koch's postulates.

How to recognize and collect the various types of pathogens is well described in the comprehensive Manual of Techniques in Insect Pathology by Lacey (1997). Collection can be carried out in nature or in crops, as with arthropods. Various pathogens can be found and further research is needed to identify pathogenicity and virulence to the target insects.

In microorganisms every isolate can be different and as many as possible isolates should be looked for to study in further screening programmes. The isolates need to be purified and stored properly. Identification on strain level will be necessary later in order to be able to recognize it and for regulatory reasons.

For isolates of fungi, bacteria and viruses there are several culture collections (such as the ATCC (USA), CABI (UK), CBS (the Netherlands), DSMZ (Germany)) where pure isolates can be ordered. There is no public culture collection of EPNs.

2.4 *Selection of Species and Strain for the Development of a Bio-Insecticide*

The first selection level regarding the selection of the optimal microorganism in order to control a pest is the "type of organism" level. Is a fungus the most suitable candidate versus a bacterium or an entomopathogenic nematode? This depends largely on the pest, its biology and on features such as feeding behaviour, mobility and habitat. The second level is the species/strain selection, once the decision on the level one "type of organism" has been made: which isolate is the most pathogenic and virulent one?

When a microbial agent is considered, there are five groups of organisms from which a selection can be made: bacteria, viruses, fungi, protozoa and entomopathogenic nematodes (EPNs). Many products have been developed for augmentative biocontrol as biopesticides based on these organisms, although products based on protozoa (mainly microsporidia) are the least common and successful. Depending on the pest insect and the knowledge on its pathogens (or of related insects) a first selection can be made, considering also the biology of the pest and the pathogen. The possibilities are different for leaf-feeding insects than for sucking insects, or for foliar insects or soil insects, or boring insects in a hidden habitat. Federici (1999a) summarized the utility of the different types of organisms against different types of insect pests. This is, of course, a theoretical exercise based on the biology of both organisms: the pest and the pathogen. Many other factors influence the best choice of a biocontrol agent for a given pest and these will be dealt with later in this study. Burges (1981a) divided the insect groups in: feeders on plant surfaces, sucking insects, boring insects, soil insects and aquatic insects. Next, he tried to assess the efficacy of pathogens against these different pest groups.

The possibilities for use of a pathogen on the dominating pest groups in protected crops are given in Table 2.1. This is an attempt to assess the likelihood of successful control based on the biological characteristics of both pest and pathogen, not on production and toxicological aspects.

The reasons for the given indications in Table 2.1 are the following:

- mites: few bacteria are known to infest mites (De Kogel and Ravensberg 2006); some fungi are known to cause diseases in tetranychid mites (van der Geest 1985) and in eriophyoid mites (McCoy 1996), but many of these fungi are from the group of Entomophthorales and cannot be cultured; only few viruses are known of tetranychid mites (van der Geest 1985); and EPNs are too large to enter and kill mites and are not very effective against foliar pests;
- aphids, whiteflies and scales: these phloem feeders are not likely to ingest bacteria; many fungi are known to infest aphids, but many of them are Entomophthorales (Latgé and Papierok 1988); whiteflies and scales can both be infested by many fungi (Fransen 1990; Evans and Hywel-Jones 1997); viruses are rare; EPNs can hardly penetrate them and are difficult to use for foliar pests;
- leafminers: few pathogens are known to infest them. The eggs and larvae are hidden in the leaf tissue, the pupae are well protected and pathogens of the adults are unknown. EPNs have been used with reasonable effects (Williams and Walters 2000);

Table 2.1 The potential of different kinds of pathogens to control greenhouse pests

Pest group	Bacteria	Fungi	Viruses	EPNs
Mites	+	++	–	–
Aphids, whiteflies, scales	–	+++	–	–
Leaf miners	+	–	–	+
Thrips	+	++	–	+
Caterpillars (defoliators)	+++	+	++	+
Caterpillars (soil, cryptic)	+	–	–	+++
True bugs	–	+	–	–

–: No chance of successful control.
+: Some possibility of control.
++: Reasonable possibility of control.
+++: Good possibility of control

- thrips: few bacteria are known, although they can be infested by them; they are known to be killed by various fungal species (Brownbridge 1995); pathogenic viruses are not known, EPNs can kill them (Wardlow et al., 2001) and also parasitic nematodes are known to regulate populations (Mason and Heinz 1999);
- caterpillars: susceptible to Bt and can be controlled depending on their biology and the chance that they ingest the pathogen; the same holds for baculoviruses, although only a few are known. A number of entomopathogenic fungi are known (Inglis et al., 2001) and EPNs can be used, especially in cryptic or soil habitats (Van Tol and Raupp 2005);
- true bugs; these active phloem feeders are hard to infest by any of the pathogens; some entomopathogenic fungi are known to kill them (Liu et al., 2002).

This table could help in setting up an exploration program for pathogens for a certain pest. When it is already well known which agent to look for or which one to use, selection level two becomes important. In this phase the screening and selection of the collected isolates (of one or more species) takes place. Differences between isolates can be large per characteristic or set of characteristics in all pathogens. In a screening/selection programme species and strains of various types of pathogens can be studied simultaneously. For instance, when a soil-inhabiting caterpillar needs to be controlled, the efficacy of a bacteria and an EPN can be studied. Finally, however, isolates of one type of organism are investigated. Our research group has screened bacteria on thrips and spider mites and in this project several strains of three bacteria species were screened simultaneously for mortality and speed of kill of the targets.

2.4.1 Strain Selection in Entomopathogenic Bacteria

The most successful microbial insecticides are based on bacteria. Worldwide, only a small number of bacteria species, however, has been used to develop these products, namely three *Bacillus* species, *B. thuringiensis*, *B. sphaericus* and *B. popilliae* and *Serratia entomophila.* Most products have been developed on the basis of strains of *B. thuringiensis* and a number of subspecies of it. Various isolates have a good activity on caterpillars (*B.t. kurstaki, B.t. aizawai*), mosquito larvae (*B.t. israelensis*) and beetle larvae (*B.t. tenebrionis*). *B. sphaericus* also kills mosquito larvae and *B. popilliae* and *S. entomophila* are pathogens of white grubs.

B. thuringiensis is the dominant pathogen in the market in terms of field of application and commercial success. Natural isolates of the subspecies are used as the active ingredient for the products. Selection of the isolates is based on a LC 50 test. Some recent products are based on transconjugate bacteria. An overview of the subspecies, the products and their use is given by Federici (1999b). For more details on Bt development and its uses the reader is referred to Charles et al. (2000).

Selection of isolates is indispensible in Bt as its activity depends on which endotoxin it makes. Each endotoxin has a specific mode of action and host range (Navon 1993). Finding isolates and screening for activity has become a field of science where standardization of bio-assays is important but also still a debate (Skovmand et al., 2000). It seems that only a specialized company, in collaboration with scientists, can successfully develop new Bt products. In 1992 more than 25,000 isolates were found (Milner 1992); less than 10 years later estimates were about 60,000 (Glare and O'Callaghan 2000). More-over, Bt's are also used to develop transgenic plants, an even more costly development in which only large multinationals are active, such as Monsanto and Syngenta.

2.4.2 Strain Selection in Entomopathogenic Fungi

The literature in which strain selection studies with fungal isolates on arthropod targets are described is vast. Few authors indicate which selection criteria they use to come to a final selection and few studies continue with the development of the chosen strain into a commercial product.

A tiered approach of selecting efficacious strains is described by Milner (1992) for the selection of *Metarhizium anisopliae* strains for locust and termite control. In Tier 1 he tested pathogenicity of the strains under optimal conditions (scarab larvae rolled in moist tissue paper at 25°C for 14 days); in Tier 2 the pathogenic strains were tested quantitatively by dipping larvae in spore solutions and incubating them as in Tier 1. The best isolates were then tested in Tier 3 in a bio-assay under simulated field conditions (in soil treated with conidia). The best three isolates were then used for field tests, in fact a Tier 4 step. Milner found the three tiered assay useful for quickly assessing the efficacy of a large number of isolates. Most authors use a similar sort of system, but few of them have described it in such a methodical approach.

The approach of Milner is a logical one for assessing the pathogenicity and the efficacy, but it can be shortened and made cheaper. Tier 1 and 2 can be combined and should be tested under more challenging conditions, for instance, at a lower humidity and for a shorter time, more strains would be eliminated earlier and fewer could be tested in the bio-assay (Tier 3).

Gillespie and Moorhouse (1990) listed nine characters for consideration when selecting fungal isolates for control of pests. Among these are pathogenicity and feasibility of production, obviously relevant characteristics, but also disease spread and fungicide resistance, of which I fail to see the relevance. The authors did not discuss the importance of these nine characteristics in any detail. On the other hand they considered that strain selection "cannot be over-emphasized as many characters can vary greatly between strains". This is well recognized nowadays with most pathogens.

As described above Fransen (1987) and Meekes (2001) investigated *aleyrodis* as a potential MPCA and selection criteria for strains were defined. The salient criteria were high spore production and virulence (Meekes 2001) which I also consider two of three of the most important characteristics on which a decision for development has to be based (see below).

Posada and Vega (2005) described a new screening method based on six parameters. They investigated the interaction between 50 strains of *B. bassiana* and the coffee berry borer. A scoring method with a minimum value for each parameter was developed to rate the potential of the studied strains. The parameters they used seem to be appropriate for the studied case, but they may not always be key features. For instance, the speed and quantity of spore production on cadavers to initiate an epizootic effect is often not relevant for commercial biocontrol products used in greenhouses. Generally, greenhouse crops do not generate the necessary high relative humidity conditions. Nevertheless, the scoring system is an interesting way of evaluating different features for selecting the most promising biocontrol agents.

2.4.3 Strain Selection in Baculoviruses

A small number of baculoviruses out of the approximately 700 known species (Moscardi 1999) are used as microbial insecticides. Two types of occluded baculoviruses, nucleopolyhedroviruses and granuloviruses, cause epizootics in insects. They cause diseases in caterpillars and in some Hymenoptera (sawflies) and some Diptera, and kill their hosts in a number of days. They have a narrow host range and are safe to most other organisms. These attributes make them suitable for the development of a bio-insecticide.

Among isolates of a baculovirus large differences may exist in virulence. Natural populations are often heterogeneous and contain many isolates and an isolate often contains a number of genotypically distinct variants (Murillo et al., 2006). Isolates can show a 50-fold difference in activity (Shapiro and Ignoffo 1970).

Selection of the most virulent isolate is done in a bio-assay on the host insect(s). Attention needs to be paid to using a standardized set up with a dose range and a suitable larval stage. The best isolates should then be tested under conditions that closely mimic the field conditions. However, few isolates have been found and producers have a low number of isolates available to select from, or sometimes only one, as in the case of the codling moth granulovirus (J. Huber, personal communication). Only five strains were available for the selection process in the development of the biological insecticide for control of *Spodoptera exigua* (Smits and Vlak 1988).

At the same time, these isolates need to be checked for production of infective particles on the host insect to select for a good isolate for the mass production, which, up to now, needs to be done in vivo. Producers of baculovirus products consider efficacy and yield of infective particles as the selection criteria (M. Andermatt, personal communication). Smits and Vlak (1988) selected the best strain on the basis of efficacy, speed of kill, and on host specificity, as this would facilitate registration. In products based on baculoviruses producers choose an isolate which consists of a mixture of genotypes in the production and in the final product. The reason for this is an increased virulence and to avoid development of resistance (M. Andermatt, personal communication).

2.4.4 Strain Selection in Entomopathogenic Nematodes

Among researchers on EPNs selection of virulent strains is considered imperative, because large variations exist between nematodes species and strains (Bedding et al., 1983; Grewal et al., 2005b; Van Tol and Raupp 2005). In the selection process of EPNs authors often compare a relatively low number of species and strains in bio-assays. This selection is usually directed towards infectivity, mortality, and the determination of a lethal dose, sometimes in combination with environmental factors such as soil temperature, soil texture or moisture content (Van Tol 1993; Westerman and van Zeeland 1989). These studies are often relatively short studies by academic scientists and are not a part of a development process as is needed to come to a commercial product.

To facilitate strain selection Bedding et al. (1983) proposed a predictive approach whereby not all available strains need to be tested. He eliminated on (1) the basis of knowledge of the biology of the nematode (temperature demand, behaviour, penetration mechanisms) and the host (movement, defence mechanism) and (2) the temperature requirement of the host's environment and a simple lab mortality test with 100 nematodes per pest insect; he called this a preliminary scan. The remaining species/strains need to be tested further in field trials. However, how to continue then was not described and they did not provide a hierarchy for the above-mentioned aspects or for testing. Nevertheless this approach is a process of strain elimination and seems valuable, quick and time- and money-saving.

Strain selection is complex since two organisms are involved, the nematode and the symbiotic bacterium. The nematode's features are crucial for successful finding and invasion of the host insect, the bacterium determines the pathogenicity towards the host. A workshop was held among EPN researchers on strain selection and a number of authors described their methodology in selecting an effective strain, including selective breeding through many generations (Simões et al., 1998). The workshop did not get any further then a case by case description. A general methodology for a selection process was not developed. Generally, three different kinds of foraging behaviour are recognized within EPNs: ambush foragers (*S. carpocapsae*), cruise foragers (*Heterorhabditis* spp.) and those with an intermediate foraging strategy (*S. feltiae*). This behaviour, however, does not give a high predictive value in screening as the "ambusher" *S. carpocapsae* showed to be the best candidate for control of a sedentary pest, the large pine weevil *Hylobius abietis* (Dillon et al., 2006). This illustrates the complexity of a screening programme. Researchers have not as yet developed a good system for selecting the optimal strain. Currently there is no consensus on a clear set of criteria for selecting strains for commercial development (R.-U. Ehlers, personal communication).

According to Georgis and Gaugler (1991) most work within the field of applied entomopathogenic nematology has been by trial and error and has not been based on any predictive screening. The authors analysed 380 treatments with nematodes and concluded that there was a high risk of failure when selection of the optimal strain under the prevailing soil conditions was not performed or not sufficiently performed. They recommended screening for better isolates (for instance low-temperature active strains), and also close examination of the environmental factors that hamper success of the nematode treatment. This last aspect should be tested not only in the lab, but under real field conditions. I can only stress this last conclusion. The problem herewith is that it is time-consuming and therefore expensive.

The number of nematode species and strains (isolates), however, is limited compared to those in fungi and bacteria. And only a limited number of species/strains can be produced by a commercial company due to practical and economic reasons. It is just too expensive to maintain and produce a large number of strains. As a group, EPN companies produce a limited number of species: five *Steinernema* and five *Heterorhabditis* species and two others (Grewal et al., 2005c: p. 480, Table 28.1) and only one strain per species. In a number of cases they even produce the same strain. These strains usually have been selected by academic researchers and then handed over to a particular company. Biosys (Georgis 2002) did not conduct strain selection by themselves, but acquired licenses to strains from several universities in the USA.

For many uses it is better to optimize the application method and conditions of the target site to favour the nematode's behaviour, then to search for a new, more effective strain. This search would cost time and money and may not provide a better candidate. Moreover, 10–20% increased mortality in small scale field trials by a better strain is often not recognized in commercial field applications due to the many variables. Therefore, it is more economical and practical to focus on the whole range of aspects around the application and to optimize those than to search for a new strain. It is recognized that soil moisture is the most relevant physical factor and this could be manipulated before and after treatment and contribute to increasing the effect of the nematodes and their persistence. Application technology needs more research and innovations in this area can stimulate a broader use of nematodes (Shapiro-Ilan et al., 2006).

In nematodes the most infective strain is not necessarily the strain which was found on its natural host, better control might be achieved by another species or strain (Bedding et al., 1983). This means that in screening all combinations need to be investigated: local strains for a native pest and for an introduced pest, as well as non-local strains for both pests. This has also been found for fungi in biocontrol (Gillespie and Moorhouse 1990), as well as in biocontrol of insects (van Lenteren and Tommasini 1999).

This poses the question of how valuable all the academic research on new strains is. It confuses the companies in further development of new products and new applications. Nevertheless, it is still essential to choose the correct species in accordance with the biology of the target insects. Strain selection is a crucial part of the EPN development, but its relevance should be evaluated against all other aspects of the development of an EPN product. It should not be overestimated in relation to the ecological and behavioural characteristics. The temperature range over which the nematode shows a good activity, however, is an key trait, as is tolerance to desiccation, which is needed in order to formulate the nematodes and have a good shelf-life. In the end, the most relevant points for a company are production efficiency in a liquid fermentation system, including formulation, and efficacy in the field, so strain selection should focus on these aspects.

3 Selection Criteria for a Microbial Pest Control Agent

3.1 The Most Important Selection Criteria

Which criteria are the most important in a screening study of a microbial pest control agent? In order to be able to answer this question, it is necessary to have in mind the foreseen application of the pathogen. Here we are selecting for a biocontrol agent that will be used in an inundative approach, an application, usually by spraying, that is meant to give a high mortality of the pest. It would be different when we want to apply a pathogen for long term control, as in classical biological control or for one seasonal application leading to an epizootic in the field. The decisive criteria for a pathogen that will be developed into a commercial bio-insecticide and that will be used for inundative application are:

(1) mortality:

- dose rate
- mode of action
- speed of kill
- host range
- sensitivity to abiotic factors
- persistence

(2) production efficiency

- yield
- costs

(3) safety:

- toxicological aspects
- non-target effects

It is hard to put these features in order of importance: many of them cannot be seen as a single factor, but must be seen as a set of parameters that together will ultimately lead to a go or a no-go decision. But mortality and production efficiency are clearly the decisive factors. With a low mortality or an almost impossible to produce organism further development into a commercial product is useless. With other factors a compromise can be accepted, for instance, a high mortality with a slow kill still can be accepted in many cases. Ecological factors, such as host plant effects, effects on natural enemies, etc., are often studied in great detail by researchers. These are, of course, necessary in order to better understand the relationships between the pathogen, the target and its ecology and could improve the efficacy if we are able to incorporate these ecological factors into the application of the pathogen. This is not always possible and therefore these studies are not pivotal for the decision whether or not to develop a strain or species into a biopesticide.

Further they are often complicated, long, and expensive and it is not always realistic for a company to investigate this in an early stage of the product development.

3.2 Mortality

It is evident that a high mortality is a highly preferred feature. The goal of the use of the pathogen is a quick and good control of the pest. The optimum would be that all stages of the pest quickly be killed, but this is rarely the case. Often only one stage of the cycle is sensitive to the pathogen which implies one or more treatments at some point in time to break through the cycle of the insect. For instance, the egg stage is often not susceptible to any of the pathogens. Bacteria, viruses and EPNs need to be ingested, or to get inside the insects, thus they are not effective on eggs. Only fungi can penetrate the cuticle into the insect, but in most cases not into the egg. Microsporidia can be transferred vertically within the egg, but usually do not give any mortality at this stage. The larval stages are often the most susceptible stage: through feeding they become infested with pathogens or in the case of fungi through penetration of the germ tube through the cuticle. Pupal stages are quite well protected in most cases, adults can be killed, but often they are also less susceptible or they do not pick up the pathogen (different feeding behaviour, living in another habitat, etc). Consequently, often the larval stage is the preferred stage that is being tested in bio-assays and field studies, and in many cases the larval stage is the eventual target stage for the biopesticide.

Mortality is caused by the pathogenicity and virulence of the pathogen, is dose-dependent, and dependent on the quality of the administered inoculum. In literature many different types of testing methods are described, often bio-assays, for the various pathogens and all kind of different target pests. For overviews and critical considerations regarding the use of bio-assays I refer to Robertson et al. (1995), Lacey (1997), Navon and Ascher (2000), and Hatting and Wraight (2007). It is beyond the scope of this study to analyse all the different types of methods, however, the goal of the testing should be clearly defined beforehand and an attempt is being made here. To do this one has to look at the final goal of the product: to control the pest effectively. In practice this means for a grower that a high percentage of the pest be killed within a short time after he has applied the product and at affordable costs. In other words the two crucial parameters are efficacy and economy. These parameters are the same for all pest control methods, including chemical pesticides. This is the point of departure in the screening programme. The isolate that gives the highest mortality, and preferably quickly, at the lowest possible dose (= costs) is the most preferred isolate for further development.

3.2.1 Dose Rate

Mortality is dose related. The susceptibility of an insect or mite is dependent of the dose of the pathogen. The dose-mortality relationship is usually the principal component in bio-assay testing. Virulence between species and strains is measured by LD 50, LC 50 and LT 50 values. The LC50 is the concentration of a given entomopathogen required to kill 50% of the insect population within a given time period. Similarly, the LD50 is the dose required to kill 50% of a population. Beside this value, the LT50, the time required to kill 50% of a population at a given dose or concentration of a pathogen, is indicative for the "speed of kill" of a pathogen and is valuable in the selection process. A dose-mortality curve should be made by testing a number of dosages and the slope of the curve gives useful information when comparing isolates. A flat curve indicates that the insect's susceptibility is less dependent on the dose, and that the pathogen strain is more virulent then a strain that gives a steeper sloped curve. The above is generally the case for all pathogens and is important in selecting the best strain. Obviously, the lower the dose, the better the chance that a product can become economically produced.

3.2.2 Mode of Action

Preferably the mode of action of the pathogen is known, but is not absolutely necessary to know it in detail for the development of a final product. Similarly, for many chemical pesticides the exact mode of action is not always precisely known. Understanding the mode of action of the pathogen can be helpful and is also needed for registration, although rather superficial descriptions have been accepted. Whether or not toxins are involved is more important to know, and if toxins are involved pesticide regulators want to know more before an approval for usage is given. More details on this will be discussed in Chapter 5.

In the case of pathogens the mode of action is often complex and several processes may be involved in the insect's death. For instance, an infection sequence of an entomopathogenic fungus may consist of the following steps: adherence to the insects' cuticle, host recognition and germination, appressorium formation, penetration of the cuticle by making a penetration peg and cuticle degrading enzymes, growth inside the insect and overcoming its immune system, possibly by release of toxins, and further development in the insect and finally killing the insect. This can be followed by sporulation outside the insect and spread of the spores for a new infestation cycle (Butt 2002).

In the commercial development of a bio-insecticide there are two reasons to have detailed information on the mode of action of the microorganism:

(1) the first is related to the virulence of the pathogen and whether virulence can be affected by the storage of the inoculum material (does the strain attenuate during storage?) and/or during the mass production (is virulence kept during production in artificial media?). Depending on the mass production system, that is whether a fungus is produced by liquid fermentation or by solid substrate fermentation, it may or may not produce a certain metabolite, and perhaps a toxin, which is part of in the mode of action process. In this case knowledge on the toxin and a method to measure its quantity in production will be useful;

(2) the second is the registration of the pathogen as an active ingredient for a biopesticide. When mortality is caused by a toxin or partly caused by a toxin, pesticide regulators want to have information on the toxin: which compound it is and in what quantity it is produced. Information will also be required on persistence on food and in the environment and on the risk for the applicator and workers. To generate knowledge on the toxin will be costly. If we can choose a strain that is efficacious without the production of a toxin, it is obvious that that is the preferred one.

3.2.3 Speed of Kill

The trait "speed of kill" is also useful. The following example illustrates this: when leaf-eating caterpillars are present in an ornamental crop economic damage is obviously increasing with time; a grower will choose a pesticide that can stop this. If he uses a Bt product, caterpillars will quickly stop feeding due to the early disease symptoms that paralyse its jaws, so the damage is not increasing. When a virus is used, the caterpillars will only die in 3–7 days, depending on the pest species and the larval stage. When a grower has the choice between a virus and a bacterium based insecticide, he will not choose a bio-insecticide based on a virus since caterpillars will only stop feeding after a number of days and then die. Differences in speed of kill are present between species of pathogens, but can also be present between strains within one species. Screening for the best available isolate should also take this intra-specific variation into account.

3.2.4 Host Range

Biocontrol scientists often search for a pathogen that is selective to a narrow host range and at the same time safe to non-target organisms. A more selective and specialized species or strain is often better adapted and as a consequence more virulent to its host(s). Selectivity must be known for both the registration process and for use in IPM systems with natural enemies and pollinators. The trade-off is that this allows only a small market for the biopesticide which may make it difficult to have an economic product at the end. Eventually, the market size determines whether a product will be successful, so selectivity is conflicting with market size. In the case a pest is serious enough and its market large enough to justify the investment in the development of a biopesticide, than we can allow selectivity to weigh more. But of course, it is clear that when a larger number of pest species can be controlled by the same product, the development costs of this product can be justified and the product will be more profitable (see Box 2.3).

In the case of fungi and nematodes selectivity needs to be considered carefully since they can have a broad host range, while baculoviruses and *Bacillus thuringiensis* are specific and non-target effects are less likely to occur. Selectivity in relation to non-target organisms and the registration will be discussed more extensively in Chapter 5.

Screening on selectivity should be focussing on two levels: (1) more in general to safety to humans, mammals and the environment and (2) more specific to effects on the pest species and other exposed organisms in the direct application area of the pathogen, such as target plants, natural enemies and pollinators. First of all, the desired pathogen should not have any harmful effects on humans, so broad opportunistic pathogen cannot be used. Secondly, it should be effective on a small group of insects with minimal effects on the non-target organisms occurring concurrently in the habitat of the area of application.

Box 2.3 The Importance of a Broad Host Range for the Market Size of a Microbial Insecticide

The product Vertalec has been developed for control of aphids, but in the field it only is effective against *Aphis gossypii*, the cotton aphid and on the peach aphid *Myzus persicae*, pests in cucumber and sweet pepper and in chrysanthemum greenhouse crops. Other aphids are less effectively controlled or not controlled at all; this is partly due to the limited host range of the Vertalec strain, partly due to the high humidity requirement of the strain. In a widely-grown crop as tomato it cannot be used effectively for aphid control.

The strain does not kill whitefly and therefore Mycotal was developed. Mycotal has a good effect on whitefly, both on *Trialeurodes vaporariorum* and *Bemisia tabaci* and its strain is less dependent on a high relative humidity. Mycotal can be used in cucumber, sweet pepper and tomato and in many ornamental crops. As a consequence the market for Vertalec is much smaller than for Mycotal. Vertalec is only registered in the UK, Switzerland, Norway, Finland and Japan. Mycotal is registered in The Netherlands, Denmark, Finland, France, Japan, Switzerland, Turkey, UK and approval is pending in Greece, Spain and Italy. Over the years 2003–2006 the sales of Mycotal were ten times higher than those of Vertalec. Due to its small market Koppert decided not to re-register Vertalec, which actually concerns its active ingredient *Lecanicillium longisporum*, strain Ve2. The high costs of updating the dossier to the current requirements and of the evaluation by the authorities do not justify to investment related to the expected future sales of the product Vertalec in the EU.

Mycotal also kills thrips (*Thrips tabaci, Frankliniella occidentalis*) and gives reasonable control, to about 60% after three consecutive applications (Ravensberg et al., 1990) in an IPM system. Some approvals allow the label to say "Mycotal has a side-effect on thrips" and this opens the possibility for more sales. In cucumber, whitefly and thrips almost always occur simultaneously and Mycotal will contribute to the control of both pests. So a broader host range leads to a more successful product.

3.2.5 Sensitivity to Abiotic Factors

Abiotic factors, such as temperature, relative humidity (RH), light (UV) and rain need to be considered in the screening process. For all pathogenic organisms, the temperature range of the pathogen and the optimal temperature in relation to the target pest's biology and occurrence are determinative. In greenhouse crops temperature can be regulated within limits, the lower temperature by heating, and to a lesser extent the higher temperature by cooling. Usually, pathogens survive low temperatures, although they may not be active then, while high temperatures can be lethal. In screening for species or strains, the optimal temperature for the pathogen for infectivity and growth needs to be considered, together with the prevailing temperatures for the pest species and the crop in seasons in which the pest is the most problematic in the crop. This means that a rather large range should be looked at and strains with a large temperature range would be more desirable. In fungi and EPNs this is more important than in viruses and bacteria. The activity of the first two organisms is strongly influenced by temperature: germination in fungi and searching behaviour and penetration in EPNs. Baculoviruses and Bt's are more passive pathogens, they need to be ingested by the insect and only become active once inside the insect.

Soil moisture is crucial for EPNs, without a water film they cannot move and find a host. This aspect is not considered important in strain selection since variability between strains is expected to be low and it has not been taken into account.

Relative humidity (RH) is a determining factor for fungi, but much less so for viruses and bacteria. In the case of screening strains of fungi, it certainly needs to be considered, as RH requirements may determine success or failure (see Box 2.3). Therefore, it is vital to screen fungal species/strains in a bio-assay system where the RH can be set. Many authors have tested infectivity of fungi at a 100% RH to find the most virulent strain. At lower RH conditions the selected strain may perform less well. Screening at 100% RH is useless and often gives the wrong outcome. A certain strain may be effective at 100% RH, but fail to give good results at RH conditions that prevail in the target crop. Screening should take place at the RH that is most prevalent in the foreseen crops.

Light and UV can have negative effects on the pathogen and its survival, especially in baculoviruses (Ignoffo 1992), but strain differences do not usually help in overcoming this problem. This is something to investigate with the product formulation (Chapter 3). Rainfall and washing-off of pathogen particles are generally irrelevant in greenhouses as overhead sprinkling is done in a few crops only. Formulation should take this aspect into account when necessary.

3.2.6 Persistence

The longer the inoculum survives on the leaves or in the soil the greater the chance that the pest comes into contact with the pathogen. When the pathogen is applied on the foliar parts of the plant contact between pest and pathogen may be immediate due to spraying, or the pest may contact or eat the pathogen later in time, even days or weeks later. The longer the inoculum is infective, the better the control, even though direct control is desired. In EPNs persistence in the soil can enhance the effects in outdoor applications. When used in greenhouses this is usually less relevant.

Persistence is also influenced by many other factors (ecological, host plant effects, target effects) which cannot be studied in the screening process. This requires fundamental and detailed research. It is interesting, but too expensive and too time-consuming. Screening for persistence is less crucial than for virulence, host range and abiotic factors. It can be improved by formulation techniques and then can be a helpful trait of the product, but it is not an essential one.

3.3 Production Efficiency

A prerequisite for a pathogen to be developed as a biopesticide is that it can be easily and economically mass-produced. This should be taken into account from the very beginning. In general, it is nowadays known whether a certain pathogen can be economically mass-produced or not. Bacteria usually are relatively easily to produce, certainly *Bacillus* spp. Fungi are more difficult, but the main ones used in biocontrol can be produced relatively well. Viruses still need to be produced on an insect host. EPNs can be produced in liquid fermentation or in vivo. More information on production of pathogens is given in Chapter 3.

New candidate strains must be checked for their ability to be mass-produced with a high yield of infective propagules. In this phase it can still be a global check; further optimization will take place once mass production is really investigated. For bacteria, this screening step should focus on the ability to produce spores and/or crystals in high numbers and in a short time. For fungi this should also focus on spore production, either in solid or in liquid fermentation. For viruses it is important to obtain an overall idea on the production on the host and of the host itself and whether this is the target pest or a related species that is easier to produce. Whether or not caterpillars show cannibalistic behaviour and consequently need to be produced separately or in groups will have an impact on production costs. EPNs can be produced in the target insects or alternative hosts such as *Galleria mellonella*, but scaling up will be expensive because of labour costs. Usually, EPNs are produced in liquid fermentation and this possibility needs to be investigated early in the process.

Differences between strains can be large. Strains that showed a high mortality of the pest in bio-assays but are difficult to produce should be discarded early in the selection process. A good example of this is given by Meekes (2001) in the screening of *Aschersonia* strains, where on the basis of global spore production strains were discarded in the first step of the screening process, even before assessing their virulence in a bio-assay.

3.4 Safety Profile

Two areas need to be considered for harmful effects, on humans and on the environment. Considering the human aspect, there are two reasons that ask for an early survey on the toxicology of the pathogens that are under research. The first is to make sure that persons who are doing the research with the pathogens are not exposed to harmful organisms and that standard laboratory safety precautions are sufficient to protect them. The second is the registration of the pathogen as an active ingredient of a biocontrol product. Products based on fungi, bacteria and baculoviruses need to be registered in the EU as pesticides according to the EU Directive EU 91/414 (EC 1991). A great number of questions have to be answered or studied to fulfil the requirements for the approval of the active ingredient. EPNs are exempt from registration in the EU, although some countries require a registration or an import permit (see Chapter 5).

In the case of fungi and bacteria the mode of action could be (partly) based on metabolites. This should be investigated in literature and/or by testing of a culture filtrate on various organisms for biological activity. In the case that mortality is caused by a toxin, further research should first focus on this in relation to the registration requirements and costs. It could become too expensive to generate all data related to the toxin. Consultation with the competent authorities is needed before continuation of the development the MPCA. If the pathogen is a potential human pathogen or closely related to one, consultation with the authorities should also follow before further time and money is invested. It could be useful to do an acute animal testing package first in order to be able to make the right decision. See Box 2.4 for an example on the possible harmful effects of a pathogen.

Considering negative environmental effects, several groups of organisms need to be looked at such as aquatic organisms (fish, daphnia's and algae), and soil organisms such as earthworms, soil micro-flora and non-target organisms in the crops, like natural enemies and pollinators. Literature and information on the mode of action, the host range and persistence can give indications as to whether or not there are any risks for these types of organisms. Further tests may be necessary when there are doubts on environmental effects.

Box 2.4 Possible Harmful Toxicological Effects of a Pathogen and the Consequences

In a research project carried out by Plant Research International, Wageningen and Koppert BV, several bacteria were investigated as possible biocontrol agents for control of thrips, mites and whitefly in greenhouse crops. The bacteria were found as endophytes and some showed promising effects on these pests. Several species and strains were further tested in bio-assays and one bacterium gave good results on thrips. Mortality was 100% in only a few days on thrips (*Frankliniella occidentalis*) and spider mites (*Tetranychus urticae*) in a broad range of relative humidity and temperature. Also production of this strain appeared to be easy.

Identification showed that it was a strain of *Serratia marcescens*, an opportunistic human pathogen. Although this pathogen was found as an endophyte, we contacted a toxicological consultant and the helpdesk of the Ctgb (Dutch Board for Authorization of Pesticides) for their opinion on this bacterium in regards to filing an application as a biopesticide. The Ctgb reported that it will not be possible to get approval for this pathogen because it is an opportunistic human pathogen and a producer of a range of toxins. Even without specific information on this strain and its mode of action, the conclusion was not to proceed with its development. Unfortunately other strains were not found with a similar virulence on these greenhouse pests and the project did not lead to the development of a new bio-insecticide due to their toxicological aspects (De Kogel and Ravensberg 2006).

4 Conclusions and Recommendations

4.1 *Considerations Leading to a "go/no go" – Decision*

In the development of a MPCA into a microbial insecticide the finding and selection of the isolate is a crucial component. Further research based on this isolate will follow on mass production, formulation, shelf-life, field trials, etc. The building of a dossier for registration will also be based on that isolate and this should start from the moment of the selection of the isolate. From this point on, costs and labour investments will be larger and more substantial. The decision whether to continue with the MPCA development, based on the initial findings, has a great impact. At this point a project can still be stopped without having lost a lot of time and money. If it is decided to go ahead, investments in time and money will become substantially larger. At the same time it is then implicitly decided (in most cases anyway) not to spend the budget and time on finding other solutions. This should be realized. Therefore a "go/no go"-decision at this stage is the most important decision taken in the development of a microbial pesticide. The generation of the information on which the decision is based needs to be done with great care and always with its purpose in mind. The information used to make the decision to develop an agent has to give good insight into the efficacy and the costs of the MPCP.

Efficacy is usually defined as high mortality, preferably with a fast "speed of kill", and comparable to mortality caused by other competing control methods, usually chemical pesticides. Costs relate to costs of production, or to costs of a unit of the final product, and to registration costs. The first are easier to estimate based on some initial production tests than the latter. Registration costs are often difficult to estimate and depend, among others, on the company's experience in this area. High costs of registration can be foreseen if metabolites need to be identified, studied and tested on animals. If many toxicological questions are expected, it is best to consult an expert in this area for an initial cost estimate of generating all necessary data for the registration dossier. The focus should be on human toxicology since authorities are mainly concerned on applicator and worker exposure (see Box 2.4). If the investment looks to be very high this could lead to the decision to stop working with this strain or even species.

I consider the initial information on efficacy, production costs and registration costs absolutely crucial in order to make the "go/no go"-decision on the continuation of the development of the biopesticide. Once the prospects of these aspects look positive, a company can decide whether to continue the development and to invest in the next research steps. Other aspects such as speed of kill, mode of action, host range and persistence influence the decision and depending on the case, they can be of greater or lesser importance. Whether a MPCA is affected by abiotic factors and how dependent its efficacy is on these abiotic factors is important, but not crucial.

In the literature few examples are found in which similar considerations are taken into account and where studies are done on the three main traits efficacy, production and registration costs. In the Lubilosa project it was recognized early in the process that the key factors for strain selection were good production features and good field performance (Prior 1992). Roberts et al. (1991), when describing the ideal microbial pesticide, list five traits which a product must meet to become successful: good efficacy, economical production and good shelf life, safety and a potential market. This is similar to my conclusion, that effectiveness, costs and the toxicological profile are the decisive factors. Academic researchers understand the requirements for a biopesticide. Efficacy is studied, but costs and safety, however, are often neglected in their research.

4.2 Importance of Parameters in the Selection Process

Above I have described how crucial the selection the MPCA is and what the main considerations should be. But which parameters should weigh the most for which pathogen? I will indicate the importance of the parameters for the different pathogens in Table 2.2 and explain briefly why that is. Three areas are considered, the first concerns efficacy, the second production efficiency and the third safety (Fig. 2.1).

The traits related to efficacy are:

- dose-mortality response; the dose is critical in fungi, EPNs and viruses since these organisms are difficult to produce and therefore expensive. Bacteria are relatively easy to produce and, thus, the dose is less critical;
- host range; in baculoviruses the host range is narrow, so apart from screening isolates on the target pest, investigations to find out what the host range is are not really useful. For the other pathogens, this is relevant and it could be useful to know other potential targets and markets;

- survival and persistence; survival of EPNs in soil is usually limited and therefore can play a lesser role in screening tests. It is laborious and expensive to test. In bacteria, baculoviruses and fungi, and at foliar applications where 100% targeting is usually impossible, persistence of infective propagules adds to the total mortality. A long persistence increases the chance of contact or ingestion by a target insect. Few studies, however, have reported on the importance of this fact and personally I do not consider it a key trait. Persistence usually ranges between a few days and a week, and is also dependent on the formulation. In glasshouses, rain and UV, usually factors that reduce persistence, are not relevant;
- abiotic factors; mainly temperature and relative humidity (RH) can limit the activity of the MPCAs. Bacteria and baculoviruses have to be ingested and only become active inside the insect. In this case, temperature affects the insect host activity more than the MPCA and selection for a certain temperature range is accordingly less relevant. For EPNs, however, it is very important to know their temperature range. If it is too cold, the nematode is not active, does not search or cannot penetrate the host, at high temperature the same happens or the EPNs even die. In fungi, the RH is the main abiotic factor taken into account as fungi need a high RH for germination and this should be studied in the selection assays.

Production-related considerations are:

- for use of a pathogen as a biopesticide it is necessary to have infective propagules that can be applied to a crop. These are usually spores for bacteria and fungi (bacterial cells and mycelial parts are also possible), occlusion bodies for baculoviruses and infective juveniles or dauer larvae (L3) for EPNs. Spore-producing bacteria are generally easy to produce and formulate, while fungi are more difficult and more expensive to produce. Baculoviruses are produced in vivo and production is both laborious and expensive. EPNs can be produced in vivo and in vitro; production is both complicated and expensive. For all pathogens it is important to investigate how easy or difficult it is to produce them. This is especially true for the production of infective propagules. In addition to ease of production, a long shelf-life is highly desirable. For example, whether a bacterium can make spores or not influences the selection because spores are stable and can be stored for a long time.

Safety-related considerations are:

- toxicological characteristics; the different types of pathogens vary in potential for harmful effects towards humans and the environment. In registration, safety to the applicator and worker in the crop are the main concern. Regarding bacteria this is critical, because many of them make metabolites and many of the ones used for biopesticides are closely related to harmful bacteria such as *Bacillus* and *Serratia* species. B.t is closely related to *B. cereus*, a food poisoning bacterium for humans. For fungi this is generally less important, although many of them make metabolites. Baculoviruses and EPNs, including their symbiotic bacteria, are specific, and safety to humans needs little attention in the selection process;

- environmental effects; the potential of harmful effects of the pathogens on non-target organisms are usually relatively small. Bt's and baculoviruses have a narrow host range and are quite specific. Fungi and EPNs can have a broader host range and that needs to be studied further. Effects on natural enemies need to be known too, not only for the registration dossier, but also for use in IPM programmes. Depending on the organism and its mode of action (metabolites) there may be concern for aquatic organisms, although serious effects are not known with the current approved biopesticides.

In Table 2.2 I have estimated the importance of the described traits in the selection process. A trait that is critical for the success of the pathogen needs to be studied with great care, such as relative humidity requirements (abiotic factors) for a fungus.

Table 2.2 Importance of traits in screening MPCAs for biopesticide development

Trait	Mortality				Production	Toxicology	
MPCA	Dose	Host range	Survival	Abiotic factors	Propagules	Effects on humans	Non-target effects
Bacteria	++	++	+/–	+/–	++	+++	+
Fungi	+++	+++	+	+++	+++	++	++
Viruses	+++	+	+	+	+++	–	–
EPNs	+++	+++	++	+++	+++	–/+	+

–: Unimportant.
+: Slightly important.
++: Moderately important.
+++: Very important.

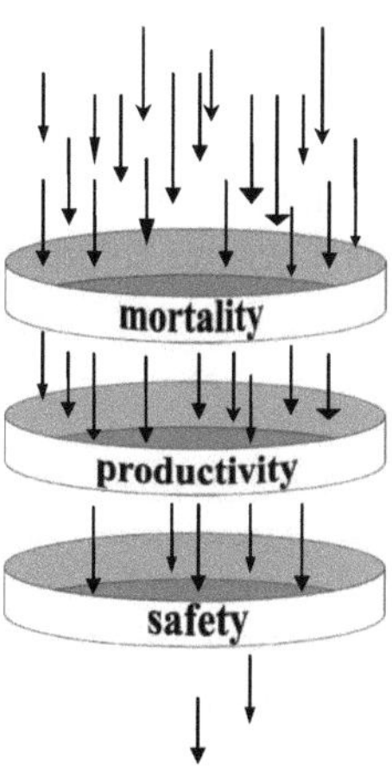

Fig. 2.1 Schematic, simplified process of sieving strains on criteria resulting in a small number of strains for further research and development

A similar table with characteristics of the different types of pathogens is given by Federici (1999a: p. 543, Table 7), but I added the importance of the various traits in a selection process for the development of a biopesticide. I included the importance of toxicological aspects.

4.3 Recommendations for Bio-Assays

Bio-assays are too often done under optimal conditions for the pathogen, such as high dosages, high relative humidity, absence of UV light, optimal administering techniques (immersion, diet incorporation, direct deposition, forced feeding) to find the optimal strain. Also, target insects or mites are often kept under unnatural and stressful conditions which could lead to susceptible animals. This kind of bio-assay set-up is not discriminating enough and could lead to false positives. Bio-assays should therefore be done under conditions that mimic the natural situation as closely as possible, only then useful data can be generated. With abiotic factors this can be easily done, more care needs to be given to biotic conditions.

Standardization of the bio-assay is imperative in order to obtain reliable results. Natural variation in bio-assays is a complicating factor and can only be tackled by repeated testing with sufficient replicates and a tightly standardized assay (Robertson et al., 1995; Hatting and Wraight 2007). Attention should be paid to possible attenuation of the pathogen strain (depending on culturing method, storage, etc.), the culture of the target organism, even to the host plant on which tests are done. Even in well-conditioned climatic rooms seasonal effects are present and susceptibility and natural mortality may fluctuate throughout the seasons. This should be realized at all times when not only bio-assays but also semi-field tests are conducted.

4.4 Final Conclusions and Recommendations

In Table 2.3 I summarize the most important steps in the selection process of a suitable candidate isolate for further development into a commercial product. Objective 4 and 5 will be treated in the following chapters. All these steps should be seen as a general approach and emphasis on selection criteria may shift a little depending on the organism and its specific characteristics and the pest problem that needs to be solved. Commercial considerations should always play a key role from the beginning.

Finally, after the selection and evaluation phase a "go/no go"-decision has to be taken based on the gathered information as described above, on efficacy, production efficiency and toxicological traits that were generated as much as possible under conditions that mimic the conditions in later phases in the development. It should be taken into account that a decision is always a compromise between finding the best strain and the best solution for the pest problem. Do not search endlessly to

Table 2.3 Recommended steps in the selection of an isolate for a microbial pest control product

Objective	Procedure
1. Collection of isolates	From culture collections, field collections, baits, crops, etc; make pure culture and test on Koch's postulates
2. Initial laboratory screen to assess efficacy	Test a large number of isolates in a bio-assay against the target species at conditions close to the application conditions. Selection is a process of strain elimination
3. Initial laboratory assessment on production efficiency	Determine growth parameters and the production efficiency of infective propagules in laboratory scale process close to the foreseen mass production conditions
4. Assessment of mode of action and toxicological properties	Determine whether mode of action is (partly) based on metabolites and study toxicological properties of the microorganism and/or its metabolites
5. Evaluation of efficacy of selected isolates in small glasshouse trials	Determine optimal dose (LD 90), application method on target species and target host plant(s) at conditions as close as possible to the grower's conditions
6. Evaluation of efficacy under commercial conditions	Determine efficacy of mass-produced propagules and with a formulated product at grower's field

Modified after Kerry (2001: p. 162, Table 5.2)

find the perfect isolate, but accept a good one or perhaps two or three for further research, and optimize one in formulation, application, etc. as far as possible. The best strain is the one that gives the best compromise between economics and good efficacy. In a commercial setting, economical considerations will be the leading ones in the decision. Consider criteria and results on a case by case study and relate constantly the economical aspects and aspects of the final field of use. Once the "go decision" has been taken for the selected strain or strains, further investments and research steps in the development process have to be taken and many of them will be addressed in the following chapters.

References

Andersch, W., 1992. Production of fungi as crop protection agents. *Pflanzenschutz-Nachr. Bayer* 45(1): 129–142.

Bedding, R.A., A.S. Molyneux, and R.J. Akhurst, 1983. *Heterorhabditis* spp, *Neoplectana* spp. and *Steinernema kraussei*: interspecific and intraspecific differences in infectivity for insects. *Exp. Parasitol.* 55: 249–257.

Brownbridge, M., 1995. Prospects for mycopathogens in thrips management. In: B.L. Parker, M. Skinner, T. Lewis (eds), *Thrips biology and management*, NATO ASI Series. Life Sciences, Vol. 276. Plenum Press, New York, NY. pp. 281–295.

Burge, M.N., 1988. *Fungi in biological control systems*. Manchester University Press, Manchester.

Burges, H.D., 1981a. Strategy for the microbial control of pests in 1980 and beyond. In: H.D. Burges (ed), *Microbial control of pests and plant diseases 1970–1980*. Academic Press, London. pp. 797–836.

Burges, H.D., 1981b. *Microbial control of pests and plant diseases 1970–1980*. Academic Press, London.

Butt, T.M., 2002. Use of entomogenous fungi for the control of insect pests. In: F. Kempken (ed), *The mycota XI, agricultural applications*. Springer, Berlin. pp. 111–134.
Butt, T.M., C. Jackson, and N. Magan, 2001. *Fungi as biocontrol agents: progress, problems and potential*. CAB International, Wallingford.
Charles, J.-F., A. Delécluse, and C. Nielsen-LeRoux, 2000. *Entomopathogenic bacteria: from laboratory to field application*. Kluwer, Dordrecht.
De Kogel, W.J. and W.J. Ravensberg, 2006. Bacterie losgelaten op plagen. *Groeten en Fruit* 35: 13.
Dent, D.R., 1998. *Removing the barriers to commercialisation: summary of the process and key issues for a mycoinsecticide for locust and grasshopper control*. LUBILOSA brochure. CAB International, Ascot.
Dillon, A.B., D. Ward, M.J. Downes, and C.T. Griffin, 2006. Suppression of the large pine weevil *Hylobius abietis* (L.) (Coleoptera: Curculionidae) in pine stumps by entomopathogenic nematodes with different foraging strategies. *Biol. Control* 38: 217–226.
EC, 1991. Council Directive 91/414/EEC of 15 July 1991 concerning the placing of plant protection products on the market. *Official Journal of the European Union L 230*, 19.08.1991: 1–32.
Enkegaard, A. and H.F. Brødsgaard, 2006. Biocontrol in protected crops: is lack of biodiversity a limiting factor? In: J. Eilenberg, H.M.T. Hokkanen (eds), *An ecological and societal approach to biological control*. Springer, Dordrecht. pp. 91–122.
Entwistle, P.F., J.S. Cory, M.J. Bailey, and S. Higgs, 1993. *Bacillus thuringiensis, an environmental biopesticide: theory and practice*. Wiley, New York, NY.
Evans, H.F. (ed), 1997. Microbial insecticides: novelty or necessity? *BCPC Symposium Proceedings*, No. 68. Coventry, April 16–18, 1997. BCPC, Farnham.
Evans, J., 2006. R & D transformed. *Agrow Magazine*, Issue 5 – July 2006: 10–12.
Evans, C.E. and N.L. Hywel-Jones, 1997. Entomopathogenic fungi. In: Y. Ben-Dov, C.J. Hodgon (eds), *World crop pests. Soft scale insects. Their biology, natural enemies and control*, Vol. 7. B. Elsevier, Amsterdam. pp. 3–26.
Federici, B.A., 1999a. A perspective on pathogens as biological control agents for insect pests. In: T.S. Bellows, T.W. Fisher (eds), *Handbook of biological control*. Academic Press, San Diego, CA. pp. 517–548.
Federici, B.A., 1999b. *Bacillus thuringiensis* in biological control. In: T.S. Bellows, T.W. Fisher (eds), *Handbook of biological control*. Academic Press, San Diego, CA. pp. 575–593.
Ferron, P., 1978. Biological control of insect pests by entomogenous fungi. *Annu. Rev. Entomol.* 23: 409–442.
Ferron, P., J. Fargues, and G. Riba, 1991. Fungi as microbial insecticides. In: D.K. Arora, L. Ajello, K.G. Mukerji (eds), *Handbook applied mycology*. Marcel Dekker, New York, NY. pp. 665–706.
Fransen, J.J., 1987. *Aschersonia aleyrodes* as a microbial control agent of greenhouse whitefly. Ph.D. Thesis, Department of Entomology, Wageningen University. 167 pp.
Fransen, J.J., 1990. Natural enemies of whiteflies: fungi. In: D. Gerling (ed), *Whiteflies: their bionomics, pest status and management*. Intercept, Andover. pp. 187–210.
Gaugler, R., 2002. *Entomopathogenic nematology*. CAB International, Wallingford.
Gaugler, R. and H.K. Kaya, 1990. *Entomopathogenic nematodes in biological control*. CRC Press, Boca Raton, FL.
Georgis, R., 2002. The biosys experiment: an insider's perspective. In: R. Gaugler (ed), *Entomopathogenic nematology*. CAB International, Wallingford. pp. 357–372.
Georgis, R. and R. Gaugler, 1991. Predictability in biological control using entomopathogenic nematodes. *J. Econ. Entomol.* 84(3): 713–720.
Gillespie, A.T. and E.R. Moorhouse, 1990. The use of fungi to control pests of agricultural and horticultural importance. In: J.M. Whipps, R.D. Lumsden (eds), *Biotechnology of fungi for improving plant growth. British Mycological Society Symposium No. 16*. Cambridge University Press, Cambridge. pp. 55–84.
Glare, T.R. and M. O'Callaghan, 2000. *Bacillus thuringiensis: biology, ecology and safety*. Wiley, Chichester.

Grewal, P.S., R.-U. Ehlers, and D.I. Shapiro-Ilan, 2005a. *Nematodes as biocontrol agents*. CAB International, Wallingford.

Grewal, P.S., R.-U. Ehlers, and D.I. Shapiro-Ilan, 2005c. Critical issues and research needs for expanding the use of nematodes in biocontrol. In: P.S. Grewal, R.-U. Ehlers, D.I. Shapiro-Ilan (eds), *Nematodes as biocontrol agents*. CAB International, Wallingford. pp. 479–489.

Grewal, P.S., A.M. Koppenhöfer, and H.Y. Choo, 2005b. Lawn, turfgrass and pasture applications. In: P.S. Grewal, R.-U. Ehlers, D.I. Shapiro-Ilan (eds), *Nematodes as biocontrol agents*. CAB International, Wallingford. pp. 115–146.

Hall, R.A., 1981. The fungus *Verticillium lecanii* as a microbial insecticide against aphids and scales. In: H.D. Burges (ed), *Microbial control of pests and plant diseases 1970–1980*. Academic Press, London. pp. 483–498.

Hall, F.R. and J.J. Menn, 1999. *Biopesticides: use and delivery*. Humana Press, Totowa, NJ.

Hatting, J.L. and S.P. Wraight, 2007. Optimising bioassay precision, with special reference to the Aphididae and Aleyrodidae. In: S. Ekesi, N.K. Maniania (eds), *Use of fungi in biological pest management*. Research Signpost, Kerala. pp. 197–237.

Ignoffo, C.M., 1992. Environmental factors affecting persistence of entomopathogens. *Florida Entomol.* 75(4): 516–525.

Inglis, G.D., M.S. Goettel, T.M. Butt, and H. Strasser, 2001. Use of hyphomycetous fungi for managing insect pests. In: T.M. Butt, C. Jackson, N. Magan (eds), *Fungi as biocontrol agents: progress, problems and potential*. CAB International, Wallingford. pp. 23–69.

Kerry, B.R., 2001. Exploitation of the nematophagous fungal *Verticillium chlamydosporium* Goddard for the biological control of root-knot nematodes (*Meloidogyne* spp.). In: T.M. Butt, C. Jackson, N. Magan (eds), *Fungi as biocontrol agents: progress, problems and potential*. CAB International, Wallingford. pp. 155–167.

Kirk, A.A. and L.A. Lacey, 1996. A systematic approach to foreign exploration for natural enemies of *Bemisia* and some current results. In: D. Gerling, R.T. Mayer (eds), *Bemisia 1995. Taxonomy, biology, damage, control and management*. Intercept, Andover. pp. 531–536.

Lacey, L.A., 1997. *Manual of techniques in insect pathology, biological techniques*. Academic Press, San Diego, CA.

Lacey, L.A., J.J. Fransen, and R. Carruthers, 1996. Global distrubution of naturally occurring fungi of *Bemisia*, their biologies and use as biological control agents. In: D. Gerling, R.T. Mayer (eds), *Bemisia: 1995. Taxonomy, biology, damage, control and management*. Intercept, Andover. pp. 401–433.

Lacey, L.A. and M.S. Goettel, 1996. Current developments in microbial control of insect pests and prospects for the early 21st century. *Entomophaga* 40(1): 3–27.

Latgé, J.P. and B. Papierok, 1988. Aphid pathogens. In: A.K. Minks, P. Harrewijn (eds), *World crop pests. Aphids. Their biology, natural enemies and control*, Vol. 2. B. Elsevier, Amsterdam. pp. 323–336.

Lisansky, S.G., 1985. Production and commercialization of pathogens. In: N.W. Hussey, N. Scopes (eds), *Biological pest control. The glasshouse experience*. Blanford Press, Poole. pp. 210–218.

Lisansky, S.G. and R.A. Hall, 1983. Fungal control of insects. In: J.E. Smits, D.R. Berry, B. Kristiansen (eds), *The filamentous fungi, Vol. 4, Fungal technology*. Edward Arnold, London. pp. 327–345.

Liu, H., M. Skinner, B.L. Parker, and M. Brownbridge, 2002. Pathogenicity of *Beauveria bassiana, Metarhizium anisopliae* (Deuteromycotina: Hyhomycetes), and other entomopathogenic fungi against *Lygus lineolaris* (Hemiptera, Miridae). *J. Econ.Entomol.* 95: 675–681.

Lomer, C., 1999. LUBILOSA Phase 3 Final Report, March 1999. http://www.lubilosa.org.

Marrone, P.G., 1999. Microbial pesticides and natural products as alternatives. *Outlook Agric.* 28(3): 149–154.

Mason, J.M. and K.M. Heinz, 1999. Potential for the biological control of *Frankliniella occidentalis* (Pergande) with a nematode *Thripinema nicklewoodi* (Siddiqi). *IOBC/WPRS Bull.* 22(1): 173–176.

McCoy, C.W., 1996. Pathogens of eriophyoid mites. In: E.E. Lindquist, M.W. Sabelis, J. Bruin (eds), *World crop pests. Eriophyoid mites. Their biology, natural enemies and control*, Vol. 6. Elsevier, Amsterdam. pp. 481–490.

McCoy, C.W., R.A. Samsom, and D.G. Boucias, 1988. Entomogenous fungi. In: C.M. Ignoffo (ed), *CRC handbook of natural pesticides, Volume 5: Microbial insecticides, Part A, Entomogenous protozoa and fungi*. CRC Press, Boca Raton, FL. pp. 151–236.

Meekes, E.T.M., 2001. Entomopathogenic fungi against whiteflies: tritrophic interactions between *Aschersonia* species, *Trialeurodes vaporariorum* and *Bemisia argentifolii*, and glasshouse crops. Ph.D. Thesis, Department of Entomology, Wageningen University. 181 pp.

Miller, L.K., 1997. *The baculoviruses*. Plenum Press, New York, NY.

Milner, R.J., 1992. Selection and characterization of strains of *Metarhizium anisopliae* for control of soil insects in Australia. In: C.J. Lomer, C. Prior (eds), *Biological control of locusts and grasshoppers*. CAB International, Wallingford. pp. 200–207.

Moscardi, F., 1999. Assessment of the application of baculoviruses for control of Lepidoptera. *Ann. Rev. Entomol.* 44: 257–289.

Murillo, R., S. Elvira, D. Munoz, T. Williams, and P. Caballero, 2006. Genetic and phenotypic variability in *Spodoptera exigua* nucleopolyhedrosis virus from greenhouse soils in southern Spain. *Biol. Control* 38(2): 157–165.

Navon, A., 1993. Control of lepidopteran pests with *Bacillus thuringiensis*. In: P.F. Entwistle, J.S. Corey, M.J. Bailey, S. Higgs (eds), *Bacillus thuringiensis, an environmental biopesticide: theory and practice*. Wiley, New York, NY. pp. 125–146.

Navon, A. and K.R.S. Ascher, 2000. *Bio-assays of entomopathogenic microbes and nematodes*. CABI International, Wallingford.

Posada, F.J. and F.E. Vega, 2005. A new method to evaluate the biocontrol potential of single spore isolates of fungal entomopathogens. *J. Insect Sci.* 5(37): 1–10.

Prior, C., 1992. Discovery and characterization of fungal pathogens for locust and grasshopper control. In: C.J. Lomer, C. Prior (eds), *Biological control of locusts and grasshoppers*. CAB International, Wallingford. pp. 159–180.

Ravensberg, W.J., M. Malais, and D.A. van der Schaaf, 1990. Applications of *Verticillium lecanii* in tomatoes and cucumbers to control whitefly and thrips. *IOBC/WPRS Bull.* 18(5): 173–178.

Roberts, D.W., J.R. Fuxa, R. Gaugler, M. Goettel, R. Jaques, and J. Maddox, 1991. Use of insect pathogens. In: D. Pimentel, A.A. Hanson (eds), *CRC handbook of pest management in agriculture*, Vol. II. CRC Press, Boca Raton, FL. pp. 243–278.

Robertson, J.L., H.K. Preisler, S.S. Ng, L.A. Hickle, and W.D. Gelernter, 1995. Natural variation: a complicating factor in bioassays with chemical and microbial pesticides. *J. Econ. Entomol.* 88(1): 1–10.

Shapiro, M. and C.M. Ignoffo, 1970. Nucleopolyhedrosis of *Heliothis*: activity of isolates from *Heliothis zea*. *J. Invertebr. Path.* 16: 107–111.

Shapiro-Ilan, D.I., D.H. Gouge, S.I. Piggott, and J. Patterson Fife, 2006. Application technology and environmental considerations for use of entomopathogenic nematodes in biological control. *Biol. Control* 38: 124–133.

Shuler, M.L., H.A. Wood, R.R. Granados, and D.A. Hammer, 1995. *Baculovirus expression systems and biopesticides*. Wiley-Liss, New York, NY.

Simões, N., N. Boemare, and R.-U. Ehlers (eds), 1998. Proceedings of the COST 819 workshop on entomopathogenic nematodes. *Pathogenicity of entomopathogenic nematodes versus insect defence mechanisms: impact on selection of virulent strains*. EUR 17776-COST 819, European Community Press, Luxembourg.

Skovmand, O., I. Thiéry, and G. Benzon, 2000. Is *Bacillus thuringiensis* standardisation still possible? In: J.-F. Charles, A. Delécluse, C. Nielsen-LeRoux (eds), *Entomopathogenic bacteria: from laboratory to field application*. Kluwer, Dordrecht. pp. 275–296.

Smits, P.H. and J.M. Vlak, 1988. Selection of nuclearpolyhedrosis viruses for control of *Spodoptera exigua* [Lep.: Noctuidae]. *Entomophaga* 33(3): 299–308.

Stenzel, K., 2004. From genes to compound discovery: unique research platform combining innovative screening technologies. *Pflanzenschutz-Nachr. Bayer* 57(1): 35–45.

van der Geest, L.P.S., 1985. Pathogens of spider mites. In: W. Helle, M.W. Sabelis (eds), *World crop pests. Spider mites. Their biology, natural enemies and control*, Vol. 1B. Elsevier, Amsterdam. pp. 247–258.

van der Pas, R.K., W.J. Ravensberg, A.C. den Braver, A.C. van Buijsen, and M. Malais, 1996. A comparison between Mycotal (*Verticillium lecanii*) and *Aschersonia aleyrodis* for the control of whitefly. *IOBC/WPRS Bull.* 19(9): 200–207.

van Lenteren, J.C. and G. Manzaroli, 1999. Evaluation and use of predators and parasitoids for biological control of pests in greenhouses. In: R. Albajes, M.L. Gullino, J.C. van Lenteren, Y. Elad (eds), *Integrated pest and disease management in greenhouse crops*, Kluwer, Dordrecht. pp. 183–201.

van Lenteren, J.C. and M.G. Tommasini, 1999. Mass production, storage, shipment and quality control of natural enemies. In: R. Albajes, M.L. Gullino, J.C. van Lenteren, Y. Elad (eds), *Integrated pest and disease management in greenhouse crops*. Kluwer, Dordrecht. pp. 276–294.

van Tol, R.W.H.M., 1993. Control of the black vine weevil (*Otiorhynchus sulcatus*) with different isolates of *Heterorhabditis* sp. and *Metarhizium anisopliae* in nursery stock. *Proc. Exp. Appl. Entomol.* 4: 181–186.

van Tol, R.W.H.M. and M.J. Raupp, 2005. Nursery and tree application. In: P.S. Grewal, R.-U. Ehlers, D.I. Shapiro-Ilan (eds), *Nematodes as biocontrol agents*. CAB International, Wallingford. pp. 167–190.

Vurro, M., J. Gressel, T.M. Butt, G. Barman, A. Pilgeram, R.J. St. Leger, and D.L. Nuss, 2001. *Enhancing biocontrol agents and handling risks*. IOMS Press, Amsterdam.

Wardlow, L.R., S. Piggott, and R. Goldsworthy, 2001. Foliar applications of *Steinernema feltiae* for the control of flower thrips. *Med. Fac. Landbouww. Rijksuniv. Gent.* 66(2a): 285–291.

Westerman, P.R. and M.G. van Zeeland, 1989. Comparison of *Heterorhabditis* isolates for control of *Otiorhynchus sulcatus* at low temperatures. *Med. Fac. Landbouww. Rijksuniv. Gent.* 54(3b): 1115–1123.

Williams, E.C. and K.F.A. Walters, 2000. Foliar application of the entomopathogenic nematode *Steinernema feltiae* against leafminers on vegetables. *Biocont. Sci. Technol.* 10: 61–70.

Zare, R. and W. Gams, 2001. A revision of *Verticillium* section *Prostata*. IV. The genera *Lecanicillium* and *Simplicillium* gen.nov. *Nova Hedwigia* 73(1–2): 1–50.

Chapter 3
Mass Production and Product Development of a Microbial Pest Control Agent

Abstract The feasibility of economic mass production of the selected strain and the development of a stable product are key factors to a successful microbial pest control product. Two phases are distinguished, the development of the production process, and the development of the product, including formulation, packaging and field testing. The critical technical and economic factors are identified and evaluated for the four types of pathogens. Various production systems are reviewed. Bacteria, fungi and entomopathogenic nematodes are generally produced in vitro. Baculoviruses must be produced in vivo. Advantages and disadvantages in terms of costs, manageability and versatility are provided for each production system. The development of a stable product that is able to deliver effective pest control requires a formulation. The four main objectives in formulation are: to stabilize the propagules; to make a user-friendly product; to protect the propagule, once applied; and to minimize risks of exposure to the applicator. Formulation considerations are presented per formulation function as well as per type of pathogen. Field testing is a key phase in the product development which links all steps in the developmental process. It provides information on the efficacy of the selected strain, on the quality of the produced propagules, on the formulation, on the optimal application strategy, on efficacy that is necessary for registration, on compatibility, on the implementation of the product in an IPM system, and on the marketability of the final product. The method of field testing is crucial and should reflect "real world" conditions. A cost price model for biopesticides is provided with cost factors involved from production to product, and from product to market. The model provides a perspective on the makeup of the end-user's price and all the costs that must be considered to achieve a profitable business. In the conclusion, production and formulation considerations and recommendations for a economically feasible production are presented. Economy of scale, full use of the production capacity, and capacity planning are essential factors to keep the costs low.

W.J. Ravensberg, *A Roadmap to the Successful Development and Commercialization of Microbial Pest Control Products for Control of Arthropods*, Progress in Biological Control 10, DOI 10.1007/978-94-007-0437-4_3, © Springer Science+Business Media B.V. 2011

1 Introduction

A key factor in the success of a MPCA is cost-efficient production and development of a product that will be stable and able to deliver effective pest control. In the phase of mass production and product development, decisions will have to be taken on the choice of production system, on downstream and formulation equipment, often even on building a production plant with the entire infrastructure. A company needs to make large investments, although the start can be done on a pilot-plant scale, but eventually it will be followed by a full scale commercial production facility. In the case that production equipment is already available, this can be used first, before new equipment is bought. Nevertheless, the research and development of the production system and of the final product is the main process in terms of money and budget and needs the full commitment of the company. This also underlines the importance of the initial phase described in Chapter 2, i.e. the definition of the problem and the selection of the potential microbial control agent as the solution for the pest. After the selection of a microbial agent, two phases can be distinguished in the development of a microbial control product:

(1) the development of the production process: a cost-efficient and reliable method of production must be developed which gives predictable yields of infective propagules (spores, bacteria, etc.) in a large quantity and of a high quality within a given period of time. This includes the choice of the equipment to be used for production, for harvesting and processing of the propagules (drying, centrifuging, sieving, etc.): the downstream process. The process further includes development of a method for storage of a stable inoculum (known as the stock culture, or "mother" culture), preparation of inoculum for subsequent steps in the production process (from small bioreactors to large ones), medium development, research on determinative process parameters and how to adjust these parameters during the production process, process quality control, harvesting and storage methods including methods for stabilizing the yield into a "technical product". The whole process must be focussed on obtaining the optimal yield quantitatively and qualitatively against the lowest possible costs;
(2) the development of the product: a final product must be developed which is stable, has a long shelf-life, and which will give the required control results when applied in the field with an appropriate delivery system at the lowest effective dose. The formulation must be developed taking into account the specific purpose of the product and the delivery system. The most efficient dose must be established in semi-field and field tests under commercial conditions. Product specifications and quality control methods need to be developed to safeguard this. The final formulation needs to be packaged appropriately.

Both phases cannot be seen as independent processes, they are complex and interconnected. The ultimate goal is to develop an affordable product that is user-friendly and will consistently produce good results. During the development of the product, registration aspects also need to be kept in mind. Registration authorities require

Table 3.1 Steps in the development of the production process and the product development of a microbial control agent

Production process	Product development
• Production method and equipment	• Formulation
• Downstream method and equipment	• Dose
• Medium development	• Application method and delivery system
• Inoculum quality	• Packaging
• Process parameters control	• Product specifications
• Process quality control	• Product quality control
• Production costs	• Product cost price

more or less the same as the producing company and the end-user: a stable and safe product that gives a certain predictable efficacy when used according to label recommendations. It is beyond this book to describe in detail mass-production systems, production manuals, formulation methods and formulation recipes. Many researchers from academia and the biocontrol industry have published details of production methods and formulation of MPCAs (Ignoffo 1973; Bedding 1981, 1984; Smits 1987; Bartlett and Jaronski 1988; Jackson et al. 1997; Burges 1998; Couch 2000; Wraight et al. 2001; Ehlers and Shapiro-Ilan 2005).

In this chapter I will identify the critical aspects of production and formulation that need to be considered in the developmental process of a MPCP (Table 3.1). These factors are of a technical and economic nature, and are vital to the success of a company. Recommendations on how to consider these factors will be given, because few papers have addressed this topic in this way. Notable exceptions are Lisansky (1985) and Jackson et al. (1992). Lisansky briefly described the whole process that leads to commercialization of a pathogen. He highlighted the decisive aspects of the process and product development. His paper is predominantly based on his work with entomopathogenic fungi. It is a complete listing that is still appropriate and useful more then 20 years later for all types of pathogens. Jackson et al. (1992) briefly described the development of a bacterium for grub control and concluded that the process is not a neat, stepwise linear programme, rather a step by step spiralling process. I am sure that many researchers and companies have similar experiences, and by identifying the conclusive factors in the production and product development I will try to give guidelines that will result in a more predictable pathway and outcome.

2 Process Development

2.1 Mass Production and Downstream Processing

The production and the downstream method should deliver a yield of infective propagules, called the technical product, which can be further formulated into a final product. The production method is the phase in which the pathogens are mass-produced, and includes the bioreactor type and the medium. Once infective propagules are produced, these will be collected by the downstream process: any large-scale processing of the ferment or solid state matrix to recover and purify the propagules until making of the technical product. "The technical" usually consists of pure propagules, sometimes with medium remnants and/or with produced metabolites. It may also include components that are added during the downstream process as a protectant. This intermediate product will be formulated further with co-formulants to make the final product.

Depending on the pathogen, processes and production systems or bioreactors may vary greatly. A company can choose a highly sophisticated technical approach or a low-budget approach with simple production systems: the production of a fungus can, for example, take place in a computer-steered bioreactor or in autoclavable plastic bags. These approaches are also referred to as the capital-intensive model and the labour-intensive model (Swanson 1997). Often investment costs versus labour costs will determine this choice, but the scale of the production and the eventual market price also need to be considered. Many production systems have been studied and developed for the pathogens discussed in this study and many researchers have reported their findings in handbooks and papers. The research generally concerns laboratory scale production, but these studies often provide valuable information that can be used in developing a mass production process. Scaling up is *the* challenge for the many biocontrol companies that are producing MPCAs. Information on details is usually proprietary. Production methods, however, of a certain pathogen whether performed by in vitro solid or liquid state fermentation or in vivo (Table 3.2), are generally known.

Below I will review the current status, type of system and kind of information needed to develop a mass-production system for the four groups of pathogens: bacteria, fungi, baculoviruses and entomopathogenic nematodes. A pivotal element is production costs in large scale productions. This is largely unreported in the literature. I will present a cost model that can be used to estimate the costs of each step in the overall product development and that provides insight into the cost factors. I have experienced many times that researchers quote production costs, but often it

Table 3.2 Economic feasibility of mass-production methods for microbial pest control agents (SSF = solid state fermentation; LSF = liquid state fermentation)

MPCA/production method	In vivo	In vitro SSF	In vitro LSF
Bacteria	+/–	+	+++
Fungi	+	++	+++
Baculoviruses	+++	–	–[x]
EPNs	++	++	+++

– : Not possible ([x]: insect cell culture).
+/–: Possible, in some cases the only method.
+: Possible, but not or hardly economical.
++: Economical method.
+++: Most economical method.

is unclear which costs are meant, from what type of scale they are calculated and what the end-user's price will finally be. The cost model will take into account all contributing factors so that estimations can be made on turnover, profit, and return on investment.

2.1.1 Mass Production of Entomopathogenic Bacteria

Entomopathogenic bacteria can be easily produced in in vitro systems, although one species, *B. popilliae*, used for the control of the Japanese beetle, can only be produced in its natural host. It is effective against this beetle, which is a problem in outdoor crops in the USA. In New Zealand, a product called Invade has been developed on the basis of *Serratia entomophila* for control of grubs in grasslands. Production of bacterial cells is carried out in vitro by liquid batch fermentation (Johnson et al. 2001). These examples will not be discussed further since they are not used in greenhouses.

Bacillus thuringiensis is the most successful biopesticide and this is largely due to its economical production (Federici 2007). Three subspecies are produced, and commercial products have been developed on the basis of these subspecies: *B.t. kurstaki* against caterpillars, *B.t. israelensis* against mosquito larvae and *B.t. tenebrionis* against larvae of the Colorado potato beetle. *B. sphaericus* is also produced and used against mosquito larvae. Production for all of these bacteria is similar and they can be easily produced in large fermentors in liquid culture. Research on Bt as a biopesticide is almost 100 years old and hundreds of researchers have published on it, including papers on mass production. Bt can be produced by (semi-)solid state fermentation, but industrial production is performed by liquid state fermentation. Semi-solid and solid state fermentation is done on a small scale in developing countries for local use (Bernard and Utz 1993; Vimala Devi et al. 2005; El-Bandary 2006). I have seen in 2006, however, that even in Cuba submerged fermentation is taking over with fermentors of a few thousand litres. Semi-solid and solid state fermentation is difficult because is takes a long time, and parameters cannot be adjusted. Liquid fermentation is therefore the preferred process in industry. Only when production takes place in vessels of over 30,000 l does Couch (2000) consider it mass production, which indicates today's scale of production. Hundreds of products have been developed by tens of companies. In the USA about 125 Bt products are registered (http://www.epa.gov/pesticides/biopesticides/product_lists/microbial_prods_by_ai.pdf).

Details of the production of Bt are given by Lisansky et al. (1993), Bernhard and Utz (1993), Couch (2000) and El-Bandary (2006). An excellent overview of "Bt in biological control" on its biology, mode of action, commercial development and on Bt transgenic crops is given by Federici (1999). The Bt development has always been dominated by large companies (Solvay, Abbott, Novo Nordisk), including chemical pesticide companies (Syngenta, formerly Novartis, resp. Ciba Geigy and Sandoz). American venture capital companies joined this area in the late eighties (Mycogen, Ecogen). They focused on genetically improved strains, but never marketed Bt products on a large scale. They have sold their strains and technology to companies which are focussing on transgenic plant development with Bt genes (Monsanto, Syngenta). Currently two companies dominate the world's Bt market, Certis USA and Valent BioSciences. Typically, production is conducted in fermentors of 40,000–120,000 l (Federici 2007). Many small companies produce Bt products, but apparently the market has become very competitive so that only two companies have large sales. The large scale productions definitely are profiting from the economy of scale and most likely that is the only way to be profitable. Downstream processes have become an integral part of Bt production and large facilities are needed to accomplish drying of the bacteria sludge by spray-drying and/or fluidized-bed drying. Formulation is often included in this process. Obviously this production has become capital-intensive because of its size and technology and it shows that competitive biopesticides can be produced in this way.

2.1.2 Mass Production of Entomopathogenic Fungi

Entomopathogenic fungi have been studied widely for use as a mycoinsecticide and many researchers have given an overview on mass production of these fungi (Quinlan and Lisansky 1983; Bartlett and Jaronski 1988; Bradley et al. 1992; Goettel and Roberts 1992; Feng et al. 1994; Jenkins and Goettel 1997; Wraight and Carruthers 1999; Wraight et al. 2001). Most of this work has been performed on a select group of Hypocreales (formerly Deuteromycetes), particularly on fungi of the genera *Aschersonia, Beauveria, Metarhizium, Isaria (Paecilomyces)* and *Lecanicillium (Verticillium)*. Two species have been the subject of most of the research: *Beauveria bassiana* and *Metarhizium anisopliae*. Both fungi have a broad host range and are relatively easy to produce. Many researchers have studied these fungi for control of various pests (Inglis et al. 2001), including their production and formulation. An extensive review has been given on production, formulation and application of *B. bassiana* by Feng et al. (1994). This fungus has been and is still produced on a large scale in many countries by solid state fermentation (SSF), liquid state fermentation (LSF) and di-phasic fermentation (Ferron 1981; Bartlett and Jaronski 1988; Bradley et al. 1992; Feng et al. 1994). *M. anisopliae*, (including the acridid-strains, formerly called *M. flavoviride*) is usually produced by solid state fermentation (Mendonca 1992; Cherry et al. 1999; Milner and Hunter 2001), but the production of blastospores by submerged production (LSF) has also been studied (Kleespies and Zimmermann 1992). Most products are based on conidiospores, and some on conidia and hyphae on a granular substrate. There is no blastospore-based product on the market. The majority of the mycoinsecticidal products have been developed with these two fungi: 58 based on *B. bassiana* and 61 based on *M. anisopliae* out of 171 mycoinsecticides and mycoacaricides (De Faria and Wraight 2007).

Entomophthoralean fungi have successfully been used in classical biological control in two cases, with *Entomophthora maimaiga* against gypsy moth in the USA and with *Zoophthora radicans* against the spotted alfalfa aphid in Australia, but they are not used in inundative programmes. Their potential is reviewed by Pell et al. (2002), but unfortunately they cannot be mass-produced in vitro and they are not further discussed here. Fungi used for control of plant diseases, plant pathogenic nematodes and weeds are often mass-produced in similar ways as entomopathogenic fungi. For an overview of fungal species products, see Butt et al. (2001: pp. 2–4). Relevant aspects of mass production considered in this chapter relate also to these fungi, regardless of their field use.

Fungi can be mass-produced by solid state fermentation, also called solid substrate fermentation, and by liquid state fermentation (LSF), also called submerged culture fermentation. A third method is an intermediate system, called a di-phasic fermentation, where both phases are involved: mycelium, produced in liquid culture, is allowed to sporulate in shallow trays (Feng et al. 1994; Jenkins and Goettel 1997). The choice of how to produce a fungus is generally led by the ability of an organism to produce a high yield of spores of a high quality in a certain system. Conidiospores are preferred over blastospores or hyphae because of their better survival in downstream processes and during storage, and for their superior persistence after application. Most fungi produce conidiospores only when produced by solid state fermentation, still, some do produce conidiospores in submerged fermentation, like *B. bassiana* and *Trichoderma harzianum*. Other fungi like *I. fumosorosea* and *L. longisporum* (*V. lecanii*) are produced by liquid state fermentation and form blastospores.

Production by solid state fermentation can be conducted by a simple method, e.g. growing on a carrier – often enriched cereal grains or parts thereof – in plastic autoclavable bags or in trays. This is often seen in developing countries or in situations where labour is relatively cheap or investments low. Well-established biocontrol companies use this method, as well as small local "cottage-industry" like organizations as the CREEs (Centros de Reproducción de Entomofágos y Entomopatogénos) in Cuba (Nicholls et al. 2002) and in Central America (Grimm 2001) and South America (Mendonça 1992). The start-up costs are low, the technology used relatively simple and the labour input is high. Nevertheless, it can be successful as proven by the Lubilosa project (Jenkins et al. 1998) and by BCP's production of *M. anisopliae* var. *acrididum* for locust control (www.biocontrol.co.za).

More sophisticated bioreactors are used for solid state fermentation by a number of companies such as Koppert (unpublished data, see Fig. 3.1), Laverlam (the former Mycotech facility) (Bradley et al. 1992), NPP (Guillon 1997) and Prophyta (Kiewnick 2001). In these reactors environmental conditions can be regulated much better than in bags or trays, for instance by means of conditioned airflow, heating or cooling. Investments are rather high, labour input is medium. Some examples of products with conidiospores produced by solid state fermentation are: Mycotal, based on *L. muscarium* and produced by Koppert; Botanigard, based on *B. bassiana* and produced by Laverlam; Naturalis, based on *B. bassiana* and produced by Intrachem (formerly by Troy Biosciences); Bio1020 and Met52, based on *M. anisopliae* and produced by Novozyme (formerly Earth Bioscience, resp. Taensa and Bayer).

LSF has also been studied for entomopathogenic fungi such as *B. bassiana* (Hegedus et al. 1990), *Hirsutella thompsonii* (McCoy et al. 1975), *M. anisopliae* (Kleespies and Zimmermann 1992; Andersch 1992), *I. fumosorosea* (Jackson et al.

2003) and *Lecanicillium* spp. (Hall 1981; Feng et al. 2000, 2002). In submerged cultures, fungi may produce blastospores (thin-walled, single-celled hyphal bodies) and/or submerged conidia and/or mycelial parts or pellets. This depends on the species, the strain and the production medium and parameters. Most of the research on submerged fermentation is aimed at blastospore production. In the case of *H. thompsonii*, McCoy et al. (1975) studied the mass production of mycelial parts as propagules. They assumed this would be the easiest and cheapest method, but the product Mycar developed by Abbott on this basis gave variable results and poor stability in storage. Later, a conidia formulation was developed with good storability (McCoy 1981). Still, poor field efficacy resulted in withdrawal from the market in 1985 (McCoy 1996). Rombach et al. (1988) investigated the production and formulation of mycelium parts of *B. bassiana* as infective material because it seemed the simplest, thus cheapest fermentation process. Another approach was taken by Andersch (1992) who produced mycelial pellets of *M. anisopliae* as the active parts of the product Bio1020. Bayer chose for the production of pellets as the foreseen formulation was a granule that could be mixed with growing substrates for plants (Reinecke et al. 1990). An other reason to choose for submerged fermentation was Bayer's perception that solid state fermentation cannot be scaled up to an industrial production method. This idea, however, did not succeed and the product with this formulation was never launched since production costs were too high. Bio1020 is currently registered as a formulation based on rice kernels with conidiospores (www.ctb-wageningen.nl). Some examples of products based on blastospores are Preferal, based on *I. fumosorosea* and produced by Certis USA (formerly Grace & Co, resp. Thermo Trilogy). Production costs and scaling-up considerations determined that LSF be used to produce this product (K. Bolckmans, personal communication). Vertalec, based on *L. longisporum* (formerly *V. lecanii*) is produced by Koppert by LSF since spore yields by solid state fermentation are low with this strain.

Generally LSF is investigated for reasons of scale and costs, or because the yield of conidiospores is too low when using solid state fermentation. Spore type and production costs generally drive the choice on how to produce a fungus, see Box 3.1 for an example. Economical mass production of stable propagules is essential for the commercial development of a mycopesticide.

2.1.3 Mass Production of Baculoviruses

Baculoviruses can be produced in vitro and in vivo. The commercial development of baculoviruses as viral insecticides started in the 1960s in the USA, based on in vivo production of the *Heliothis* NPV virus (Ignoffo 1973). About 20 years later, interest in virus products in Europe increased and even then in vitro production was investigated and compared with in vivo production because of cost concerns (Gröner 1987). Today, mass production of baculoviruses for biopesticide products still occurs in vivo, in most cases on the original host, sometimes on a factitious host. In vitro production is performed in research and for purposes such as protein production, but not yet for biopesticide production. Many authors have reviewed the development and production of baculoviruses in general (Ignoffo 1973; Granados and Federici 1986; Huber and Miltenburger 1986; Shieh 1989; Weiss et al. 1994; Black et al. 1997; Hunter-Fujita et al. 1998).

In vivo production of baculoviruses has been described in detail by Shapiro (1986). He identified the production of the host: moths, eggs and larvae (labour, medium, containers and facility costs) as the chief cost factor. Automation and optimization of the artificial media (simplification) for caterpillars are areas for improvement. Production costs are higher where caterpillars are cannibalistic and need to be reared singly in contrast to in groups. Finding an alternate host that is not cannibalistic is in some cases a possibility for reducing costs. In virus production, inoculation and harvest mechanization could lead to further cost reduction. Essential aspects in the production are the optimal inoculation dose and larval stage and the optimal harvesting time, giving the highest yield of infectious occlusion bodies (OBs) or polyhedral inclusion bodies (PIBs) per larva. A difficulty with in vivo production is contamination with microorganisms, mainly bacteria, and degeneration of the virus (Black et al. 1997). Using inoculum from a well-defined mother stock can overcome this risk of loss of activity by mutations, and companies are usually using this method. A general problem is that process control of in vivo production of viruses is limited and therefore many authors suggest that production in vitro would be much better. The possibility for this method is discussed below. Many authors have given detailed descriptions of in vivo production methods of various baculoviruses in different insects (e.g. Ignoffo and Couch 1981; Lewis 1981; Smits 1987; Shieh 1989; Cherry et al. 1997; Grzywacz et al. 1998; van Beek and Davis 2007).

An advantage of the production of baculoviruses and their hosts is that it can be started in a simple set-up without high initial investments. This gives a company the possibility of a step by step approach, implementing improvements continuously over the years. Likewise, production systems in developing countries also rely on relatively simple technology. Techniques have been developed for high-density rearing of various caterpillars (Wood and Hughes 1996). Besides, produced virus can be stored frozen for many years without loss of activity, allowing the producer to produce all year round and to build up stock. Failures in production are therefore not as problematic as with EPNs or fungi. The large scale production of the insect host is pivotal. Optimization and automation can reduce costs. Other problems such as diseases and premature virus infection can be devastating. Quality control parameters should be developed to detect problems, like virus degeneration, in an early phase. Production in vivo is not a sterile process and it is not as controllable as production in vitro. Microbial contamination can be a problem (Hunter-Fujita et al. 1998), leading to variable yields and discarding of batches. Generally, registration requirements do not allow high bacterial contamination levels, and for this reason too, contamination needs to be kept under control. Guillon (1995) considers quality control the main cost factor, which underlines the importance of this aspect in baculovirus production. Downstream processes such as harvesting dead larvae and virus purification can be laborious, but are relatively easy in terms of technology. Product can be stored easily, unformulated as well as formulated. For high value crops this in vivo technology has proven to be feasible in the western world. In low value crops baculoviruses are also used, even on a scale of over 2 million hectares in soy. Production, however, is too expensive for these crops, therefore diseased larvae are collected in the field and virus is extracted from these caterpillars and prepared for application (Moscardi 1999). This is called field production of baculoviruses.

The number of baculovirus biopesticides is about 35, based on 15 different virus species (Copping 2009). Baculoviruses are used to control lepidopteran pests and dipteran pests (mainly sawflies in forests, the latter are not discussed further here). Production systems have been developed for codling moth, fruit tortrix, and gypsy moth and for *Heliothis* spp., and various virus products have been registered and are marketed. Examples are Carpovirusine, Granupom, Madex, CYD-X and Virosoft for control of codling moth, Capex for control of the fruit tortrix, Gemstar for control of *Heliothis* species and Gypscheck and Disparvirus for control of the gypsy moth *Lymantria dispar* in North American forests (Copping and Menn 2000). There are two products used for a greenhouse pest, Spod-X, produced by Certis USA, and Virex, produced by Biocolor, Spain, both against the beet army worm *Spodoptera exigua*. Virus products are able to compete in the market with chemical pesticides, for example in the case of codling moth in apples and pears in Europe. The amount of active ingredient needed per ha is a significant factor. For codling moth the dose is 3×10^{12} OB/ha, for the beet army worm the dose is 3×10^{11} PIB/ha (polyhedral inclusion bodies). Recently, a baculovirus has been found of *Chrysodeixis chalcites*, the tomato looper (Van Oers et al. 2005). This caterpillar is the predominant lepidopteran pest in Dutch greenhouses and this virus may be developed into a product.

2.1.4 In Vitro Production of Baculoviruses

In vitro mass production of baculoviruses for biocontrol purposes has been studied for decades by academic researchers (Stockdale and Priston 1981; Weiss and Vaughn 1986; Shuler et al. 1995; Wood and Hughes 1996; Black et al. 1997; Szewczyk et al. 2006) and by industry, by chemical companies such as Hoechst (Gröner 1987), American Cyanamid (Gard 1997), Dupont and Zeneca (Harris 1997), as well as by biocontrol companies such as Biosys (Weiss et al.1994). For two important reasons, in vitro production of baculoviruses could be advantageous over in vivo production:

(1) economy of scale. As described above, in vivo production is economically feasible up to a certain scale of use of a virus product or, when labour is not expensive. If there were a large demand, say over several 100,000's ha, scaling up would give production in fermentors the advantages of economy of scale. Product costs for growers could become lower and products more competitive with traditional chemicals;
(2) baculoviruses kill their host relatively slowly. This is a drawback of the products. The possibilities of improvement by genetic modification of strains aiming at an accelerated speed of kill are reviewed by Inceoglu et al. (2001). Recombinant strains, in which the "speed of kill" trait is improved, for instance in toxin-expressing recombinant baculoviruses, however, give much lower

virus yield per larvae than the wild-type. Larvae infected with improved baculoviruses die quicker and the production of occlusion bodies is therefore too low. Yields can decrease from 20 to even 95% (Sun et al. 2005; Burden et al. 2000) depending on the virus species and the recombinant virus. This will render in vivo production no longer economically feasible.

Production in cell lines may not have this disadvantage and may be the solution for the production for genetically improved strains. The lower yield problem may be overcome by production in insect cell lines. Many authors (e.g. Wood and Hughes 1996; Inceoglu et al. 2001; Szewczyk et al. 2006) believe that in the near future these strains will be used as pest control products and that cell line cultures will become the appropriate production method.

The production of insect cells has been studied extensively and an overview of this research is given by Vlak et al. (1996). Many authors have been optimistic about the potential of baculovirus production in insect cell lines by submerged fermentation. A variety of baculoviruses can be produced on selected Lepidoptera cell lines, and the formation of occlusion bodies (OBs) in high titres is possible (Goodman et al. 2001). Baculovirus production in cell cultures is well studied for pharmaceutical purposes. The baculovirus expression vector system is able to produce proteins efficiently in insect cells and this could be used to develop a method for large scale production of biologically active proteins for human therapeutic and vaccine use (Ikonomou et al. 2003). Much progress has been made in this area and some companies are able to produce insect cell lines in 600 l fermentors. They claim that scaling up to 30,000 l is feasible and that they have overcome the challenges with oxygen and nutrient supply and high shear forces, and can produce high titres of viruses on serum-free medium (Cox 2004). Nevertheless this technology is complicated and demands highly skilled biotechnology experts. Moreover, there were in 2003 already over 5,000 patents on insect cells or baculovirus production. This complicates production and could make it expensive due to license fees (Cox 2004).

Another problem is the instability of the recombinant virus in such a production process. Deletions in the virus genome often occur and after a number of passages through cell lines in the cascaded process of scaling up, production of occlusion bodies per cell decreases. The quality of the final yield may be completely unsuitable for the control of the pest because of the loss of biological activity caused by this "passage effect" (Pijlman et al. 2001). Recently a patent (WO 2005/045014) has been published suggesting a method to overcome this problem. A two step method was developed by which the initial inoculum is produced in insects, and the second step is using this inoculum for the production in cell lines with only a limited amount of passages. The inventors claim to be able to produce infectious OBs of the *Heliothis armigera* SNPV in a 10,000 l fermentor with about 2.5×10^{11} OBs/l and that this method can be used to develop a formulated product. The use of this technology is theoretically possible for agricultural use, but more research is needed on the stability of the recombinant baculoviruses. The prospects of the future use of recombinant baculoviruses are reviewed by Szewczyk et al. (2006). Technically, much progress has been made and many examples show that an accelerated speed of kill can be reached. The safety of these recombinant viruses still needs to be investigated, although it is not believed to be a great concern. Registration and use of such strains, however, will be much more difficult or even impossible in some countries due to the political or societal attitude towards genetically modified organisms.

Production costs are not considered by Szewczyk et al. (2006) in their future outlook, but from a commercial point of view this seems to be the major factor that will not allow for in vitro production of baculoviruses for the near future. The scale on which in vitro production is currently possible is too small, and it is too expensive due to high costs of specialized reactors, media and patent license fees. An economic analysis of baculovirus production in vitro has been made by Rhodes (1996). He concludes that it is economically feasible, but due to high investments, mainly in bioreactors, it can only be profitable if the market is over 1 million hectares. Such a large area would most likely apply to a low value crop where only low prices can be afforded for pesticides. This makes it a highly unlikely scenario and means that in vitro production is still far away as a commercial production system for viral insecticides. In two EU projects (AIR1-CT92-0386 and FAIR-CT96-3222, see www.biomatnet.org) the large scale production of baculoviruses in insect cell cultures was studied, but due to the instability of the virus through a number of passages and due to the fact that available cell lines do not produce high enough titres, commercial production is still not possible and not much progress has been made in this field for the last 10 years (J. Vlak, personal communication). Biocontrol companies keep an eye on the trends but are not actively engaged in research of in vitro technology. Andermatt Biocontrol AG does not expect in vitro production in the next 10 years (M. Andermatt, personal communication). Despite the optimistic views of many researchers and the industry, including major agrochemical companies, over the last 20 years, in vitro production of baculoviruses for viral insecticides seems still far away. Commercial development of genetically engineered baculoviruses and production thereof in insect cell lines is even further away from successful commercial use (Summers 2006).

2.1.5 Mass Production of Entomopathogenic Nematodes

There are three methods by which EPNs can be commercially mass-produced: in vivo or in vitro either in solid or in liquid culture. Each method has its advantages and disadvantages relative to costs of production, investments, technical expertise, economy of scale and quality of nematodes. An overview of the three production methods is given by Gaugler and Han (2002) and by Shapiro-Ilan and Gaugler (2002).

For research purposes EPNs are usually produced in larvae of the greater wax moth *Galleria mellonella*. Most nematode species reproduce well in this insect. In vivo production has been studied in many different nematode species on many different host insects: *G. mellonella, Tenebrio molitor* and many other lepidopteran and coleopteran larvae. Inconsistency of infection of hosts, variable yields, costs of the insects and labour make mass production in insects not attractive for an industrial production (Shapiro-Ilan and Gaugler 2002). Still, some companies used in vivo mass production in the past and the odd one may still use this method. It requires little capital investment, but it is labour-intensive. Scaling-up is difficult due to labour-intensive, space- and time-consuming production and harvesting methods (Gaugler and Han 2002). Attempts have been made to improve in vivo production on *G. mellonella* by automation of the production and the harvesting technology (Gaugler et al. 2002). This system could be used in places where labour is less expensive and capital investment limited and for small business companies in niche markets. Also, in vivo production is suitable for production of nematode species or strains that cannot be produced in vitro, or for research purposes where large numbers of nematodes are needed for field trials. For large volume production though, in vivo mass production is not suitable, mainly due to the lack of economy of scale.

Production of nematodes without a host insect became possible after Bedding (1981, 1984) developed a low cost solid-phase system using animal offal homogenate as a medium in a three-dimensional inert carrier in plastic bags. This was a major breakthrough and opened the door for economically commercial mass production. A number of companies (Biotech Australia, De Groene Vlieg and Koppert Biological Systems, both from the Netherlands, and Bionema, Sweden) started to produce and sell EPNs for soil pests, mainly in greenhouses (Ehlers 2001a). These nematodes were *Steinernema feltiae* and *Heterorhabditis megidis* against larvae of Sciaridae and larvae of the black vine weevil *Otiorhynchus sulcatus.* In vitro solid culture does not require high technology inputs and large investments, but is labour-intensive and the system lacks economy of scale. Stability of the production process is also a problem, as well as the variable quality of the animal offal used as medium and, consequently, yields. Koppert stopped solid culture mass production of *H. megidis* and *S. feltiae* after about 7 years of commercial production. The main reasons were the inconsistency of the cultures, production runs were long (2.5–3 weeks) and contaminations often occurred. Further, scaling up did not lead to substantially lower cost prices. Labour costs for preparation of the bags, inoculation and harvesting were the most limiting factors. Foam and waste removal were other relevant costs that did not profit from economy of scale effects. For similar reasons, Biotech Australia, later taken over by Ecogen-Australia, stopped their production (Gaugler and Han 2002). For Biosys, cost prices of EPN products which could not compete with prices of chemical insecticides were the main incentive to study in vitro liquid culture technology (Georgis 2002). Few companies use the solid culture method nowadays, although it still has its role. Andermatt Biocontrol AG produces *H. megidis* in this way since production of this species in liquid culture is not cost-effective due to low yields. BioLogic USA uses this method and it is also used in developing countries (Gwynn 2006).

In vitro production in liquid culture was developed step by step by several researchers (Ehlers 2001b). It is a complex rearing process which demands medium development, understanding of the biology of the nematode and the bacterium, and bioreactor development, including understanding and control of the process parameters. Two excellent overviews of the history of these developments are given by Ehlers (2001b) and by Gaugler and Han (2002). Collaboration between public researchers and research within companies contributed largely to further developments. Biosys, a venture capital company founded in 1983 in the USA, had a large research group and they were able to produce three species of nematodes in large fermentors of 15,000 l in 1989. Some years later, production even succeeded in 80,000 l fermentors. Biosys patented their liquid production technology (1989; WO 04602). An overview of these developments within Biosys is given by Georgis (2002). A similar situation was seen with Ecogen where many researchers studied this process; a brief historical overview on this within Ecogen is given by Ehlers (2007). The large amount of capital spent on research by both venture capital companies contributed enormously to the progress in liquid culture production. EPNs could now be produced for lower prices and for bigger markets, although still there is no comparison with the prices of chemicals pesticides. Many authors have reported on the production technology in liquid culture of EPNs over time and research is still being continued. For more details the reader is referred to Ehlers (1996, 2001b, 2007), Ehlers and Shapiro-Ilan (2005), Shapiro-Ilan and Gaugler (2002) and De la Torre (2003).

Gaugler and Han (2002: p. 293, Table 14.1) compared the different production systems in terms of investment, direct costs, R&D efforts, waste and space requirements and concluded that there is a place for all three systems. But in order to expand the use of nematodes and to compete with chemical pesticides, the only way to produce them is in large bioreactors where overall production costs will be the lowest (Ehlers 2001b; Ehlers and Shapiro-Ilan 2005). This is shown by the production of EPNs by Becker Underwood, formerly Microbio Ltd, currently the largest producer in the world, which uses liquid culture systems up to a scale of 30,000 l fermentors. Prices in the market have gone down in recent years, indicating that they have managed to lower their production costs. The other two companies currently producing in liquid culture are Enema and Koppert Biological Systems, ranking below Becker Underwood in scale of production. These three companies are the world's leading EPN producers, indicating that liquid culture is successful and that the other productions systems are not economically feasible for large scale production. Apparently, economy of scale determines the method used, although for liquid culture production investments are quite high and great technological expertise is needed within the company. A number of EPN species can be produced in liquid culture such as *Steinernema carpocapsae, S. feltiae, S. glaseri, S. kraussei, S. riobrave, S. scapterisci*, and *Heterorhabditus bacteriophora, H. indica* and *H. megidis*, although with different yields/ml. Some of these species, like *H. megidis,* cannot be produced (yet) economically in liquid culture due to the variability in recovery and yields (Ehlers et al. 1998). *H. marelatus* was produced by solid state production by BioControl Systems Inc, USA, until 2007. Production of *S. scarabaei* is not yet possible; this could be a valuable species for control of white grubs (Koppenhöfer and Fuzy 2003). There are still many challenges in the mass production of the various EPN species and research will need to continue in order to make production more reliable and economical (Ehlers and Shapiro-Ilan 2005). More research is particularly needed on the process parameters (Ehlers 2001b; De la Torre 2003) that influence the bacteria as well as the nematodes in liquid cultures. Expertise from other disciplines like biotechnology, fermentation technology, and from large-scale production of bacteria could be helpful. Another significant factor is phase variation of the symbiotic bacteria, shifting from primary to secondary forms, which leads to unpredictable yields. More fundamental research is needed to be able to control phase variation in these bacteria (Ehlers 2001b).

Eventually, the goal is to achieve a predictable and stable production method. Process quality control and insight into biotic and abiotic variables is required. Further, we need to know more about the relationship between production and shelf-life, and between production and field performance. Producing high yields of low quality nematodes is futile when their shelf-life is short and their field performance variable. To improve acceptance of EPNs as alternative pest control solutions, we need to be able to ensure better efficacy. It is not understood why only a part of the applied nematodes are active or pathogenic in the field and others are inactive or show delayed activity (Grewal and Georgis 1994). Costs could go down due to improved production. The continuous scale up of bioreactor volumes will further reduce production costs when accompanied by improved process stability and downstream processing (Ehlers 2001b). If the numbers applied in the field could be decreased and good results could still be obtained this would create even more market opportunities for EPNs. Further research is needed on these intertwined aspects of production and field performance.

2.2 Other Important Aspects in Mass Production

An efficient and low-cost medium for the production is an essential part of the process. Each organism needs a medium tailored to its demands and the growing conditions, and research is needed to develop such a medium. The choice of equipment is determinative. This depends on how an organism will be produced, a chief part of the cost factor of a product. Below the relevant considerations for these two aspects are given. Stability and quality of the inoculum used in the mass production over a certain period is an absolute necessity in order to have good production and an effective final product. Relevant considerations are briefly presented. Quality control related to the production process will be discussed in Chapter 4. Production process parameters need to be established and routinely checked. Depending on the system and the pathogen, a production run may take a few days up to more than 2 weeks, as in the case of nematode production. Any fault, deviation or contamination that may lead to an unsuccessful yield or to discarding of the batch should be detected as early as possible. This will save time and money. Knowing the system in detail is of paramount importance. In the design of a bioreactor and a production process, possibilities for parameter control and sample taking needs to built in. A production manual which includes parameter behaviour and checks needs to be established and a logbook be kept reporting all the events. In the end, gained experience and knowledge on the production process and the predictability of the yield, quantitatively and qualitatively, are extremely valuable. Analysing failures will give useful information and could improve the understanding of the system and help preventing future failures. A stable production process is indispensible.

2.2.1 Medium Development and Optimization

The production efficiency of an organism greatly depends on the nutrients that are available to it and whether they can be utilized by that organism. When production takes place in in vivo systems, the producer can only have a limited influence on the nutritional value of the hosts. Insects used as hosts should be of good quality, and by means of medium composition this can be influenced. Relationships between host propagation and pathogen production are both direct and indirect and optimization is more an art than science. This is seen in the production of baculoviruses in insect larvae. Medium composition is optimized in order to get large and healthy caterpillars, but the relationship between the medium and the virus production is hardly studied and largely unknown.

In in vitro production systems the relationship is more direct, and can be studied in detail. In fungi and bacteria this has been studied by many researchers. Methods are available, for instance, to find the optimal C:N ratio. The necessity of other minor nutrients can also be studied experimentally. In case of defined artificial media for LSF of fungi, bacteria and EPNs, medium optimization is possible. Still, this is a complex matter since there are generally two phases in the production, a vegetative phase and a generative phase which should lead to the production of spores or other propagules. In EPN production two organisms are produced more or less simultaneously in the same medium that makes optimization complex. The amount of propagules produced per millilitre as well as their quality is important, and whether they are suitable for surviving the downstream process and the formulation process. Further, do they have a long shelf-life and a good virulence in the field? Media components and production conditions are known to influence all of those factors.

An overview of effects that can be obtained by manipulation of the medium for various fungi is given by Magan (2001). Fungal spore survival under harsh conditions can be improved by adding certain sugars in the medium. For instance, trehalose and glycerol improve germination in *B. bassiana* and *M. anisopliae* at low water activities (a_w value = proportion of unbound water) (Magan 2001). Additives to standard media could give higher yields of blastospores in *M. anisopliae*, as demonstrated by Kleespies and Zimmermann (1998), but at the same time, spore resistance to high temperatures and low humidity decreased, as did virulence. Carbon-nitrogen ratios may influence yield, germination rate and survival rate in *I. fumosorosea* (Jackson et al. 1997) and virulence in *M. anisopliae* (Shah et al. 2005). Nutrient rich media may cause attenuation in many entomopathogenic fungi (Butt et al. 2006). Improvement of one trait does not necessarily mean improvement of another trait and the effects of production changes should be studied on all relevant traits with great care. In the production of Bt's, medium composition should support both the vegetative growth phase and the production of spores. Medium optimization has to be carried out for each strain. The salient criterion for an efficient production process is the amount of endotoxins that are produced. Exotoxins may be produced depending on the medium and culture conditions, but exotoxins are a serious concern to regulatory authorities. Medium and production parameters need to be standardized in order to prevent production of these metabolic toxins (Glare and O'Callaghan 2000).

For solid state fermentation of fungi, an inert carrier or an organic solid substrate, often a cereal product, is used. The latter is also defined as solid substrate fermentation. When an inert carrier is used, conditions come close to LSF production. Media components can be added according to the demands of the fungus. Various inert carriers can be used such as clay, perlite, glass beads, or synthetic material like polyurethane foam. When an organic carrier, such as cereal grains or bran, rice, hemp, wood chips, or bagasse is used, it fulfils two roles at the same time, both as a nutrient source and simultaneously as a carrier. The carrier function is to create three-dimensional structure that allows gas and heat transfer and it is important for the needs of a fungus related to spore formation. For instance, in *Aschersonia* spores are formed in pycnidia on the surface of the carrier and not inside the material. Creating a large surface with the help of the structure of the carrier improves the production efficiency. In *Lecanicillium* sporulation takes place wherever mycelium is able to grow, but it needs space since this fungus makes conidiophores. An open structure of the carrier improves the efficiency. Sporulation does not take place inside a cereal grain, but on the outer surface only. Fragmented cereal or rice kernels or bran are better in this case. In some uses of fungi, spore production takes place in the field and production is aimed at high production of fungal mass, in this case whole cereal kernels are used. An example is the use of *Beauveria brongniartii*, where mycelium overgrown barley kernels are applied into the soil (Keller 2000). Examples of products that have been developed on this base are Melocont-Pilzgerste from Kwizda Agro, Austria, and Beaupro from Andermatt BioControl.

Medium optimization for production of EPNs is complex since a bacterium and a nematode are produced on the same medium, subsequently. In LSF, the medium is predominantly developed for the bacterium, while the bacterium reverses the medium for the nematode. The medium composition is important, but according to Ehlers (2001b) it should not be overestimated. Process stability and mean yields are more meaningful parameters.

Medium optimization and growth kinetics of the organism in a particular medium/carrier matrix should be evaluated in the production of pathogens because they determine production efficiency in terms of production time, yield of propagules, quality of propagules and costs. Even shelf-life and field efficacy can be influenced by it. The methods that can be used in the downstream process, and the formulation type depend on propagules' properties to some extent. Accordingly, it is necessary to profoundly study this aspect of the mass production: primarily, considering production efficiency, and secondarily, considering medium costs. Advantages and disadvantages of media composition in mass-production systems are given in Table 3.3. PhD studies are devoted to this subject of optimization, but in the biocontrol industry this is too expensive and takes too long. During the development of a MPCP, collaboration with a university would give the opportunity to study the fundamental requirements of the production of a certain pathogen. Eventually, medium composition will be a compromise between production efficiency and medium costs.

Table 3.3 Advantages and disadvantages of medium type and composition in mass-production systems (SSF = solid state fermentation; LSF = liquid state fermentation)

Production system	Advantage	Disadvantage
In vivo	• On natural host or (closely) related host • No attenuation	• Limited influence on composition • Non-sterile, contamination risk • Downstream complicated • Quality and yields variable
In vitro: *SSF*		
– Inert carrier	• Sterile • Defined artificial medium	• Downstream easy – complicated • Attenuation possible
– Cereal (like) carrier	• Some influence on medium • Semi-sterile to sterile	• Variable quality • Downstream complicated
In vitro: *LSF*	• Sterile • Defined artificial medium • Downstream easy	• Can be expensive • Risk for attenuation • Unnatural

2.2.2 Mass-Production Equipment

The fundamental question to be asked at the start of product development is which type of bioreactor should be used for the mass production of a pathogen. Many factors play a role and it depends on the possibilities of production of the organisms (Table 3.2), but decisive is the expected scale of the production for the coming years in relation to investment costs. A well-performed marketing study will give estimates of the expected market size, but nothing more than an estimate. Looking ahead for more than 5 years is difficult and investments should not be based on long term projections. It is wiser to choose a system that can be extended or scaled up once the market demand increases. It should also be kept in mind that registration requires detailed information on production and product and that, when scaling up, no major changes occur that would need additional registration information to be submitted and evaluated.

It is beyond the scope of this book to elaborate in detail on the various types of bioreactors and to consider which one to choose for which organism. It is a case by case study depending on the organism, the required propagule for the final product and the scale of production. For solid state fermentation, non-sterile or sterile reactors could be used, non-stirred or stirred or rotating. For LSF sterile reactors are needed. Solid state fermentation is used for many purposes, such as for production of enzymes, metabolites and for the production of (Asian) foods like soy and tofu. To some extent it is also used for the production of entomopathogenic fungi. Many different types of bioreactors have been designed. An overview of techniques and bioreactors is given by Roussos et al. (1997). The literature on solid state fermentation is vast, including on production of entomopathogenic fungi and on nematodes. A careful study of this literature is useful before a new production system is built. Scale and costs and the biology of the pathogen are decisive. It is essential to study the organism in detail in order to know its nutritional and environmental requirements and to adapt the production system to meet these requirements. Process control and management of the crucial parameters are vital. There is no standard recipe for the production method of an entomopathogenic fungus, bacterium or nematode. Each case needs its own approach, although much can be learned from other cases. Fine-tuning is always needed and must be continued over the years.

What companies use is also strongly influenced by their expertise and available equipment. Experience tells us that, particularly in the case of solid state fermentation reactors, many companies develop their own equipment, focusing on their first and main microbial organism. Koppert built their own solid state fermentation reactors, with a tray system, for production of *aleyrodis* and *Lecanicillium muscarium*. The former one needs light to sporulate which made it necessary to build "windows" in the reactor, making it more vulnerable to contamination (see Fig. 3.1). Other companies have built and even patented their equipment and technology, like Mycotech (now part of Laverlam) (1995: WO 95/10598) and Prophyta (2004: EP 1502946). Mycotech developed a packed bed reactor for the production of *B. bassiana* (1995: WO 95/10598); Prophyta produces a biofungicide based on conidiospores of *Coniothyrium minitans* (Kiewnick 2001). NPP (now part of Arysta LifeSciences Europe) used a packed bed, non-sterile reactor, designed and patented (WO 94/18306) by INRA Dijon (Durand et al. 1997), for the production of *B. bassiana* for corn borer control. Another reason for these in-house developed bioreactors is that solid state fermentation reactors for pathogen production are hardly available on the market, in contrast to various types of LSF reactors which are standard in the biotechnical and pharmaceutical industry. Sterility, removal of metabolic heat, maintaining water content, and transfer of gases must all be carefully managed in the production of pathogens in SSF reactors. This is similar for LSF reactors, and additional critical aspects are the shear forces. When developing or buying equipment, care should be taken that these parameters can be controlled.

Recovery of the propagules from the mass production system is an intrinsic part of the process. In most cases propagules need to be separated form the medium and this process should not negatively influence the quality of the spores, dauer juveniles, etc. The percentage recovery should, of course, be as great as possible. Harvesting often means a sudden change in conditions and care should be taken to avoid potential damage to propagules by osmotic shock, low or high temperatures, desiccation, shear forces or high pressure. Harvesting methods depend on the production system whether from submerged or solid fermentation. Downstream equipment is usually not a standard type of equipment available to the biocontrol industry and needs to be developed in-house or bought from other industries. Typical apparatuses are centrifuges, separators, sieving and milling equipment, and machinery to dry spores or medium/carrier with spores or mycelium, such as a spray-dryer, a fluidized-bed dryer, a freeze-dryer or a simpler forced-air blower. This equipment is needed to concentrate progagules from the production phase, and in the formulation process. Biocontrol companies have to investigate which equipment is suitable for their processes and develop the appropriate protocols. This is a case by case study. Advantages and disadvantages of various production systems is given in Table 3.4. Mass production is a multi-disciplinary process. On an industry scale both advanced biological and technical knowledge and skills are necessary.

Fig. 3.1 Mass-production equipment (*left*: 1,000 l fermentor for liquid state fermentation; *right*: 1.3 m^3 bioreactor for solid state fermentation)

2.2.3 Inoculum Stability

The first aspect in producing a microorganism is storing the strain in a way that it keeps its relevant characteristics, mainly its virulence. It is well known that repeated sub-culturing could lead to a reduced virulence or loss of other properties, such as reduced sporulation (Butt et al. 2006) and a decrease in production of cuticle-degrading enzymes (Nahar et al. 2008). Further there is a risk of contamination. Every production batch should lead to yielding propagules of the same quality and obviously the inoculum for the production is the basis for this. Loss of virulence can often be restored by passing it over the target host again, but in a mass-production system this is cumbersome and costly. The solution is to make a master stock from which many sub-cultures are made, also called working seed stock that can be stored for a long period at the appropriate conditions. This is often at low temperatures, deep-frozen or by cryo-preservation. A strain should also be properly identified and deposited in a public collection culture as a back up. This is also a registration requirement. Every production batch can now be started using one of the stored samples. Depending on the organism, a new master stock needs to be made every 2–3 years. A newly made master stock needs to be rigorously checked on identity, on virulence via a bio-assay and on microbial contaminants. Re-identification of the culture by a specialized laboratory further ensures the identity and that the quality of the stock material is still maintained. This makes the whole procedure of keeping stable inoculum relatively easy and cheap. Further, the number of culturing steps in producing inoculum for scaling up to the final mass production should be kept as low as possible to minimize the risk of any loss of characteristics. Any new batch needs to start with inoculum from the original master stock.

Table 3.4 Advantages and disadvantages of various mass-production methods (SSF = solid state fermentation; LSF = liquid state fermentation)

Method	Advantage	Disadvantage
In vivo	• Only economic method for baculovirus production • Exceptionally used for bacteria (*B. popilliae*) • Low start-up costs • No high technical expertise required • Low developmental costs • Easy to scale up when labour costs are low	• Labour-intensive • High risk for contamination • No effect of economy of scale • Host and baculovirus production must be kept separate • Little influence on "medium"
In vitro SSF: bioreactors	• Can be used for many fungi and bacteria • Medium to high start-up costs • Technical expertise required • Sterile technology • Control of parameters possible	• Start-up costs medium to high • Labour requirements medium • Effect of economy of scale limited • Production facility of medium technology required • Limited influence on medium composition • Limited homogeneity
In vitro SSF: bags	• For fungi and EPNs • Low start-up costs • Low technical expertise required • Cheap to scale up to mid-range level • Low quality facility will suffice	• Labour-intensive • Risk for contamination, particularly in drying steps • Labour-intensive downstream methods • No effect of economy of scale
In vitro LSF	• Versatile; fermentors can be used for fungi, bacteria and EPNs (viruses in cell cultures) • Sterile technology • Homogeneous process • Large effect of economy of scale • Labour requirement low • Good ability to control conditions • Medium optimization possible to high degree	• High start-up costs • Great technical expertise required • High quality equipment and facility needed • Fungi may not produce conidia • Large scientific input required, particularly in virus/cell line production

Modified after Lisansky (1993)

Box 3.1 Production of Lecanicillium muscarium in SSF and LSF

In 1989 Koppert started to mass-produce *Verticillium lecanii* strain Ve6 (renamed as *Lecanicillium muscarium*) for Mycotal, in SSF home-built bioreactors. In the past, Tate & Lyle produced it in LSF as they did with the *V. lecanii* strain Ve2 (renamed as *L. longisporum*) for Vertalec (Quinlan 1988). Tate & Lyle's biopesticide department became independent as Microbial Resources Ltd in 1984. In 1986 all registrations and technical know-how of Microbial Resources were acquired by Novo Industria AS of Denmark (S. Lisansky, personal communication). In 1988 Koppert bought the IP rights of these products and started to produce them. Koppert choose for SSF in order to improve the products and their efficacy. Tate & Lyle produced Mycotal in LSF, the main reason for this was the scale-up possibility which was considered too difficult with SSF technology. LSF was a "second choice" based on spore quality: conidiospores have a better shelf-life and field persistence; on the other hand economy of scale is much better with LSF (S. Lisansky, personal communication). Koppert wanted to produce conidiospores in order to have a better shelf-life and better efficacy and persistence in the field. At the same time, there was SSF production equipment available as well as experience with the production of *Aschersonia aleyrodis*. The bioreactors were designed with a tray-system. The fermentors had a surface of 28 m^2 (52 trays of × 0.55 m^2) and each held 100 kg dry medium. Koppert built 12 fermentors in 1990 and had drying and sieving-milling equipment built. The fermentors were equipped with an inoculation system of ultrasonic nozzles on each tray, and with internal cooling systems. Pre-conditioned air can be lead through the reactors in order to keep the RH high and for supply of oxygen and transfer of CO_2 and metabolic heat. Everything was built to keep the system sterile. A large autoclave was designed and built in which one reactor, equipped with wheels, could be sterilized, including the medium. Investment for the reactors was around €120,000 in 1990. A production run (including medium preparation, autoclaving and downstream processing) takes 10 days and mean yields are about 8.10E14 spores/reactor. Annual full capacity is about 25 runs with 10 reactors per run, giving 2.10E17 spores (Approximately 20,000 kg Mycotal (10E13 spores/kg)).

In 2005 our group started to investigate again the LSF technology for Mycotal. The production of blastospores reaches 2.10E10 spores/ml in a 120 l fermentor, and scaling up to 1,000 l fermentors gave a similar yield per ml. A complete production run takes 7 days. To have the same capacity on an annual basis, a LSF fermentor capacity of 1,000 l only needs 10 runs (2.10E16 spores/fermentor). This would only occupy a quarter of its annual capacity. A new set of SSF bioreactors, including autoclave and downstream equipment, would cost around €750,000 in 2007, a 1,000 l fermentor around €300,000, including all equipment. This clearly shows the economy of scale, the difference is about a factor 10: 2.5 times the equipment price, and 4 times the capacity. If variable costs are considered, SSF demands about 10 times more labour to produce 20,000 kg Mycotal. This example shows that the costs for a yearly supply of Mycotal for SSF are much higher than for LSF when investment and running costs are considered (see Table 3.5). Thus, LSF has a considerable advantage over SSF in terms of economy of scale. The decision on how to produce a pathogen, however, is also influenced by quality of spores (see Feng et al. 2002), shelf-life, efficacy, production of metabolites, contaminants, and registration issues. A final decision on whether to produce by SSF or LSF can only be made based on all aspects.

Table 3.5 Parameters of the production of *L. muscarium* spores in SSF and LSF (SSF = solid state fermentation; LSF = liquid state fermentation)

Production parameters	SSF	LSF
Run time	10 days	7 days
Runs per year	25 × 10 = 250	40
Total spore yield	8 × 10E14/unit	2 × 10E16/1000 l
Investment (reactor, incl. downstream equipment)	€750,000/set of 10 units	€300,000/1000 l
Spores per ml or gr	1 × 10E10 sp/gr	2 × 10E10 sp/ml
Medium costs	€0.06/gr	€0.16/ml
Fermentor labour	4 h/unit	2 h/1,000 l
Downstream labour	4 h/unit	8 h/1,000 l
Annual capacity	250 × 8.10E14 = 2 × 10E17 conidiospores	40 × 2.10E16 = 8 × 10E17 blastospores

3 Product Development

3.1 Formulation

As a general rule, pesticides are formulated products containing an active ingredient or active ingredients and co-formulants or additives, components with specialized characteristics that are needed to make and keep the pesticide efficient in many aspects. Biopesticides are no exception to this rule and formulation may even be more consequential as they contain living organisms, and as a result, small particles, in contrast to chemical pesticides which contain molecules. There are many different types of formulations depending on the active ingredient(s), the production process, the target(s) and the application method. This is true for biopesticides as well. In the chemical industry, formulation technology has a long history and many different types of formulations have been developed. Most information is, however, proprietary and confidential. The knowledge of formulating chemical pesticides is not available to biocontrol workers, but neither is it very appropriate. Of course, there is some similarity in requirements for both types of products, and in biocontrol we could learn from the chemists, but mostly, formulation is a new research area for biopesticides and there are no "off the shelf" solutions available. New technologies have to be developed for each of the different pathogens. First and foremost we need to ask the question "why do we need formulation?" The four main objectives in formulating a microorganism, virus or nematode are:

(1) to stabilize the propagules that are collected from the production process by means of the downstream process so that they ultimately can be packaged, stored and shipped to the end-user;
(2) to make a user-friendly product that can be applied economically by the end-user and can be effectively delivered to the target;
(3) to protect the propagule, once applied, against harmful environmental influences, thereby maintaining and even improving its persistence at the target site;
(4) to minimize the risks of exposure to the applicator during loading, mixing and applying the product as well as to the worker in the crop, and to the consumer in the case of food crops.

The solutions for these four objectives are not always compatible and the final formulation is a compromise between all these demands. The different types of pathogens vary in their formulation demands concerning storage and application aspects; some may be the same for some pathogens, while others may be very different. Formulation is specific to a certain pathogen and determined by the features of that pathogen and its specific use: it demands a case by case approach. The mode of action is the most essential feature that determines the formulation requirements. Further, the type of the propagules, and whether they are used for a foliar or a soil application, will influence formulation choices. Formulations for foliar applications demand different considerations than formulations for soil applications. Formulations for the latter application can be simpler. In Table 3.6 the most significant formulation considerations are given per type of pathogen.

3.1.1 First Function of Formulation: Stability

The first function of the formulation is to keep the propagules in a state of low metabolic activity or non-metabolic activity and, at the same time, maintain the infective propagules viable and virulent for as long as possible during storage in the final product package: this is called shelf-life. The protection of propagules could already be necessary during the downstream process, especially when drying is a part of that process. Thus formulation considerations can start even before the infective propagules have been collected, and before formulants are mixed with them. Storability features too can be influenced during production by the composition of the medium and culture conditions. This has been shown for fungi such as

Table 3.6 Formulation considerations per type of pathogen

	Bacteria	Fungi	Viruses	EPN
1. Stability				
Maintain virulence	++	++++	++	++++
Standardization of potency or content	+++	++	++++	+
Maintain physical stability	+	++	++	+++
Maintain chemical stability	– – –	– –	+	+++
Minimize number and growth of contaminants	++	– – –	++	– –
Packaging	++	+++	++	++++
2. Efficacy and user-friendliness				
Spray characteristics	+++	++++	+++	– – –
Application method and equipment	++	++++	++	++
Efficient targeting	+++	++++	+++	–
Convenience of use	+	+++	+	++
Enhancement of propagule activity	++	++++	+++	– –
3. Persistence				
Protection	+++	++++	++++	– –
Stimulation of growth of propagules	– – –	+++	– – –	– – –
4. Registration and safety				
Minimize exposure	++++	++++	+	– – –
Minimize microbial contaminants	+++	+++	+++	– – –

+ to +++++: Increasing amount of attention needed during formulation.
– : Not relevant.

M. anisopliae, B. bassiana and *P. farinosus* (Hallsworth and Magan 1994) and *T. harzianum* (Agosin et al. 1997). Virulence can be influenced by the nutrients in the production as was demonstrated in *M. anisopliae* by Shah et al. (2005) and by repeated sub-culturing in vitro (Nahar et al. 2008). Degeneration of entomogenous fungi on nutrient rich media has been described for many species (Butt et al. 2006) and needs to be monitored in commercial productions. The lipid content in dauer juveniles (DJs) can be influenced by the production medium and eventually this may have an effect on shelf-life and infectivity (Patel et al. 1997). These examples demonstrate that formulation is not merely adding components to the propagules once they have been collected from the production. Ultimately, formulation considerations start from the beginning of the production process and formulation features can be influenced by medium components and culture conditions.

Formulation can also be part of the downstream process. Protectants and other compounds may be added before or during drying. This depends on the method used and on the type of formulation. For instance, when granules are made by spray-drying, protectants may be added in the drying process, as well as formulants with other functions that play a role in spraying characteristics or in enhancing persistence. Standardization must be realized in formulation. Each product batch must contain similar numbers of viable propagules within a narrow range. This often requires counting, germination tests and/or bio-assays whose results are used to set the standard content or potency of the active ingredient. A difficult issue in setting the standard is whether over-formulation can be done to compensate for loss of viable propagules over time. An example: if a shelf-life of 1 year is aimed for, but survival of propagules is only 80%, over-formulation can be done to 125% so that after 1 year 100% is still viable, which is in absolute numbers the required amount. Registration generally demands stability over the shelf-life period within a narrow range, but the possibility of over-formulation is not well laid down in the regulations. Death of 20% of the propagules is likely to be accompanied by a quality decrease in the still viable ones. This is an aspect that has hardly been studied in pathogens. Apart from the quality concerns, costs also play a role. Generally, 10% over-formulation seems acceptable, but there is no information available on what companies actually do in this respect.

The formulation needs to be physically stable, i.e. no clump-forming in wettable powders or irreversible sedimentation of propagules in suspensions should occur, nor separation of carriers in the case of emulsions. Besides protecting the active ingredients from degradation, formulation also serves to inhibit growth of possible contaminants. Other unwanted microorganisms may be present in low numbers and need to be killed or limited in their presence and growth as much as possible. Not only could they harm the active ingredient during storage, they could interfere with the biological activity once applied, or pose a risk to the applicator and consumer. Also registration requirements reply to product stability with regard to physical-chemical stability as well as to biological stability.

3.1.2 Second Function of Formulation: Efficacy and User-Friendliness

The second function of the formulation is related to its use in the field, where the formulation should enhance the desired effect on the target pest. Pests can acquire pathogens by direct contact and by secondary pick-up from the spray deposit via contact or *per oral*. Another possible route is pick-up after a primary growth cycle on the leaf or secondary growth after cycling through a dead insect. The latter is not influenced by formulation. A primary consideration is the spray characteristics in case of a foliar application that is aimed at an optimal coverage of the target pest and/or the target's host plant. Next, good contact with the target insect should be aimed for, as well as enhancement of the activity of the organism such as germination and growth on the insect and on the foliage. Stimulants could also be added to the formulation to enhance oral uptake or to modify the insect's behaviour through which more propagules may be picked up. In the case of a soil application, an even distribution matters more than spray characteristics. Spray characteristics are a well-studied area in chemical products, but this is an almost neglected area in biopesticides. Droplet size spectrum and droplet behaviour during spraying (evaporation and drift) and when droplets hit the plant surface (bouncing off, spreading and drying) can be manipulated by the formulation and the spraying equipment. Propagules should be easily dissolvable and well dispersed in the spray solution and be evenly distributed over the droplets, and to the plants and targets. These spray characteristics can be influenced by additives such as surfactants, spreaders, stickers, humectants, etc. The list of compounds for these functions is huge. Finding the right additives requires an extensive range of investigations on the target and target crops. Few papers refer to these parameters as part of a biopesticide formulation, but they are a relevant part of successful targeting the pest. Bateman and Alves (2000, Fig. 1) illustrated where spray application effects can be manipulated by the formulation of biopesticides. Co-formulants can influence propagule distribution in the spray droplet spectrum, evaporation of the spray solution during atomization, and effects on the deposit on the leaves. The latter can be retention, spreading, drying, survival and availability of active ingredients for the target insect. Particularly persistence, adherence, protection, propagule growth and feeding stimulants are important for mycoinsecticides (see Fig. 3.2). Evans (1999) reviewed the principles of dose acquisition and transfer of pathogens. He emphasized the connection between the biology of the microbial agent and the biology of the target, as well as the delivery system. Bateman and Alves (2000) studied the effects of equipment type in relation to the numbers of spores per droplet and the impact on delivery to the target. Their work underlines the role of these factors for biopesticides. A formulation should be appropriate for a variety of crops and field conditions.

Other aspects are convenience of use and compatibility with the end-users' equipment and practices. The formulation should be an acceptable quantity with regard to logistics, storage, handling, and to measuring the desired quantity for an application. An unmanageable quantity that needs cold storage is a big disadvantage that will

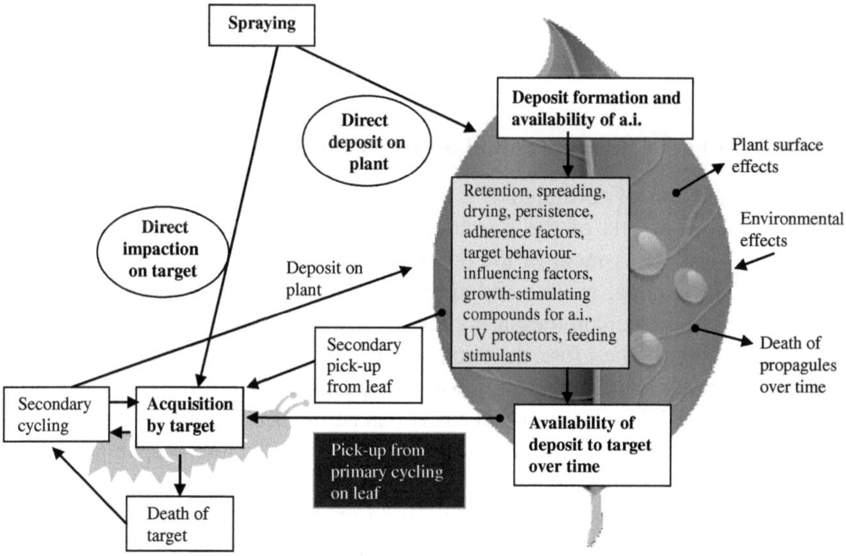

Fig. 3.2 Spray application, deposit and pick-up processes of the active ingredient. The *shaded box* represents effects which can be manipulated by the formulation and which are particularly important for mycoinsecticides. The *dark box* represents effects only relevant in mycoinsecticides (modified after Bateman and Alves 2000)

cause failure of the product in the market. An example from my own experience in the mid 1980s is R350, a fungal nematicide based on *Arthrobotrys irregularis*: it needed cold storage and the rate of application was 1,000 kg/ha. Further, the formulation consisted of mycelium-overgrown cereal kernels that formed one big mass that needed to be separated manually into small pieces that could be spread over the soil. Besides efficacy problems and costs, the difficulties with the formulation and with logistics caused this product to fail in the early days of biopesticides.

3.1.3 Third Function of Formulation: Protection and Persistence

The third function of formulation is to protect the propagule from harmful environmental effects and increase its persistence on the target site. Protection is needed against desiccation, high temperatures, de-activation by sunlight, and washing off by irrigation or rain, particularly when used outside the greenhouse. Good persistence increases the chance of secondary pick-up and pick-up after a growth cycle of the pathogen on the leaf. The addition of growth stimulants to the formulation may give rise to the formation of new propagules which increase the chance of secondary pick-up. This is particularly useful with fungi as illustrated by the following example. Additives in the formulation of Mycotal allow the fungus *L. muscarium* to grow on the leaf surface and form new conidia. These conidia are formed in clusters and covered with a sticky slime that prevents desiccation and improves adherence to insects. This slime dissolves in water and as a result is not present anymore on spores that have been sprayed onto the plant. Newly formed spore clusters are produced in slime heads and enhance secondary pick-up by passing insects such as thrips (Hall 1980).

3.1.4 Fourth Function of Formulation: Registration and Safety

The fourth function of formulation relates to registration demands and concerns possible harmful effects of the formulation on, especially, the applicator. In the toxicological risk assessment of the product, exposure to the microorganisms is a concern during loading, mixing and spraying. Especially dusty formulations may pose a hazard to the person measuring the needed amount and during pouring this into the spray liquid. Therefore, the trend in agrochemicals is to change from powder formulations to liquids and granules. Most biopesticides are formulated as a powder, and small particles as spores can easily can be inhaled and thus pose a possible risk. This may cause sensitization by or allergenicity to the microorganism. Liquid formulations are difficult to make with microorganisms, because water activity has to be low to keep the propagules inactive and to prevent contaminants from developing. However, currently an increasing number of oil suspension formulations is being developed, which minimize inhalation risks during loading and mixing. The aspect of microbial impurities is also a concern in registration and needs attention during formulation.

3.1.5 Formulation in Academic Research

Many authors have recognized the importance of a good formulation (Couch and Ignoffo 1981; Rhodes 1990; Guillon 1993; Hofstein and Fridlender 1994; Jones et al. 1997; Jones and Burges 1997; Couch 2000; Boyetchko et al. 1999; Brar et al. 2006; Jackson et al. 2010) and many aspects of formulation have been studied by scientists and by the biocontrol industry. A comprehensive overview of formulation of microbial pesticides is given by Burges (1998). This book covers formulation of beneficial microorganisms and EPNs, including usages as insecticides, herbicides, fungicides and seed treatments. It discusses the principles of formulation and specific features per group of microorganisms. This review is very helpful for any product developer. Jones and Burges (1998) recognized four basic functions of formulations; Couch (2000) listed six criteria for Bt's that need to be satisfied for commercial formulations. All criteria are similar to the first three described above. These authors, however, do not mention the aspects related to registration, but these are surely relevant and need to be considered seriously. For an overview of the different types of formulations, such as dry products (wettable powder, granules, etc.) and liquid suspensions (water or oil-based, emulsions, etc.), I refer to Jones and Burges (1998). They also included the advantages and disadvantages of equipment that can be used for various specific application methods for each group of entomopathogens. A more recent overview is given by Lisansky (CPL 2006) with a specific focus on commercial formulations. He also provided a number of tables with the advantages and disadvantages of wettable powders, flowable concentrates, granules and high-potency formulations.

In academic research, formulation has received much attention. Often, investigations have focused on a specific aspect of the formulation. Rarely are all aspects covered. In Burges (1998) the state of the art for the mid 1990s is presented for the insect pathogens covered in this book. I will review developments since then per type of microorganism and relevant to bio-insecticides for greenhouse applications, particularly for foliar applications. The details of formulated commercial products, however, are rarely published or patented and in most cases details are kept as trade secrets. I will also discuss possibilities to improve efficacy of biopesticides by using adjuvants, by other enhancing agents, and new developments in formulation.

3.1.6 Formulation of Entomopathogenic Bacteria

Standardization, spray characteristics and protection from environmental factors must be considered in formulations of bacteria. The formulation of bacteria seems to be the least demanding when the four pathogen types are compared, at least when it concerns bacterial spores and proteineous substances, as in *Bacillus* species. Spores formed by these bacteria are long-lived and resistant to heat and desiccation. There is a big difference between these dormant stages versus metabolically active stages. For bacterial cells as in the case of *Serratia* spp. and *Pseudomonas* spp., formulation is much more complicated. As only Bt's are used in greenhouses, these will be considered here. The good stability and easy handling of Bt formulations surely are part of their success. Shelf-life is generally 2 years at room temperature. Points of attention here are the concentration of the spores, the drying technology, and the determination of the potency standard (Skovmand et al. 2000). Most formulations are wettable powders or water dispersible granules, some are oil-based or water-based or mixtures thereof; detailed information is given by Burges and Jones (1998a) and Couch (2000). Every conceivable type of formulation has been developed with Bt. Brar et al. (2006) review recent advances in formulations of Bt's. They summarize screening factors for which additives need to be chosen and they give a list of different types of additives and their function that are used in microbial formulations. Formulation requirements and formulation types are different depending on the usage of a Bt product, in agriculture, forestry or in aquatic habitats.

Advances in formulation technology like encapsulation and controlled release formulations offer possibilities to improve efficacy. Brar et al. (2006) also review possibilities to better protect Bt's from harmful factors such as sunlight, rainfall, dew, extremes in temperature, pH and foliage effects. They recommend studying the whole process from fermentation to formulation since many factors are interconnected. The large Bt producing companies have developed successful and convenient formulations, but details of commercial formulations are proprietary. Improvements are always possible and new formulations appear on the market. In most Bt products the potency has been doubled in recent years in order to make the products more cost-effective and to comply with the trend to reduce spray volumes. To increase the effect of Bt's, feeding stimulants have been studied, but satisfactory improvements have not been found and feeding stimulants are generally not included in formulations nor added as tank-mixes (D. Avé, personal communication).

3.1.7 Formulation of Entomopathogenic Fungi

Formulation of fungal propagules is the most challenging among the four types of pathogens discussed in this study. This is related to their mode of action which is primarily by adhesion to the outer parts of an insect and penetration through the cuticle. This infection process is described in detail by Butt (2002). The infection process is complicated compared to bacteria and baculoviruses with which infection takes place after take-up *per oral*, or compared to the active search and penetration behaviour of EPNs as part of the infection process. Formulation of fungi poses challenges related to targeting, persistence, protection against harmful environmental factors, growth stimulation and enhancing agents. Generally, formulations of fungal products contain spores, either conidiospores or blastospores, sometimes mycelial particles.

The literature on formulation of fungal spores is vast. Burges (1998) reviewed the developments in formulation of mycoinsecticides and provided an excellent overview of the various formulations and additives that stabilize and protect the propagules and enhance the biological activity. Vega compiled a bibliography called "references dealing with formulations of fungal entomopathogens", see the website http://www.sipweb.org/fungi.htm. There are about 150 references, updated to 1999. Most products are based on aerial conidia, some on conidia and hyphae on a granular substrate, and some on blastospores (De Faria and Wraight 2007). Most entomopathogenic fungi produce large amounts of aerial conidia which are responsible for dispersal and infection. These spores are thick-walled and relatively stable. In solid state production these spores are also formed in high numbers. These characteristics make them the preferred propagule for formulating final products. Some fungi are difficult to produce efficiently on solid media and submerged produced blastospores are used for formulation. Blastospores may also be more infectious as is shown in *I. fumosorosea* against *Bemisia argentifolii* (Jackson et al. 1997) and in *M. anisopliae* against *Locusta migratoria* (Kleespies and Zimmermann 1994). Wraight et al. (2001) reviewed stabilization and formulation in relation to production of the different kind of propagules of entomopathogenic and herbicidal fungi. Feng et al. (1994) reviewed formulations of *B. bassiana* based on conidiospores, blastospores and mycelium. Jackson et al. (2010) reviewed the ecological considerations in producing and formulating fungal entomopathogens with emphasis on interactions between the propagule and the insect cuticle, and the role of a surfactant.

Formulation-related aspects can be improved during production and this area has received considerable attention in entomopathogenic fungi. Hallsworth and Magan (1994) demonstrated that inoculum quality (in this case virulence and shelf-life) can be improved by the culture conditions (low water activity) and nutrient composition in conidia of *B. bassiana, M. anisopliae* and *P. farinosus*. Jackson et al. (2003) studied blastospore production in *I. fumosorosea* and showed that blastospore tolerance to desiccation can be positively affected by supplementing certain medium components. Ypsilos and Magan (2004) showed that modifications in physico-chemical culture conditions such as water stress, and washing treatments of blastospores of *M. anisopliae* have the potential to improve spore quality in terms of drying-resistance and storage stability. Tolerance to desiccation is an integral factor in the process of drying blastospores and for long shelf-life and this should be considered early on in the production. Many features are influenced in the production and this needs to be studied and used in the product development. Formulation additives are another tool to improve the propagules' stability and activity. Tolerance to desiccation can be further improved by formulation additives. Blastospores of *I. fumosorosea* mixed with starch-oil mixtures improved survival of spores during freeze-drying and improved storage stability (Jackson et al. 2006).

Oil-based formulations have been developed for conidiospores of *M. anisopliae* var. *acridum* for locust control. This proved to be a major breakthrough, enabling the usage of this fungus in hot and dry environments with ULV applications (Bateman et al. 1993). This formulation technology has several advantages that made it so successful: shelf-life is quite long for a mycoinsecticide (>18 months at 17°C); good persistence in the field; enhanced biological activity; and usage by conventional equipment at ultra low volumes (Bateman 1997). The precise role of the oil in the infection process is not well understood. Bateman and Alves (2000) suggested that secondary pick-up from plants plays a greater role than direct impaction of spray droplets in the case of locusts. This would mean that persistence of spores was improved by the oil as well as the pick-up and adherence to insects. These aspects relate to efficacy, user-friendliness, and costs, and resemble the use of a chemical, making acceptance of this product easy. This technology was developed in the Lubilosa Programme (www.lubilosa.org) and the product Green Muscle, based on *M. anisopliae* var. *acridum*, is formulated in mineral oil.

Oil formulations have been developed with mycoinsecticides (Botanigard ES, Naturalis, Green Muscle) based on mineral and paraffin oils. It is believed that oil formulations increase spore adherence to insects because they improve contact between spores and the hydrophobic cuticle of insects. In oil-based formulations sprayed with water as a carrier, the assumed mode of action is that the mixture once sprayed with a pressure of 10–15 bar forms an emulsion with fine oil droplets. Under high pressure and considerable shear forces an inverted emulsion may be formed with oil droplets that contain water and spores. These oil droplets may enhance adherence to insects once the water has evaporated. In these droplets the spores may germinate and grow out of the droplet and infect an insect. In this way, spore germination and adherence are enhanced, resulting in greater insect mortality and reduced dependency on a high relative humidity and the availability of free water. Where an invert emulsion does not occur, the result might be protection of spores from dehydration under a thin film of oil (Auld 1993) and a better adherence to insects, still resulting in improved efficacy. These effects were seen using a vegetable oil-based adjuvant with Mycotal, which contains hydrophilic conidia of *L. muscarium*, for control of thrips (Van der Pas et al. 1998). Many other studies indicate that oil formulations improve the efficacy of mycoinsecticides; a review is given by Inglis et al. (2002). On the other hand, Ugine et al. (2004) showed that there was no increased secondary acquisition of conidia of *B. bassiana* by *Frankliniella occidentalis* adults and larvae between unformulated conidia and two formulations with conidia. These were the WP and the ES (emulsifiable suspension) formulations of Botanigard. It was tested, however, under optimal conditions for the fungus, at 25°C and a 100% RH. Formulations are developed to enhance product efficacy under sub-optimal conditions, e.g. lower RH conditions. Testing under optimal conditions is not likely to show enhancement of efficacy by a formulation. It also may explain why the oil-based formulation (ES) did not increase conidia acquisition. *B. bassiana* spores posses a hydrophobic cell wall (Inglis et al. 2002), so any adherence effect of oil is much less relevant, even under optimal conditions for the fungus.

Vegetable oil formulations could also be developed to improve shelf-life of vulnerable fungal material. This has been studied with mycelium of *Lagenidium giganteum* in water-in-oil emulsions, together with prevention of sedimentation of cells (Vandergheynst et al. 2007). Oil thickeners based on silica nanoparticles proved to be effective in preventing sedimentation to a large extent in the formulation as well as in the mixture with water. The authors assume that shelf-life was improved by protecting propagules from desiccation in the water-in-oil matrix. In biocontrol of weeds with pathogenic fungi various kind of oil formulations have been tested, including invert emulsions, oil suspension emulsions, and many adjuvants. An overview is given by Green et al. (1998).

Many compounds have been shown to protect spores and to enhance biological activity. This area of research offers interesting ideas that could be useful in developing formulations with entomopathogenic fungi. Other non-agricultural areas where microorganisms are used for pest control should also be considered. Examples are the control of mosquitoes and weeds in aquatic environments, insect control in forestry where specialized formulations, often oil-based, have been developed for low volume aerial applications, and control of termites, ants or cockroaches in domestic usages.

Some formulations consist of the production substrate including the spores. These are usually fungi grown on rice or cereal kernels where the spores are not separated from the production substrate and where fungus-overgrown granules are applied to soil. Examples are Melocont from Kwizda (Austria) and Engerlingspilz form Andermatt Biocontrol AG for control of white grubs in meadows, lawns and orchards. The production matrix is here the final formulation and these products are not formulated with any additives.

A small number of products have been formulated with mycelium. The product Laginex AS (Agraquest, USA) for control of mosquitoes, based on the fungus *L. giganteum*, is formulated as an aqueous suspension with mycelial parts since production of oospores gives uneconomic yields (Vandergheynst et al. 2007). The product is no longer on the market due to an unstable formulation with a short shelf-life and a limited market (D. Edgecomb, AgraQuest; personal communication). A formulation of *M. anisopliae*, based on mycelial pellets, was developed by Bayer, but it was never marketed due to economical constraints (Zimmermann 2005). The product Mycar, based on mycelium of *H. thompsonii* was put on the market, but it was withdrawn because results were insufficient (McCoy et al. 1975). In mycoherbicides mycelial formulations are also used. An example is the fungal herbicide Chontrol Paste, from MycoLogic Inc., Canada, which contains mycelium of *Chondrostereum purpureum* for inhibition of re-sprouting of tree stumps (www.epa.gov.). A definition of formulation types and an overview of mycoinsecticides and mycoacaricides, including the type of the commercial formulation, are provided by De Faria and Wraight (2007).

3.1.8 Formulation of Baculoviruses

Key aspects in the formulation of baculoviruses are standardization, microbial contamination, protection against harmful environmental factors, improvement of the speed of kill by faster and greater uptake of viruses, as well as enhancement of the pathogenicity of the virus. Shelf-life is of minor importance because products can be stored for many years when kept at approximately –20°C, or 1 year at refrigerated temperatures, and even for a few weeks at room temperature. Products can also be re-stored from one temperature to the other to some extent. Spray characteristics should ensure an even distribution on foliage for optimal oral uptake by caterpillars.

A main issue in virus formulations is microbial contaminants. Purification of the crude technical product is needed for product stability and registration requirements, and various methods can be used (Huber and Miltenburger 1986). On the other hand, it is well known that remainders of the dead insects enhance stability of the virus during storage and that these particles also aid in the protection against UV radiation (Huber 2005). This creates a formulation challenge for producers who will have to find a compromise between these two aspects. At the same time, the formulation needs to be set at certain potency, and counting occlusion bodies (OBs) or polyhedral inclusion bodies (PIBs) is difficult and unreliable. Particles in the formulation can be hard to distinguish from the OBs or PIBs and inactive ones are also counted. Bio-assays are needed to standardize the product.

Many researchers have focused on formulation improvements such as adding feeding stimulants, stickers for rain-fastness, and protectors against UV. It is generally accepted that exposure to sunlight rapidly inactivates baculoviruses. Accordingly, most formulation studies focus on protection from harmful wavelengths. Overviews of formulation methods and research are given by Burges and Jones (1998a) and by Hunter-Fujita et al. (1998); both papers include extensive lists of wetting agents, stickers, phagostimulants and UV protectors. Encouraging results have been obtained in apple using CpGV with various additives against the codling moth (Ballard et al. 2000). The use of these compounds (molasses, sorbitol, ά-farnesene) led to less codling moth damage, although the exact mechanism was not clear. Phagostimulants, built into a granular formulation of SfMNPV, increased infection rate in *Spodoptera frugiperda* in maize and improved persistence on crop foliage compared to an aqueous spray application (Castillejos et al. 2002). Sunscreens to block damaging UV wavelengths have been studied widely (Burges and Jones 1998a). Optical brighteners can give protection against UV radiation and at the same time enhance baculovirus activity (Dougherty et al. 1996). The results, however, are considered insufficiently effective for open field applications. The use of adjuvants based on lignin and particle films were studied as putative solar protectants, but showed no significant improvement in field trials (Arthurs et al. 2008). The situation for feeding stimulants and stickers is similar and the use of these compounds is not recommended in practice (M. Andermatt, personal communication). On the other hand, Lasa et al. (2007) showed that the use of an optical brightener, i.e. a UV blocker, in the formulation increased mortality in greenhouse trials with SeMNPV against larvae of *Spodoptera exigua* in Spain. Mortality increased in late instars larvae, whereas persistence was not improved by the additive. The effect on mortality was most likely due to an increased rate of virus acquisition, suggesting that the optical brightener leucophor AP increases pathogenicity of this virus and/or acts as a feeding stimulant. The effect as a UV blocker was not seen. Apparently, UV radiation is much less harmful in a plastic-covered greenhouse than in the field.

The formulation ingredients of current baculovirus products are not known as the information is proprietary. Formerly, virus products were often formulated as wettable powders, particularly those used in forestry (Young and Yearian 1986). Oil formulations have been studied, but oils are often repellent or act as an anti-feedant to the target pests (Cherry et al. 1994). Nowadays, most formulations are aqueous suspensions that are developed for spray applications in orchards and greenhouses and are user-friendly.

3.1.9 Formulation of Entomopathogenic Nematodes

Nematodes require formulation for similar reasons as do microorganisms, although they present unique features. Shelf-life and user-friendliness are integral to an optimal nematode formulation. Persistence in the soil cannot be influenced by adding formulation ingredients, and registration is not an issue of great importance. The DJs are relatively large survival stages that are able to move actively. They also present challenges to oxygen and moisture requirements and to temperature conditions during storage, and are more sensitive in this respect than the microorganisms. Generally, nematodes are applied to the soil which poses less formulation considerations then when pathogens are applied onto foliage where spray characteristics and protection from environmental factors are necessary. On the other hand, when nematodes are applied through irrigation systems, blocking of filters and small drippers, either by the DJs itself or formulation lumps, needs to be avoided. DJs can be stored in refrigerated and aerated water to supply sufficient oxygen and to prevent settling. For storage and transport of products, formulation is needed. The main goal is maintenance of quality for as long as possible. In formulation, the stage of DJs of nematodes is used in which they become metabolically inactive and can withstand a relatively high degree of desiccation and low temperatures. Further, the growth of contaminants must be prevented. Species, and even strains, differ in their biology, and storage and formulation needs to be studied case by case (Strauch et al. 2000). Temperature should be kept as low as possible to ensure a long shelf-life. Once temperature allows activity and mobility, DJs use up oxygen and their energy resources, which will negatively influence shelf-life and ultimately quality and efficacy, although it can vary depending on the species (Patel et al. 1997). Once DJs have moved inside the package, the formulation ingredients protect them less efficiently and quality rapidly decreases, even when the temperature is restored to a level where they can not move anymore. In this respect nematodes are much more sensitive than non-moving propagules of pathogens. High temperature can be lethal. The moisture content of the formulation demands a balance between the level that keeps the DJ's from desiccation and the level that prevents them from moving. Moisture content should also be low to prevent contaminants from developing.

Formulations need to be tailored towards the species and the type of formulation, and requires studying the physiological chemistry of a nematode and its ecology and behaviour. A historical overview of formulation with EPNs is given by Georgis and Kaya (1998). The authors also identified significant steps in the process of formulation (Table 9.6, p. 298). A more recent overview of the various types of formulations is presented by Grewal (2002). Many types of commercial formulations have been developed and marketed, including water-dispersible granules. This last formulation was not successful and was taken from the market (Georgis 2002). The dominant and successful formulations are clay powder formulations and gel formulations.

Liquid formulations may have several advantages over solid matrix formulations. The latter have a relatively short shelf-life of about 3 months, are difficult to handle, and the entire package needs to be used up at once because of the non-homogeneous nature of the package contents. High temperature effects are irreversible due to the change of the EPN-matrix mixture when DJs have crawled out of the carrier. On top of that, active nematodes produce heat and gases and contaminants may start to develop, resulting in a negative spiral effect with regard to product quality. Aqueous formulations, however, have the disadvantage of sinking of EPNs and lack of oxygen. Wilson and Ivanova (2004) investigated neutral density liquid formulations based on colloidal silica suspensions which could overcome these problems. Survival and virulence with *H. megidis* and *S. feltiae* DJs was improved compared to nematodes stored in aerated water solutions with Ringer's solution at 4°C and at 15°C for *S. feltiae*. Still they found a decline over time of surviving DJs. This is not acceptable in commercial formulations. The number of DJ/ml was at maximum 10,000 per millilitre which would mean a 5 l volume for a 50 million nematode package which is generally about 200 ml with clay formulations. Both aspects are not acceptable for a commercial formulation and efforts to improve these points were not sufficient (unpublished data of our group). Although a liquid formulation has advantages such as easy measuring off the desired quantity, less or no visible residue on plants, and no blockage of nozzles, this idea has not been developed into a commercial formulation due the disadvantages mentioned above.

3.1.10 Foliar Applications of Entomopathogenic Nematodes

There is an increasing interest in using EPNs for control of foliar insect pests such as leafminers, thrips and caterpillars. To be successful, DJs need sufficient time to find and penetrate the target insect. They are only able to move in a water layer and desiccation within a short period of time after application limits their effect. Formulations are mainly developed to have a long shelf-life. For enhancement of EPNs' effectiveness on foliage through good mobility and survival, formulation requirements are different. This formulation should aim at maintaining the presence of the water layer long enough for the nematodes to find and penetrate the host, usually a matter of a few hours. This can be reached by tank-mixing an adjuvant or by developing a new formulation.

Piggott et al. (2000) studied the formulation of DJs in a polyacrylamide polymer and its effect on leafminers. This formulation improved survival of DJs on leaves because of slower drying of droplets compared to water and gave a significant increase of pest mortality, compared to DJs in water, and water plus a wetting agent. This formulation is marketed by Becker Underwood as Nemasys F for use against thrips and leafminer. The exact formulation is proprietary and not known, however. Enema and Koppert developed similar gel formulations. These formulations have a similar shelf-life as the clay formulations, and can be used for foliar as well as for soil applications. The gel formulations also have the advantage that there is no visible residue on plants, unlike the clay formulations.

The use of formulation additives to enhance EPNs effectiveness on foliage has been studied by Schroer et al. (2005a) against diamondback moth larvae. Many different compounds such as polymers, surfactants, oils, waxes, and mixtures thereof were tested in combination with *S. carpocapsae* DJs in bio-assays on cabbage leaf discs. The authors found that a mixture of a surfactant and a polymer gave the best results, suggesting that a good deposition and distribution of DJs over the foliage is important as well as a prolonged DJ survival. In leaf bio-assays Schroer and Ehlers (2005) demonstrated that the effect of the surfactant-polymer formulation is mainly due to support of the DJ's performance on the leaf surface rather then to improving DJ survival. Survival was prolonged, but diamondback moth larvae mortality did not increase over time. The results indicate that insect penetration by the nematode on the leaf occurs within the first hour after application. Field trials under tropical conditions did not show any additional effect of the polymer-surfactant or surfactant formulation. This was attributed to humid, rainy conditions during the trial (Schroer et al. 2005b). The authors conclude that the use of anti-desiccants only slightly improves the effectiveness of EPNs on foliage and they recommend not using them due to higher costs and the small additional effect.

Studies in our group confirm that the most significant factor in foliar use of nematodes is contact between target and nematode, and that conditions in the first hours are determinative. Prolonged survival by means of anti-desiccants does not seem to be very useful. The above-mentioned gel formulations have an anti-desiccant effect and give reasonable control of foliar pests. The use of a spreading agent may improve spray characteristics and contact between DJ's and target insects. Using a separate adjuvant seems to be the most practical. This is because EPN products generally are used for soil applications and do not need this kind of formulation features, so building in features for foliar applications would be unnecessary. The use of adjuvants with EPNs offers a potential improvement in the use of EPNs against foliage-feedings insect pests and needs to be further studied in field trials focusing on initial contact between nematode and host.

The application method for foliar treatments with nematodes needs careful investigation. As DJs are large particles, the suitability of equipment (pressure, nozzle size), droplet size, survival, distribution and canopy structure are determinative. These aspects have been studied by several authors, including the use of adjuvants, but optimal application methods have not yet been developed (Lello et al. 1996; Shapiro-Ilan et al. 2006).

3.1.11 Entomopathogens and Adjuvants

Efficacy of entomopathogens can be enhanced by certain additives in the formulation or by tank-mixing adjuvants. Additives are defined as compounds incorporated into a formulation and their role has been mentioned several times above. Adjuvants are defined as separate products that can be tank-mixed with (bio)pesticides in order to enhance their efficacy. Adjuvants are increasingly used to improve the biological activity of pesticides. An overview of why, where and what kind of adjuvants are used with chemical pesticides is given by Stock (1997). The main reasons are spray characteristics such as atomization, foliar retention, foliar coverage, foliar penetration and enhancement of the mode of action of the active ingredient. Adjuvants should have no biological activity on their own. This is also a requirement related to regulations of adjuvants. Adjuvants are needed because there are limitations to the amount of additives that can be built-in into one pack-formulations (products) and some additives are not compatible with the active ingredient and can only be tank-mixed. Another reason is that an additive may not be compatible with all the foreseen usages of the product; they could be phytotoxic to some crops, but useful in some other crops. There is a limit to a multi-purpose formulation.

The considerations for developing and using adjuvants with biopesticides are given in Table 3.7. The range of adjuvants is increasing; the adjuvant industry is growing, with their own newsletters (e.g. Adjuvant Newsletter) and conferences (the 9th International Symposium on Adjuvants for Agrochemicals was held in 2010) and specific adjuvant market reports. Specialized consultants are active in this field, such as Alan Knowles (Form-AK, UK) and Hans de Ruiter (Surfaplus, the Netherlands). The market of adjuvants is estimated to be about US$1.5 billion in 2005 (Knowles 2006). In biopesticides, similar considerations for the use of adjuvants play a role and adjuvants may enhance the biological activity of pathogens or improve their persistence on the target site.

Adjuvants are sometimes recommended in combination with biopesticides. An example is Nu-Film-17 (Nufarm) that was tank-mixed with Madex as a UV protectant and sticker. Results were disappointing and Nufilm is no longer recommended (M. Andermatt, personal communication). Examples with fungi are AddQ, added to AQ10, a mycofungicide based on *Ampelomyces quisqualis* to control powdery mildew (Kiss et al. 2004), which improved efficacy. And, NuFilm added to Naturalis (*B. bassiana*) brought the level of control of whiteflies up to the chemical standard level (Mayoral et al. 2006).

Some adjuvants are specifically developed for usage with biopesticides. Our group developed a vegetable oil-based adjuvant Addit to improve the activity and persistence of Mycotal (Van der Pas et al. 1998). Initially, we attempted to develop an oil formulation with the spores of *L. muscarium* in order to increase the efficacy and persistence on the foliage. Due to technical problems such as sedimentation and clogging of the spores over time, we abandoned this idea. Another reason was the need to register a new formulation, while launching an adjuvant only needs a limited registration or none at all in some countries. The product Addit is recommended to

Table 3.7 Considerations when using tank-mixed adjuvants with biopesticides

Adjuvant aspect	Features that need to be considered
Formulation functions	Improve spray characteristics, enhance mode of action, increase persistence, improve secondary pick-up, etc.
Efficacy	Should not have direct efficacy on its own
Costs	Additional end-user's costs should be acceptable
Quantity	Logistics, storage, handling
User-friendliness	Easy to mix (dilution, emulsion) and apply
Undesired effects	Phytotoxicity, growth of sooty moulds, visible residue; harmful effects on natural enemies, etc.
Registration	Often needed, sometimes exempt; often easier than a new formulation registration

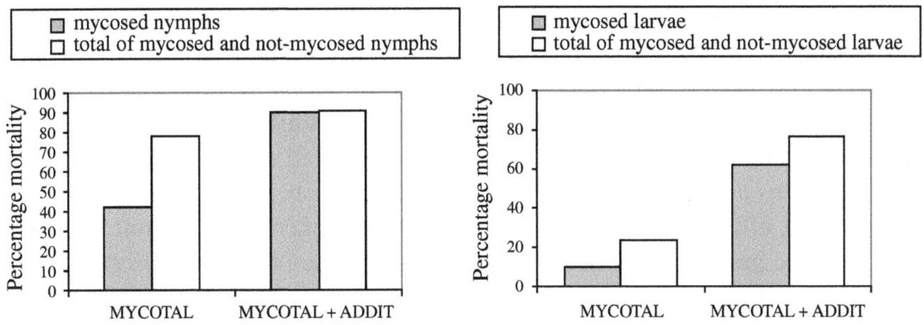

Fig. 3.3 Mortality of Mycotal and Mycotal + Addit on whitefly nymphs (*Trialeurodes vaporariorum*) (*left graph*) and thrips larvae (*Frankliniella occidentalis*) (*right graph*) 7 days after 2nd application on cucumber (RH 55–80%) (Van der Pas and Ravensberg, 1995, unpublished data)

be tank-mixed with Mycotal at a dose of 2.5 ml/l of the spray liquid. It increases the mortality of whitefly and particularly of thrips (Fig. 3.3). It shows mycosis earlier and more pronounced than in the control, so that mortality is easier to assess by the grower and most importantly, its activity is less dependent on a high relative humidity. The main effect of the oil is the improved secondary pick-up of spores resulting in a quicker death and a higher mortality overall (Van der Pas et al. 1998).

As a result of the development of Addit, the usage of Mycotal has increased, demonstrating the value of an adjuvant. The addition of arachid oil also improved the efficacy of *V. lecanii* against powdery mildew on cucumber and reduced the humidity dependence (Verhaar et al. 1999). Curtis et al. (2003) studied the effect of nutrient and non-nutrient additives to conidiospores of an Australian *V. lecanii* strain (FI40) derived from an aphid. They found that nutrient additives in the spray solution, such as whole egg powder and glycerol, increase growth on leaves and infection in aphids. A period of over 15 h of high humidity was still needed to get high mortality. Oil-based additives did not improve mortality rates. Our group found that an oil adjuvant did not improve the efficacy of Vertalec (blastospores), in contrast to Mycotal (conidiospores). Whether this is related to the spore type, the target insect or to strain differences is yet unclear. With Bt, the use of cotton oil as a tank-mix improved the efficacy in sweet corn on three lepidopteran pest species considerably, although the mechanism is not fully understood (Hazzard et al. 2003). The use of adjuvants for foliar applications of EPNs is discussed above.

The quantity and the cost are the issues in using adjuvants. Often, improvements on efficacy can be realized by tank-mixing an adjuvant, but in many studies the quantity required is prohibitive. A typical example is the study of Ballard et al. (2000). They found that adding 10% (v/v) cane molasses or 10% (w/v) sorbitol or 0.08% ά-farnesene to CpGV treatments with 250 l/ha in an apple orchard significantly improved levels of codling moth control. Damage reduction was 41% with pure CpGV and 74% with pure CpGV plus molasses. In this case, 10% molasses is 25 l, costing about €25 (purchase price (in 2007)), without any sales margins. Adding sorbitol gave a significant increase of control, but it would mean using 25 kg/ha, costing about €200 (purchase price, in 2007). Efficacy was also improved by adding ά-farnesene in a quantity of 200 ml, but the additional cost is about €55 (purchase price in 2007). One treatment with a CpGV product costs a grower about €60/ha in 2007 in the Netherlands. From a practical point of view, this renders the use of these compounds as adjuvants in terms of logistics and costs impossible. If the price of an adjuvant increases the cost of the overall treatment considerably, a grower chooses another product. An adjuvant must be cheap and easy to use.

Undesired effects of adjuvants need also to be taken into account. Molasses may stimulate growth of sooty moulds. Oils may cause phytotoxic effects. Optical brighteners and other compounds may cause visible residues which can be a problem on ornamentals and on shiny fruits like eggplant and sweet pepper. Besides, harmful effects on natural enemies may occur. All these aspects need to be considered and studied.

The use of biopesticides is often criticized for its lack of reliable efficacy. Adjuvants offer a potential improvement, and more research should be dedicated to investigating the combination of biopesticides with both available and new adjuvants. This is a relatively quick and affordable approach by which efficacy of microbial products can be improved. Adjuvants are regulated in most countries, but dossier requirements for registration are much less than those of biopesticides.

3.1.12 New Formulation Technologies

Formulation technology offers a promising field for improvements in terms of efficacy and user-friendliness. This will improve the uptake of biopesticides in the market. New research, new knowledge and new products open ways to develop more sophisticated formulations. The agrochemical industry is constantly trying to improve their formulations and new developments are taking place that could be interesting for the biocontrol industry. The latest are micro-encapsulated formulations, granular formulations and replacement of harmful solvents agents by vegetable oil or water. Some examples: Syngenta and Zeneca are looking at micro-encapsulation techniques (Anonymus 2004; Perrin 2000), Bayer is developing vegetable oil/additives dispersion formulations to optimize coverage and penetration and to improve spray characteristics and rain fastness (Anonymus 2006) and Dow AgroSciences is developing small droplet sized water-based formulations (Anonymus 2007). Incentives are the wish to create more user-friendly products, using less active substances and less to no organic solvents, better efficacy, reduced exposure and safer formulations for human and the environment. Formulation research for biopesticides can learn from these developments as these are going towards safer compounds and formulation concepts that may be useful for biopesticides. There are some examples with biopesticides: granular formulations such as dry flowables and water dispersible granules have been developed for Bt products. Aqueous suspensions have been developed for organic use: Serenade ASO and Madex.

Many new products and new concepts have been developed by the additive and adjuvant industry. Better understanding of the role of surfactants, dispersants, emulsifiers, spreaders, and other additives is resulting in new and safer formulations. These new concepts are used for built-in formulations as well as for tank-mixed adjuvants (Knowles 2006). In the past, adjuvants were mainly studied for use with herbicides, while current studies include usage with insecticides and fungicides. Since many of these developments are focusing on safer additives for humans and the environment, they may also be useful for formulating living propagules for biopesticides. Some examples of new co-formulants that can be used in biopesticides are compounds such as natural and modified alcohols, fatty acids, starches, sugars, polymers, vegetable oils, waxes, etc. Examples of companies involved in the production of these compounds are Akzo Nobel, Clariant, Degussa, Helena, Seppic, Croda and Uniqema.

3.1.13 Enhancement of Entomopathogens Through Formulation

Improvements can be made by manipulating biological characteristics of a pathogen and technical aspects of the formulation. Biological improvements can be made in various ways. A review of potential improvements through biological features is given by Charnley et al. (1997), but few of these have been incorporated in products. A better understanding of production aspects related to virulence and quality of propagules and a longer shelf-life is needed. This requires more fundamental research. Nutrition can influence virulence and quality of propagules (Butt et al. 2006). Fungi produce cuticle-degrading enzymes and bioactive secondary metabolites (Butt 2002) and a better understanding of when and which ones are produced and what their role is in the infection process could help improve fungal products. We also need to know more about these enzymes and metabolites and how to exploit them better in the production process and in the formulation. Once more fundamental knowledge becomes available, it will be possible to devise strategies to improve strains. If features such as a broader host range, improved speed of kill, reduced dosages, and increased resistance to harmful environmental factors could be improved, biopesticides may become better. Other possibilities to improve strains are somatic hybridization and genetic modification, but the latter is not accepted in Europe.

A novel approach for getting better results is to search for organisms or strains of organisms that are able to grow and reproduce on foliage or roots. Koike et al. (2004) found that strains of *L. muscarium* with an epiphytic character are able to colonize the leaf surface and give a higher mortality of whitefly over time than strains without this characteristic. This feature could be improved by designing a formulation that stimulates epiphytic growth. Similar examples are epiphytic bacteria like *Pseudomonas syringae* that colonize the leaf surface and thereby give an effective protection from frost injury (Hirano and Upper 2000), and root colonizing antagonistic fungi that are able to protect plant roots from pathogenic fungal diseases as demonstrated with *T. harzianum* (Harman 2006).

Further, there is a wide range of recent technological innovations that could be used to improve pathogen formulations. Shelf-life of biopesticides is critical and new packaging technology may help improve shelf-life. New materials such as oxygen absorbers, carbon dioxide absorbers, and water absorbing gels could stabilize the environment in the package. Packaging under inert gases may prevent degradation of the propagules and result in a longer shelf-life. Additives with a certain function could be used to improve efficacy. These could be compounds that enhance virulence (enzymes such as chitinases, proteases and lipases, or stimulators such as chitin and activity enhancers) or that increase survival (UV blockers, etc). Other additives could be used as growth stimulants or compounds that improve secondary pick-up of spores. The latter may be achieved by compounds with sticky features such as oils and gums; by behaviour modifying substances such as attractants (sugars), repellents (garlic oil), alarm pheromones (β-farnesene for aphids) or by low doses of insecticides that increase insect movement and thereby increase the chance for pick-up of spores. This approach proved to be effective when 1% of imidacloprid, a sub-lethal dose, was used in combination with *L. muscarium*. The chemical dramatically increased aphid movement and as a result secondary pick-up of spores increased and led to high mortality (Roditakis et al. 2000). Similar effects have been reported in trials studying mortality of vine weevil larvae with *M. anisopliae* and low, sub-lethal doses of imidacloprid and fipronil (Shah et al. 2007). New technologies, such as the Exosect technology (www.Exosect.com), may improve the efficacy of biopesticides through an improved adherence of particles, i.e. spores, by formulating them with electrostatic wax powder or magnetic metallic powders. Combinations of functional additives may give synergistic effects.

Technical innovations in the downstream and formulation processes offer another type of potential improvements. Protection of propagules by encapsulating techniques has been studied with bacterial cells of *Pseudomonas fluorescens* and mycelium of the nematophagous fungus *Hirsutella rhossiliensis*. Hydrogel capsules improved storage possibilities and efficacy (Patel et al. 2005). The hydrocapsule technology has been further developed by ARS, Inc., USA (www.ars-fla.com/hydrocapsules/html), and applications for biopesticides have been suggested. Propagules can be protected during formulation processes with different drying and coating technologies. New equipment and technologies together with new compounds enlarge the possibilities by methods such as spray-drying, agglomeration, fluidized-bed granulation, freeze-drying, encapsulation and (film-)coating. With these processes, various types of formulations can be made, starting with either particles or liquids, and resulting in dry flowables, water dispersible granules or capsules. A new formulation technique has been developed to coat propagules with water-soluble compounds and then to dissolve the particles in oil. This can be applied to the targets, such as locusts, where the formulation provides protection of the pathogens against heat and solar radiation (patent US 2004/0038825).

According to Hynes and Boyetchko (2006) there is a high priority for formulation development with a multi-disciplinary approach to optimize biopesticide-pest interaction. They made a plea for a better understanding of the ecology of the interaction. This would certainly be a new area for investigations that may lead to better formulations. The authors also observed that researchers in formulation have focused on individual biopesticides rather than on developing general principles on groups of pathogens. This concurs with my experience. The ecology of the interactions should be studied more, and this should be done in institutional research because of its fundamental character. If pathogenicity factors could be elucidated further, then efficacy could be improved by better adapted formulations, and if application rates could be reduced, biopesticides would become more efficacious and more competitive with chemicals. This would be highly advantageous to both the producer and the user. Budgets should be made available for this kind of academic research in order to improve biopesticides.

3.2 Product Packaging

Product packaging is usually considered from a marketing perspective only. However, it is intrinsic to the success of a microbiological control product. The final formulation needs to be packaged in an appropriate package. The packaging material needs to be compatible with the formulation and its shelf-life. Registration requirements demand evidence hereof by means of shelf-life studies in the final packaging. Other general requirements are that the packaging be well-designed and constructed, that its content cannot escape, that it be strong and solid and will safely withstand normal handling, and that opening and closing be easy. Minimizing product contact and creating convenience to the user are other desirable elements. The material of the package needs to protect the formulation in order to maintain its shelf-life. This is important with biological products, particularly fungi, which are less stable than chemical products. Biocontrol products are more vulnerable to changes in moisture content (clumping or desiccation), oxidation and microbial contaminants. It may be necessary to keep out daylight and it needs to be air-tight to maintain the formulation's moisture level. Products could be packaged in an inert atmosphere (N_2 or CO_2) although I have not yet seen examples with MPCPs. In the case of EPNs, gas exchange is required, however.

Shelf-life, as defined by registration requirements, is the period that the product in the final packaging remains stable under certain conditions. Storage stability of unformulated technical product, often called the technical grade active ingredient (TGAI), and formulated product may differ, as well as the storage stability of formulated product in bulk containers versus in the final packaging. This will differ per pathogen, per type of propagule and per formulation type. Sometimes bulk storage of the TGAI or the formulated product has better storage stability than the end-use product. This will affect operational decisions such as whether, what, when, and how to store and package. In-house and distributor storage of the end-use product uses up part of the shelf-life period which is unfortunate for the end-user and this period should be kept as short as possible. Therefore, in the process of packaging, the timing of making the end-use product is related to the product's shelf-life. The actual shelf-life length determines this decision, especially when shelf-life is relatively short. For instance, DJs of EPNs are usually stored in aerated vessels before formulation. They are formulated upon demand and from this point on packaged in the commercial packaging. The shelf-life period starts here. In the case of some fungi our group has found that bulk storage is better than storage in the final package. Some formulations and pathogens have more flexibility considering this than others; thus, it needs to be approached on a case by case basis. Further, the moment of packaging depends on the frequency of production runs and the demand of the market.

Other relevant considerations to the packaging are user-friendliness, size and costs. It is easy to underestimate the costs of packaging, especially the labour component. Whether this requires manual labour or the use of costly specialized packaging machinery, these costs need to be anticipated. Other costs are cold room storage and stock management. Size and weight of the packaging should be manageable. Handling and using the package should be easy as well as (repeatedly) opening and closing, including measuring off partial amounts of the product. Marketing is an indispensible part of the game. The packaging of the product including the label must look professional instead of amateurish. The user's first impression comes from the product's appearance, i.e. the packaging. Here we are competing with the chemical industry and the image that the product produces in the customer's mind. On the other hand, the biological industry wants to emphasize that it is a biological product and not a chemical one. Biopesticides are innovative products and creative packaging offers an opportunity to deliver this message. This is a challenge for the product developer and the marketer. Rarely is the biopesticide package employed to emphasize the innovative and environmentally-friendly character of the product. The biocontrol industry has largely ignored this element. Improvements call out to be made in this area. In the development of product packaging, commercial as well as biological aspects need to be considered. Product quality and cost of packaging are key factors. Packaging is an imperative element of overall product quality in terms of efficacy, handling, economics and image.

3.3 Field Testing

Field testing the MPCA serves a number of purposes. After the strain has been selected in a bio-assay, its effectiveness needs to be studied under more natural conditions. At the same time, it is necessary to evaluate the quality of the propagules from the mass production and to develop the formulation. Further, testing is needed to determine the dose and the best application method and strategy. All these factors are interlinked and form a complex matrix. Finally, field testing needs to deliver proof of efficacy of the final product for registration purposes and for marketing purposes.

In the development of pathogens for control of greenhouse pests, testing can be done on small plants in a greenhouse setting, ranging from small plots to large blocks in commercial greenhouses, respectively from small plants to mature crops. The advantage of working in greenhouses is that testing can be done more or less year-round and climate conditions can be controlled within certain limits. There is no need for outdoor testing and thus much less need to consider variable climatic conditions and influence of rain and UV.

Considerations for testing microbials against greenhouse pests are provided by Burges (2007), illustrated by examples with Bt and *L. longisporum*. He focused on experimental design, infestation with pests, and spraying methods. Evans (1999) described the parameters involved in dose acquisition and relevant considerations needed on field dose rate determination, and on the importance of persistence of primary inoculum and the role of secondary inoculum. To evaluate the use of entomopathogens against pests, many methods can be used, depending on the microbial agent, the target pest and the crop and the environment. An excellent overview providing methods and tools for experiments with entomopathogens is given in the Field Manual of techniques in invertebrate pathology of Lacey and Kaya (2007). Several chapters focus on evaluation considerations with each of the pathogen groups, while others focus on the use of pathogens in specific systems, like greenhouse, mushrooms, nursery, lawns, etc. Particularly useful is the chapter by Chapple et al. (2007) that discusses the application of MPCPs as particulate products in relation to spray application, equipment, and dose transfer to the crop canopy. Most of their considerations are theoretical. Still, product developers need to be aware of the application variables in order to optimize formulation and application.

A vast amount of papers has been published on testing entomopathogens under semi-field conditions and, to a lesser extent, under commercial conditions. Most of these focus on a particular aspect such as strain comparison, product comparison, determination of dose, formulation aspects, influence of abiotic factors, etc. The challenge for the product developer is to integrate in a testing programme all the aspects that need to be studied to bring about an economic, effective, and user-friendly product. This testing programme should not take too long and should be affordable in terms of time and money. This requires a very good understanding of the crops, targets and the microbial product that is under development, including the delivery system. It is necessary for the product developer to have a good idea of what the ultimate product should become. Each case demands a specific approach and a good understanding of the relevant issues. Organisms with a *per oral* mode of action, Bt's and baculoviruses, are quite different from organisms with a contact mode of action (entomopathogenic fungi) and demand another approach. EPNs have a searching and penetration mode of action and clearly require different testing methods. An overview of testing methods for EPNs is given by Koppenhöfer (2007). Most entomopathogens are sprayed on the foliar part of the crop, but sometimes they are applied to the soil. EPNs in general are soil-applied.

For all pathogens field testing is required in the development of the final product and to arrive at end-users' recommendations. A list of the essential considerations is given below:

(1) to demonstrate the suitability of the selected strain(s) in a situation mimicking as closely as possible the commercial crop. The strain selection has taken place with the help of bio-assays and must be seen as an initial phase in the selection process. Ideally, all strains are compared under more natural conditions, but that is impossible for reasons of space, time and costs. Nevertheless, this should be done with the strain or a small number of strains that were selected

in the first selection phase. Testing on whole plants, or even better in a crop system, should prove that the selected strain is also effective under these conditions or that further selection between strains should be conducted. This can initially be done with propagules from laboratory cultures, but must be followed by mass-produced propagules and the formulated product. It is not possible to give detailed information on the testing methods as they vary greatly between entomopathogens and crop systems. Information can be found in Lacey and Kaya (2007), and particularly in Burges (2007) for greenhouse pests;

(2) to identify whether mass-produced propagules are effective and of good quality. Mass production must deliver infective propagules that are able to kill the target. Production on artificial media or in artificial systems may influence the effectiveness of the propagules. In the process of optimizing the mass production, many variables are studied, usually focussing on improving yield. Downstream processes may be hard on the propagules and negatively influence them. Field testing is needed to see whether the output still performs as is expected;

(3) to improve the formulation. As described above, the formulation should, amongst others, optimize application to the target and protect the pathogen after application (Jones and Burges 1998). This is true for the products based on bacteria, baculoviruses and fungi which are foliar-applied. It plays no role in the case of soil applications of EPNs, but it is relevant however for formulations which are suitable for foliar use of EPNs. Testing formulations can be done in bio-assays and in field tests or in an in-between assay, the "plant-to-tray" assay. Our group often uses this test when testing formulations, particularly when we are looking at field persistence, thus at the protecting abilities of the formulation. The formulation is then sprayed on the plant and after a variable number of days leaf discs are taken and put in the bio-assay trays on water-agar. Insects are then transferred onto the leaf discs and efficacy is assessed after the appropriate amount of time. In this way deposit of the formulated product ages under natural conditions on the plants and many tests can be done with the same deposit over time. The plant does not have to be infested with insects, only the bio-assayed leaf discs. This proves to be a simple and easy assay. But it is also important to test the formulation on plant or crop level, assessed by mortality. First, to test direct effects of the formulation such as coverage, and targeting. Second, to test protection of the propagules, and field persistence. Persistence can be strongly influenced by the relative humidity on the plant and therefore whole-plant testing is required. There may be other plant-related effects such as anti-fungal compounds or micro-flora effects, but these are more difficult to assess and generally less relevant;

(4) to determine the optimal dose and spray volume. The most important factor determined by field testing is the optimal dose. The goal is to find the highest mortality with the lowest possible amount of formulated product using the optimal application method. Application equipment should be similar to equipment used by the grower. In these tests the optimal spray volume needed to deliver the dose in the best way should also be studied, and this may vary depending on the crop and how it is cultivated. Establishing the dose is crucial to the economics

of the product and it should answer the question: "can a product be developed that satisfies the grower and the producer in terms of results, and economics?" The initial data on the dose-mortality relationship originating from bio-assay tests should be followed by semi-field tests and finally confirmed by testing in commercial greenhouses with the final formulation;

(5) to determine the optimal application. This refers to the method of application, the timing of the application, the frequency of the application and intervals between applications. It actually concerns the application strategy for the product. The method of application is usually spraying with the equipment that is available to the grower. It is unlikely that growers will buy specific equipment for a new product; in that case he will choose for a conventional product. To my knowledge there is no commercial equipment specifically designed for the application of microbial products available on the market. Adoption of a MPCP will only be done if it does not require extra efforts or costs for the grower. The timing of the treatment is relevant with respect to the level of the pest and the age structure of its population and the position of the pest on the plant and leaves. Insects can be all over the plant or only on the growing top, on the underside of leaves only, in flowers, etc. This varies per pest and can be dependent on the crop. Often treatments have to be targeted at the most susceptible insect stages. Pathogens may have a slow killing activity and this may demand repeated applications with certain intervals. The optimal interval needs to be established as well as the number of consecutive treatments to control the population. This should lead to the final label recommendations;

(6) to demonstrate efficacy for registration purposes. In order to get approval for a product, efficacy has to be proven with a certain number of tests performed under commercial crop conditions and compared with a reference product. In the EU these tests are only accepted by the registration authorities when they are done according to accepted guidelines, often EPPO guidelines, and executed by GEP (good experimental practice) certified organizations. The testing has to be done with the final formulation and following label recommendations. Selectivity tests have to be performed also, showing that the product does not cause any phytotoxic effects to a variety of plants;

(7) to demonstrate efficacy for marketing purposes. Field tests will be used to show to growers, but also to distributors and sales people, that the product works in commercial greenhouses and that it fits in with all the other cultural practices. Actually, these are all demonstration trials that are not only needed to convince growers to use the product, but it is just as compelling for the company who developed the product to convince its own sales people that the product works when applied by growers on a large scale. There is always the unexpected and the unforeseen. Feedback from others, who have no experience with the product, is more than useful and can be used to improve the product and its application strategy before the product is definitively launched on the market.

From the above it is clear that a good field testing method is imperative. It should be a test that can be performed easily, quickly, cheaply, year-round and give reliable

Table 3.8 Purposes of field testing in product development

- Demonstrate effectiveness of selected strain(s)
- Demonstrate efficacy of mass-produced propagules
- Improve the formulation and demonstrate its effectiveness
- Determine the optimal dose and spraying volume
- Determine the optimal application strategy, incl. equipment, timing, repeated treatments and interval
- Prove efficacy for registration purposes
- Demonstrate efficacy for marketing purposes

and reproducible results. Our research group has found that cucumbers can be used for several pathogens and many pests develop well on it. It germinates easily, grows quickly in various growing substrates, and it does not require specific climate conditions or fertilizers. Diseases are usually not a problem, although powdery mildew is always a risk and resistant varieties should be used. Cucumber forms a big plant in a few weeks, mimicking a commercial crop. It needs only a little labour and can be easily pruned. It also shows an intermediate sensitivity to phytotoxic effects thereby proving to be useful in testing formulation types.

Field testing consists of a series of tests, focussing on the issues that need to be studied step by step. It should be done in a logical order, as presented in Table 3.8, although frequently, one will have to go back a step and adapt or change something, e.g. in the formulation, and repeated testing will be needed to optimize a certain aspect. After this series of testing the product has been developed and the label recommendations established. With this in hand registration trials can be initiated. Field testing should give answers to all the issues mentioned above and should confirm that the final product with the right recommendations for use is an effective product.

3.4 Product Specifications and Quality Control

During the development of a product the product characteristics need to be established, ultimately resulting in product specifications. Every product batch must conform to these "specs" and product batches that do not meet the "specs" must be improved or discarded. Products need to be routinely checked on these criteria, testing methods need to be developed and a quality control protocol needs to be designed. Product specification implies a number of aspects given in Table 3.9. For

Table 3.9 Product specifications and quality control aspects

- Number of propagules or CFU's per gr or ml product
- Viability
- Virulence (efficacy)
- Purity (microbial contaminants)
- Physical parameters (moisture level, particle size, etc. depending on the type of formulation)

every aspect a limit has to be set, including a tolerance range, and what measures should be taken if the limit within the range is not met. Procedures should comply with ISO working methods including tracing and tracking possibilities. Product "specs" need to be checked before product is sold. This is directly after packaging, but also after certain storage times. Product "specs" need to be set in such a way that the product still performs at the end of the shelf-life. Quality control needs to be carried out a number of times during various phases of the product. "Specs" refer to all types of intermediate products too. If "technical product" is produced and stored, "specs" need to be developed. The same holds for bulk-stored formulated product and the packaged final product. Details of quality control will be discussed in Chap. 4.

4 Production Economics and Final Product Costs

4.1 A Cost Price Model for Biopesticides

Decisive for the success of a biopesticide is its price in the market where it needs to be a good alternative in terms of costs and efficacy. If one of these two features does not meet the user's expectations, the product will not be used or used again. "Can the product be produced for an acceptable price in the market?" This essential question needs to be answered in an early stage of the developmental process. Which factors determine the costs of the product? And which factors are the main ones? I want to address these questions in general for the development of a MPCA into a biopesticide. Insight into the makeup of the end-user's price of a product highlights where attention needs to focus to render it more economical, if possible. It is also important for decisions relating to the building of a production plant, to the choice of how to produce a pathogen, to registration costs, etc. In fact it is the basis of a business plan that will lead to a "go/no go"-decision. A business plan also may be necessary to raise money.

4.1.1 Cost Factors from Production to Product

The manufacturing costs must be regarded as specifically product linked and calculations must be made like-wise. However, depending on the economic area and the structure of the manufacturer this could be evaluated in different manners. For instance, capital costs could be spread over various products depending on their stage in the market. For better understanding I will regard all costs as product-specific costs. The production costs can be divided into fixed and variable costs or into direct and indirect costs. These categories are not exactly the same, for instance, variable costs could also be sub-divided into direct and indirect costs, but for ease of use I will use the terms fixed and variable here. Fixed costs are usually expenditure for the production facility: the land, the building and the utilities, and production equipment. They also include overhead costs of the management of the business and of the plant. Production equipment includes bioreactors, downstream and formulation equipment, and packaging machinery. Facility and equipment costs are usually calculated by means of a depreciation factor, which may include a mark-up for interest and maintenance. Research and development costs can be seen as an investment because they are made before the product is launched and are therefore regarded as fixed costs. When a company has a certain permanent R & D staff, costs are seen as fixed, and part of them can be attributed to a certain product. After the product launch, research is often needed to improve the product or its use and then it is considered as a variable. In the case of biopesticides, registration costs may be a large cost factor and they can be accounted as fixed or as variable costs. Here too, they should be seen as an investment, thus as fixed costs. Later they may be some registration maintenance costs or costs for use extensions, and then they could be seen as variable. In the development phase I regard them as fixed costs and both R & D and registration are factored in as a depreciation factor. For the depreciation time a period of 5 years is taken as a generally accepted period for equipment, R & D, and registration, and 30–40 years for buildings. These periods are mainly determined by the national tax rules.

Variable costs usually include the costs of raw materials (medium and formulation ingredients) energy, fuel, water, steam, production labour costs, packaging costs and packaging labour costs. Costs for quality control and waste removal are considered variables since they depend on the production level. These fixed and variable costs differ from plant to plant, and are strongly influenced by the types of processes and products. Besides the above-mentioned production costs, other aspects should be considered such as a certain mark-up for production failures, product leftover when not sold within the shelf-life period (together called production buffer), extraordinary cases, etc. A small mark-up is also needed for a profit margin on the production (Table 3.10).

The total of the fixed and variable production costs and these mark-up factors determine the full product cost price. The costs can be calculated per amount of propagules, per product unit or packaging unit or per hectare, depending on what is the easiest for comparisons. The costs per hectare are often used in order to estimate the competitiveness in the market.

4.1.2 Cost Factors from Product to Market

The final market price further includes selling expenses and marketing costs, together called sales costs, and sales profit margins for producer and distributor. Sales costs generally comprise indirect costs, which include overhead costs, and direct costs. Indirect sales costs are administration and secretary costs, public relations and advertising costs. Overhead costs may include general administration (purchasing, financial, human resources, and ICT departments), general maintenance, management, office costs, insurance, etc. Direct sales costs are salary costs, car costs and telephone costs of sales people, travel and accommodation costs and shipping costs (Table 3.11). All these costs can be calculated on a product-specific basis or, what is often done, a certain percentage is used as a sales margin on top of

Table 3.10 Cost price model for a biopesticide: costs factors involved from production to product

Fixed costs: facility, equipment, R & D, registration, etc.	Fixed costs: overhead costs
• Land and buildings • Utilities • Production equipment • Downstream equipment • Packaging equipment • Repair and maintenance • R & D • Registration • Insurances, etc	• General management • Financial administration • Human resources • ICT • Facilities • Purchasing department • General costs (insurance, legal costs, etc)
Variable costs	Mark-up
• Raw materials • Utilities (energy, water, steam, etc) • Production labour • Packaging costs • Packaging labour • Quality control • Waste removal	• For production buffer • Extra-ordinary cases • Production profit margin

the full product cost price as a method to cover sales and marketing costs. Finally, the full cost is calculated for which the product will be sold to a distributor. If the product is directly sold to the end-user, the sales margin charged by the producer needs to be higher to cover the expenses which would normally be carried out by the distributor.

Often products are sold in the market by a distributor, who also makes various sales and marketing costs, and needs to make a profit on its activities. Distributor

Table 3.11 Cost price model for biopesticide: costs factors involved from product to market

Indirect sales costs	Indirect sales costs: overhead costs
• Management • Administration, secretariat • Public relations • Advertising, publicity • Registration • Insurances, etc • Miscellaneous	• General management • Financial administration • Human resources • ICT • Facilities • Purchasing department • General costs (insurance, legal costs, etc.)
Direct sales costs	Mark-up
• Direct salaries • Car, telephone • Travel, accommodation • Shipping	• Sales profit margin

Table 3.12 Example of a cost model for a biopesticide, per unit of product

Production phase		
Variable costs	30	
Fixed costs		
Facility, etc. costs	50	
Overhead	10	
Mark-up	10	
Full product cost price	100	
Sales phase		
Sales margin	60	37.5% of distributor price
Full cost price	160	
Distributor phase		
Sales margin	80	33% of market price
Market price	240	

tasks are usually storage, logistics, and billing and advising the end-user. A certain sales margin is required by the distributor and the percentage depends on his efforts. Roughly this can vary between 20 and 40% in the area of crop protection, calculated from the market price. The market price is made up of direct and indirect production costs, direct and indirect sales costs and sales margins for the producer and the distributor. A fictive example is given in Table 3.12 based on our group's experience with fungal products in a capital-intensive production system with solid state fermentation. The full product cost price is set at 100; this makes it easy to see the makeup by the various factors. From this example it can be seen that direct production costs (variable costs) constitute 10–15% of the end-user's price and that fixed costs and sales margins are a considerable part of the final product's market prize.

In research and product development generally the focus is on costs of raw materials, medium ingredients and equipment. However, they only constitute a low percentage of the market price. This model provides a perspective on the subdivision of these costs and their contribution to the end-price. Labour costs are generally a relatively high percentage (over 50%) of the variable costs, while raw materials and packaging material constitute the remainder. The full production cost price is for a large part determined by depreciation costs, a factor which is often overlooked or at least the significance of it. On the other hand, once the plant has been built and equipped, costs can only be influenced by reducing medium and labour costs.

This model is only an example and depending on the kind of product and its production method, the importance of each cost factor may vary greatly. It also depends greatly on the company, on the source of capital resources and how costs are internally calculated. Costs may vary substantially country by country, and companies may have different views on profitability. Other reasons for producing biopesticides may also be involved. Nevertheless, this cost model may help researchers and product developers to evaluate and consider the importance of certain costs in the process of product development. Medium costs are often mentioned in publications as a limiting factor, but actually in many situations they are only a small percentage of the product costs. Burges and Jones (1998b) state: "There are too few costs analyses in research literature". Costs of storage, marketing and application are lacking. Most of the information is treated as proprietary information within the biocontrol industry. Academic researchers sometimes report on production costs, but often it is not clear which factors are actually included. My experience is that they usually refer to medium costs, and only sometimes include labour costs. Unfortunately these cost factors only represent two of the factors involved and do not give a reliable basis for a complete cost calculation.

4.2 Economic Analyses of Biopesticides in the Literature

In the literature there is a limited amount of papers presenting details on production economics, I will briefly review a number of useful studies that illustrate different cases. In the Lubilosa project, an economical analysis for the product Green Muscle, based on the fungus *M. anisopliae,* has been made by Cherry et al. (1999). It concerns the mass production of spores by solid state fermentation in the "bag-system" in a purpose-build facility in Benin. The authors gave a breakdown of production costs illustrating the main factors that determine the costs, which are salary costs and capital deprecation (60%). Salaries are the largest single item, accounting for 30% of the costs. Generally, there is a concern that medium costs are a constraint, but in this case rice was used as a medium and considered an expensive product in Africa. Still the authors found that medium costs are only 7% of total costs. Sales costs are not included, which is understandable since Green Muscle is not sold to individual farmers, but mostly to sponsors and governments. This route is different for most biopesticides which are sold to individual growers.

Grimm (2001) presented a study on the economics of a small-scale production plant for entomopathogenic fungi in Nicaragua. He demonstrated the economic feasibility of the proposed plant over a 10 year period, with calculations including all the cost factors, when products are produced for 20,000 ha/year. Makeup of product costs showed that overhead costs and labour costs are the main cost factors in the production of mycoinsecticides.

An interesting and detailed desk study is presented by Swanson (1997) who compares the economic feasibility of two production systems for a mycoinsecticide against locusts in Madagascar. He compared the capital-intensive production model with the labour-intensive production model. The first one is usually exploited by biocontrol companies in the industrialized countries, the second one in developing countries where labour is relatively cheap. Swanson used data from actual operating production facilities, the Mycotech facility as the capital-intensive model and the IITA in Benin production site as the labour-intensive model. He demonstrated that economic feasibility over a 10 year period depends on the production capacity and the product price per hectare, and that at a higher capacity the capital-intensive model is profitable, even in Madagascar where labour is cheap. This publication is the only one where two different production methods are compared on their economic feasibility and therefore presents a valuable study for biopesticide developers. Swanson recognized a number of caveats that may affect his analysis such as irregular demand, insufficient performance, calculating with predetermined quantity of output, and the static nature of technology. These challenges, however, are unavoidable in such an economic analysis and pose a certain risk that entrepreneurs face when developing a MPCP.

The three cases above describe low-tech production of fungi in developing countries.

Total investment to build the plants was US$96,000 in Benin, US$264,000 in Nicaragua and US$650,000 in Madagascar. An economic analysis of a large high-tech plant for the production of *Bacillus thuringiensis*-based bioinsecticides was given by Rowe and Margaritis (2004). Here the investment was approximately US$13 Million. The assumed output of the facility is about 7% of the world's Bt market. The authors gave a detailed overview of installed equipment and total capital costs and a breakdown of the operating costs. They illustrated how profitability depends on production scale and sale price for various fermentation systems. Rowe and Margaritas concluded that it is essential to have a reliable estimate of the selling price in order to calculate the potential profitability of a production facility.

In EPN production only one author reported on a breakdown of costs for the production of *H. bacteriophora* in a 300 l fermentor (Gaugler and Han 2002, Fig. 3.2). Media costs were 8%, while formulation and packaging were 16 and 2%, labour 22% and toll manufacturing 40%, additional costs 12%. Friedman (1990) provided a detailed review of commercial and technology development in EPNs. He illustrated the effects of economy of scale of in vivo and in vitro solid and liquid production of nematodes and showed that liquid production is the best for large scale production when analysed from the perspective of scale-related costs.

4.3 Relevant Considerations for an Economic Analysis

The reviewed papers are valuable examples of economic analyses of biopesticides developments and will help to estimate feasibility and profitability in new cases. This can be done together with the model above which is an attempt to illustrate the cost makeup of a biopesticide and to give perspective on the importance of cost factors involved. The model can be used for both the capital-intensive and labour-intensive models. In a feasibility study, a sensitivity analysis should be made to identify the main cost factors. In a capital-intensive model this is often the capacity ratio. If the production capacity is not fully used, this has a large effect on the fixed costs, particularly, because depreciation is calculated over much less output, making the full product cost price considerably higher. This is especially relevant for products with a short shelf-life and with a strongly variable or seasonal demand. Capacity planning is very important to have an optimal capacity usage and to keep costs down. In a labour-intensive model the reciprocal factor is labour, more precisely productivity per man hour. Other factors that could influence the sensitivity analysis are medium costs, certainly once a plant is running at full capacity. Registration costs are often regarded as a decisive factor. The main problem is that they are hard to estimate beforehand; nevertheless they should not be underestimated to prevent great miscalculations.

To review and compare economics of the various biopesticides, based on fungi, bacteria, baculoviruses and EPNs is extremely difficult and probably impossible, given the wide range of production systems, yields, application rates and crop values. Even within one group of pathogens this would be difficult. The examples from literature provided above demonstrate this also. Nevertheless, the model helps to indicate the main cost factors and to consider where attempts to decrease costs can be made. From the model a rough estimate can be made from direct production costs to end-user's price. In an early phase of the research, variable costs can often be estimated based on labour input and medium costs. Then one could assume a factor 7–8 for the market price, or a factor 3, based on the total of the fixed and variable costs. This assumption could be used as a rough indication. When it is known what end-user's price is acceptable in the market, profitability can be analysed. This should be seen as the starting point for a feasibility study. Depending on what level of profitability and over what period of time this can be reached, a company can decide whether to develop the product or not. Clearly, production economics and final product costs need be to analysed on a case by case situation and depend on the type of product, the company and the market.

5 Conclusions and Recommendations

5.1 Production Considerations and Recommendations

A challenge in the process development is the extrapolation and realization from laboratory scale techniques to commercial and large scale technology. Investigators often assume that laboratory production parameters are valid on production scale. This miscalculation is often made. Scaling-up always presents new problems and unforeseen deviations that need to be tackled. For instance, experimental production in shaking flasks may give misleading data. Production research should be started as early as possible in the foreseen bioreactors. Optimization of each parameter also needs to be investigated at the production scale. The production process needs to be tailored to the specific pathogen and its biology. Several companies have patented their bioreactors and/or production processes (medium, carrier). Up to 2000, about 215 patents had been filed on the general topic of biopesticides (Montesinos 2003), but how many are related to mass production is not mentioned. The costs of filing a patent are high and the relevance is limited. Minor changes may avoid breach of the patent and, more importantly, checking on patent breach is also difficult. Keeping production details confidential should usually suffice with regard to concerns about competitors.

Economy of scale could provide a way to decrease costs in three areas: capital, labour and materials. Improvements in production technology should be a continuous research objective as well as reducing production costs. In designing a production plant and a production process, the possibilities and advantages of scaling-up should be part of the business plan in case the market demand increases. In in vitro production of nematodes, capital costs (depreciation, interest) are the main cost factors and may account for over 60% of the total costs, and they should, therefore, be the focus for cost savings, according to Gaugler and Han (2002). Production costs are a function of scale and they illustrate this (Fig. 14.4, p. 306) by the production of *H. bacteriophora*: a 10 times larger fermentor reduces the production costs more then 4 times. Changing from low technology to high technology may give a more controlled technology with low fault equipment and more fault tolerant processes. This will also result in better control over process parameters and more predictable yields. A high investment is then the trade-off. In the Lubilosa project, "after much debate", the capital-intensive approach with high technology of large scale SSF was chosen over a cottage-type industry production because of product quality and unit production costs (Douthwaite et al. 2001).

The developments of a mass-production system and of a formulated product are complicated and expensive processes. Production efficiency and cost-effectiveness are the key-factors. It is obvious that only a competitive product can become successful in the market and that companies will survive and be profitable when their costs are covered by selling enough products. The mass-production system is the engine that drives everything. Capital is needed to build it, and once running, it needs to run efficiently and at low cost. In Lisansky's words "any biopesticide company planning to remain in the business should be actively addressing its process technology and production facilities as a matter of high priority" (Lisansky 1993). The production should be reliable and failure-free; instability will cause many problems and increase product costs.

Comparisons between the production systems of the four different types of pathogens are challenging to make due to all the different biology and technology. The production of pathogens of the same group is comparable to some extent and systems may be quite similar, as for instance for EPNs. The use of deep tank fermentors seems the most versatile in production of biopesticides. Comparisons between SSF and LSF are difficult to make, but may be necessary to make cost estimates for a certain pathogen and to decide in which system it could be produced most efficiently. Federici (1999) attempted to compare the different types of pathogens based on product characteristics, such as mass production and cost-effectiveness, but for a commercial evaluation this is insufficient and not very useful. Only general requirements for an economical mass production therefore can be given (Table 3.13). A mass production of a certain pathogen needs to be approached and studied on a case by case basis.

5.2 Formulation Considerations and Recommendations

Shelf-life is often mentioned as a weak point of biopesticides. The agrochemical industry requires a shelf-life of a minimum of two up to even 4 years for a product to be able to go through the whole supply chain, at ambient temperatures, with a range from 0°C up to over 40°C for extended periods in tropical countries (Rhodes 1990).

Table 3.13 Considerations and requirements for an economical mass production

- Technology level: in vivo, in vitro: SSF or LSF
- Production equipment: availability and versatility
- Possibility for economy of scale for capital, labour and materials
- Ability to control process parameters; fault tolerance of the process
- Length of the production cycle, the shorter the better
- Reproduction factor and yield
- Medium composition and costs: defined or a natural source
- Downstream equipment: availability and versatility
- Virulence and storability of the propagules: technical product

In biopesticides, some companies require a long shelf-life too, but if this objective is rigidly strived for, this could mean a failure in developing a product. Jaronski (1986) stated that an 18-month period at room temperatures is a valid goal for mycoinsecticides and that this has been achieved with formulations of *B. bassiana* and *M. anisopliae*. Biosys' aim was a shelf-life of 2 years at room temperature with EPN products. They did not achieve this goal. Jones and Burges (1997) reviewed the requirements of biopesticides regarding product stability and shelf-life and they gave suggestions on how to improve this through strain selection, production, downstream methods and additives. They rightfully concluded to start investigating this aspect early in the product development.

In the greenhouse industry distributors and growers are accustomed to handling short shelf-life products like predatory mites and parasitoids. Biopesticides with a shelf-life of 6 months at refrigerated storage are acceptable. A short shelf-life is also less of a problem when the application period is known in advance and to which production can be adapted. In EPNs this is often the case, for instance, for control of larvae of the black vine weevil or white grubs. In EPNs, shelf-life is usually 3 months at refrigerated storage and the producing companies and distributors are coping well with this shelf-life period. Ideally, companies would like biopesticides to have a shelf-life of 2 years at room temperature. This would improve their position when compared with chemical pesticides on user-friendliness. Much research is still needed to achieve this.

Direct contact is often required in order to target the pathogen and this demands careful spraying with well-adjusted equipment. Biopesticides do not profit from extra features such as systemic and trans-laminar distribution of the active ingredients which is the case with many chemicals. An improvement in targeting biopesticides with more appropriate equipment is also an area where general principles need to be generated by institutional organizations.

User-friendliness is also a fundamental aspect which should not be underestimated. If there are extra requirements to what a grower is used to with chemicals, adoption of the biological will be less easy. An example from Mycotal: previously, users had to pre-soak the product for at least two hours before spraying and they complained often about it. This requires planning of the application early in the morning or late in the afternoon (as recommended since the relative humidity is higher then) and this was too complicated. Once this was no longer necessary due to a small formulation change, product acceptance was improved. Further, spraying early or late in the day makes it more inconvenient for the grower and this can be a disadvantage for a microbial product. These aspects may seem irrelevant, but in practice they should be avoided as much as possible.

The formulation of entomopathogens is an absolute prerequisite for a commercial product. Hofstein and Fridlender (1994) considered formulation as the critical stage in product development. A formulation may offer a unique selling point and add value to the final product. The development of a formulated product, including the application recommendations, is a complex process, intertwined with aspects of the production and the downstream processes. Formulation research needs to look at production, storage stability, efficacy, user-friendliness and costs, and to consider their importance continuously during the product development. A multi-disciplinary approach is needed and knowledge and products be included from other industries such as the adjuvant industry, packaging industry, coating industry and others. Cooperation with academic research may provide insight in the fundamental aspects and may study more generic principles that could improve formulation. Formulation needs a case by case approach and it can influence the products' success to a great extent. In industry, hands-on experiences and know-how are valuable in order to improve formulations for new products.

Therefore formulation studies are essential through the whole process of production and product development. Registration considerations should be included too in designing a formulation. Lisansky (CPL 2006) recommended that formulation staff be involved in process development and product design from the end of fermentation to the final product (unlike in chemical industries). I recommend including involvement early in the mass production and in registration. In the biopesticide companies this is usually in the hands of a small team anyway, but they should be aware of all these aspects. On the other hand, the product developer needs to decide what an efficient formulation is within a certain timeframe and budget. Research in this area should be pragmatic focussing on key demands. The challenge is to find the balance between what is acceptable and what may give real improvements. Based on knowledge, experience and considerations of the four main functions of a formulation, the formulation expert must make a decision. The complex interactions and the demands of the formulation mean that the selection of the best formulation will ultimately be a pragmatic compromise with a main focus on efficacy, ease of use and costs.

5.3 Final Conclusions

In optimization of production there is a maximum to the yield of propagules per ml. In many cases substantial improvements are hard to make even after 20–30 years of research in this area since one is confronted with the biological limits of the organism. Another factor that influences product costs strongly is the application rate. In many pathogens, the slope of the dose-mortality regression curve is not

Table 3.14 Key factors determining a successful economical mass production and formulated product

• Investments
• Economy of scale
• Stability of production process
• Production yield and application dose
• Capacity usage
• Product shelf-life
• Level of efficacy
• User-friendliness
• Full product cost price and end-user price
• Turnover, profitability and return of investment

steep indicating that higher application rates do not contribute significantly to a higher mortality. A small factor matters less in efficacy, but matters more in production. Even a factor of two can be pivotal. This relationship between production and application rate needs careful study. Formulation is part of this relationship and an area of improving the product's performance and reducing its costs. These areas are all interconnected and offer potential for improvements and lowering costs for the end-user.

In addition to the technical issues of the development of the mass production and of the final formulated product, a key issue is the upfront investment and the market price for the end-user. Further, R & D costs, and registration investments contribute largely to the final cost price. Depreciation is a large part of the costs. The size of the market and the potential turnover determines the market price too, since costs and depreciation can be divided over more products. Once production is running, the focus should be on variable costs, stability of the process, and optimal capacity usage (Table 3.14).

A sharp business plan is indispensible to the decision-making process. The development of a mass production and a formulated product are interconnected and technical and economical issues should be leading. The entire process is multi-factorial and multi-disciplinary and scientists and industry should work together from an early phase on. Many different aspects need to be investigated and specialists are required. These developments need the commitment of all involved, including the commitment for financial resources. The technical issues of the development of the mass production, the final formulation, and the application strategies must follow the commercial aspects. For the product to remain competitive, improvements in these areas are a never-ending process. To be successful with biopesticides, both production efficiency and product efficacy are the keys.

References

Agosin, E., D. Volpe, G. Muñoz, R. San Martin, and A. Crawford, 1997. Effect of culture conditions on spore shelf life of the biocontrol agent. *Trichoderma harzianum*. *World J. Microbiol. Biotechnol.* 13: 225–232.

Andersch, W., 1992. Production of fungi as crop protection agents. *Pflanzenschutz-Nachr. Bayer* 45(1): 129–142.

Anonymus, 2004. A microscopic formula for success. *Agriscience News*, Issue 2, 12 January 2004: 1–3.

Anonymus, 2006. New O-TEQ formulation for Bayer insecticides. *Crop Protection Monthly*, February 2006: 17.

Anonymus, 2007. Dow unveils "revolutionary pipeline". *Agrow* 515, March 9, 2007: 1–3.

Arthurs, S.P., L.A. Lacey, and R.W. Behle, 2008. Evaluation of lignins and particle films as solar protectants for the granulovirus of the codling moth, *Cydia pomonella*. *Biocontl. Sci. Technol.* 18(7): 829–839.

Auld, B.A., 1993. Vegetable oil suspension emulsions reduce dew dependence of a mycoherbicide. *Crop Prot.* 12(6): 477–479.

Ballard, J., D.J. Ellis, and C.C. Payne, 2000. The role of formulation additives in increasing the potency of *Cydia pomonella* granulovirus for codling moth larvae in laboratory and field experiments. *Biocontl. Sci. Technol.* 10: 627–640.

Bartlett, M.C. and S.T. Jaronski, 1988. Mass production of entomogenous fungi for biological control of insects. In: M.N. Burge (ed), *Fungi in biological control systems*. Manchester University Press, Manchester. pp. 61–85.

Bateman, R., 1997. The development of a myco-insecticide for the control of locusts and grasshoppers. *Outlook Agric.* 26(1): 13–18.

Bateman, R.P. and R.T. Alves, 2000. Delivery systems for mycoinsecticides using oil-formulations. *Aspects Appl. Biol.* 57: 163–170.

Bateman, R.P., M. Carey, D. Moore, and C. Prior, 1993. The enhanced infectivity of *Metarhizium flavoviride* in oil formulations to desert locusts at low humidity's. *Ann. Appl. Biol.* 122: 145–152.

Bedding, R.A., 1981. Low cost in vitro mass production of *Neoaplectana* and *Heterorhabditis* species (Nematoda) for field control of insect pests. *Nematologica* 27: 109–114.

Bedding, R.A., 1984. Large scale production, storage and transport of the insect-parasitic nematodes *Neoaplectana* spp. and *Heterorhabditis* spp. *Ann. Appl. Biol.* 104: 117–120.

Bernard, K. and R. Utz, 1993. Production of Bacillus thuringiensis insecticides for experimental and commercial uses. In: P.F. Entwistle, J.S. Cory, M.J. Bailey, S. Higgs (eds), *Bacillus thuringiensis, an environmental biopesticide: theory and practice*. Wiley, New York, NY. pp. 255–267.

Black, B.C., L.A. Brennan, P.M. Dierks, and I.E. Gard, 1997. Commercialization of baculoviral insecticides. In: L.K. Miller (ed), *The Baculoviruses*. Plenum Press, New York. pp. 341–387.

Boyetchko, S., E. Pedersen, Z. Punja, and M. Reddy, 1999. Formulations of biopesticides. In: F.R. Hall, J.J. Menn (eds), *Biopesticides: use and delivery*. Humana Press, Totowa, NJ. pp. 487–508.

Bradley, C.A., W.E. Black, R. Kearns, and P. Wood, 1992. Role of production technology in mycoinsecticide development. In: G.F. Leatham (ed), *Frontiers in industrial mycology*. Chapman & Hall, New York, NY. pp. 160–173.

Brar, S.K., M. Verma, R.D. Tyagi, and J.R. Valéro, 2006. Recent advances in downstream processing and formulations of *Bacillus thuringiensis* based biopesticides. *Process Biochem.* 41: 323–342.

Burden, J.P., R.S. Hails, J.D. Windass, M.-M. Suner, and J.S. Cory, 2000. Pathogenicity, speed of kill, and productivity of a baculovirus expressing the itch mite toxin Txp-1 in second and fourth instar larvae of *Trichoplusia ni*. *J. Invertebr. Pathol.* 75: 226–236.

Burges, H.D., 1998. *Formulation of microbial biopesticides: beneficial microorganisms, nematodes and seed treatments*. Kluwer, Dordrecht.

Burges, H.D., 2007. Techniques for testing microbials of arthropod pests in greenhouses. In: L.A. Lacey, H.K. Kaya (eds), *Field manual of techniques in invertebrate pathology*, 2nd edition. Springer, Dordrecht. pp. 463–479.

Burges, H.D. and K.A. Jones, 1998a. Formulation of bacteria, viruses and protozoa to control insects. In: H.D. Burges (ed), *Formulation of microbial biopesticides: beneficial microorganisms, nematodes and seed treatments*. Kluwer, Dordrecht. pp. 33–127.

Burges, H.D. and K.A. Jones, 1998b. Trends in formulation of microorganisms and future research requirements. In: H.D. Burges (ed), *Formulation of microbial biopesticides: beneficial microorganisms, nematodes and seed treatments*. Kluwer, Dordrecht. pp. 310–332.

Butt, T.M., 2002. Use of entomogenous fungi for the control of insect pests. In: F. Kempken (ed), *The Mycota XI, agricultural applications*. Springer, Berlin. pp. 111–134.

Butt, T.M., C. Jackson, and N. Magan, 2001. Introduction-fungal biological control agents: progress, problems and potential. In: T.M. Butt, C. Jackson, N. Magan (eds), *Fungi as biocontrol agents: progress, problems and potential*. CAB International, Wallingford. pp. 1–8.

Butt, T.M., C. Wang, F.A. Shah, and R. Hall, 2006. Degeneration of entomogenous fungi. In: J. Eilenberg, H.M.T. Hokkanen (eds), *An ecological and societal approach to biological control*. Springer, Dordrecht. pp. 213–226.

CPL, 2006. *Biopesticides 2007*. CPL Business Consultants, Wallingford.

Castillejos, V., J. Trujillo, L.D. Ortega, J.A. Santizo, J. Cisneros, D.I. Penagos, J. Valle, and T. Williams, 2002. Granular phagostimulant nucleopolyhedrovirus formulations for control of *Spodoptera frugiperda* in maize. *Biol. Control* 24: 300–310.

Chapple, A.C., R.A. Downer, and R.P. Bateman, 2007. Theory and practice of microbial insecticide application. In: L.A. Lacey, H.K. Kaya (eds), *Field manual of techniques in invertebrate pathology*, 2nd edition. Springer, Dordrecht. pp. 9–34.

Charnley, A.K., B. Cobb, and J.M. Clarkson, 1997. Towards the improvement of fungal insecticides. In: H.F. Evans (ed), *Microbial Insecticides: Novelty or Necessity? BCPC Symposium Proceedings* No. 68, Coventry, April 16–18, 1997. BCPC, Farnham. pp. 115–126.

Cherry, A.J., N.E. Jenkins, G. Heviefo, R. Bateman, and C.J. Lomer, 1999. Operational and economic analysis of a West African pilot-scale production plant for aerial conidia of *Metarhizium* spp. for use as a mycoinsecticide against locusts and grasshoppers. *Biocontl. Sci. Technol.* 9: 35–51.

Cherry, A.J., M.A. Parnell, D. Grzywacz, and K.A. Jones, 1997. The optimization of in vivo nuclear polyhedrosis virus production in *Spodoptera exempta* (Walker) and *Spodoptera exigua* (Hübner). *J. Invertebr. Pathol.* 70: 50–58.

Cherry, A.J., M.A. Parnell, D. Smith, and K.A. Jones, 1994. Oil formulation of insect viruses. *IOBC/WPRS Bull.* 17(3): 254–257.

Copping, L.G., 2009. *The manual of biocontrol agents*, 4th edition. BCPC Publications, Alton.

Copping, L.G. and J.L. Menn, 2000. Review: biopesticides: a review of their action, applications and efficacy. *Pest Manag. Sci.* 56: 651–676.

Couch, T.L., 2000. Industrial fermentation and formulation of entomopathogenic bacteria. In: J.-F. Charles, A. Delécluse, C. Nielsen-LeRoux (eds), *Entomopathogenic bacteria: from laboratory to field application*. Kluwer, Dordrecht. pp. 297–316.

Couch, T.L. and C.M. Ignoffo, 1981. Formulation of insect pathogens. In: H.D. Burges (ed), *Microbial control of pests and plant diseases 1970–1980*. Academic Press, London. pp. 621–634.

Cox, M.M.J., 2004. Commercial production in insect cells. One company's perspective. *BioProcess International* (Supplement), June 2004: 2–5.

Curtis, J.E., T.V. Price, and P.M. Ridland, 2003. Initial development of a spray formulation which promotes germination and growth of the fungal entomopathogen *Verticillium lecanii* (Zimmermann) Viegas (Deuteromycotina: Hyphomycetes) on capsicum leaves (*Capsicum annuum grossum* Sendt. var. California Wonder) and infection of *Myzus persicae* Sulzer (Homoptera: Aphididae). *Biocontl. Sci. Technol.* 13(1): 35–46.

De Faria, M.R. and S.P. Wraight, 2007. Mycoinsecticides and mycoararicides: a comprehensive list with worldwide coverage and international classification of formulation types. *Biol. Control.* 43: 237–256.

De la Torre, M., 2003. Challenges for mass production of nematodes in submerged culture. *Biotech. Adv.* 21: 407–416.

Dougherty, E.M., K.P. Guthrie, and M. Shapiro, 1996. Optical brighteners provide baculovirus activity enhancement and UV radiation protection. *Biol. Control* 7: 71–74.

Douthwaite, B., J. Langewald, and J. Harris, 2001. Development and commercialization of the Green Muscle biopesticide. *Impact*, IITA publication 1. p. 22
Durand, A., R. Renaud, J. Maratray, and S. Almanza, 1997. The INRA-Dijon reactors: designs and applications. In: S. Roussos, B.K. Lonsane, M. Raimbault, G. Viniegra-Gonzalez (eds), *Advances in solid state fermentation. Proceedings of the 2nd International Symposium on Solid State Fermentation FMS-95*, Montpellier, France. Kluwer, Dordrecht. pp. 71–92.
Ehlers, R.-U., 1996. Current and future use of nematodes in biocontrol: practice and commercial aspects with regard to regulatory policy issues. *Biocontl. Sci. Technol.* 6: 303–316.
Ehlers, R.-U., 2001a. Mass production of entomopathogenic nematodes for plant protection. *Appl. Microbiol. Biotechnol.* 56: 623–633.
Ehlers, R.-U., 2001b. Achievements in research of EPN mass production. In: C.T. Griffin, A.M. Burrell, M.J. Downes, R. Mulder (eds), *Developments in entomopathogenic nematodes/bacterial research*, EUR 19696-COST 819. European Community Press, Luxembourg. pp. 68–77.
Ehlers, R.-U., 2007. Entomopathogenic nematodes: from science to commercial use. In: C. Vincent, M.S. Goettel, G. Lazarovits (eds), *Biological control: a global perspective*. CAB International, Wallingford. pp. 136–151.
Ehlers, R.-U., S. Lunau, K. Krasomil-Osterfeld, and K.H. Osterfeld, 1998. Liquid culture of the entomopathogenic nematode-bacterium complex *Heterorhabditis megidis/Photorhabdus luminescens*. *BioControl* 43: 77–86.
Ehlers, R.-U. and D.I. Shapiro-Ilan, 2005. Mass production. In: P.S. Grewal, R.-U. Ehlers, D.I. Shapiro-Ilan (eds), *Nematodes as biocontrol agents*. CAB International, Wallingford. pp. 65–78.
El-Bandary, M.A., 2006. *Bacillus thuringiensis* and *Bacillus sphaericus* biopesticides production. *J. Basic Microbiol.* 46(2): 158–170.
Evans, H.F., 1999. Principles of dose acquisition for biopesticides. In: F.R. Hall, J.J. Menn (eds), *Biopesticides: use and delivery*. Humana Press, Totowa, NY. pp. 553–573.
Federici, B.A., 1999. Bacillus thuringiensis in biological control. In: T.S. Bellows, T.W. Fisher (eds), *Handbook of biological control*. Academic Press, San Diego, CA. pp. 575–593.
Federici, B.A., 2007. Bacteria as biological control agents for insects: economics, engineering, and environmental safety. In: M. Vurro, J. Gressel (eds), *Novel biotechnologies for biocontrol agents enhancement and management*. Springer, Dordrecht. pp. 25–51.
Feng, K.C., B.L. Liu, and Y.M. Tzeng, 2000. *Verticillium lecanii* spore production in solid-state and liquid-state fermentations. *Bioprocess Eng*. 23: 25–29.
Feng, K.C., B.L. Liu, and Y.M. Tzeng, 2002. Morphological characterization and germination of aerial and submerged spores of the entomopathogenic fungus *Verticillium lecanii*. *World J. Microbiol. Biotechnol.* 18: 217–224.
Feng, M.G., T.J. Poprawski, and G.G. Khachatourians, 1994. Production, formulation and application of the entomopathogenic fungus *Beauveria bassiana* for insect control: current status. *Biocontl. Sci. Technol.* 4(1): 3–34.
Ferron, P., 1981. Pest control by the fungus Beauveria and Metarhizium. In: H.D. Burges (ed), *Microbial control of pests and plant diseases 1970–1980*. Academic Press, London. pp. 465–482.
Friedman, M.J., 1990. Commercial production and development. In: R. Gaugler, H.K. Kaya (eds), *Entomopathogenic nematodes in biological control*. CRC Press, Boca Raton, FL. pp. 153–172.
Gard, I.E., 1997. Field testing a genetically modified baculovirus. In: H.F. Evans (ed), *Microbial Insecticides: Novelty or Necessity? BCPC Symposium Proceedings* No. 68, Coventry, April 16–18, 1997. BCPC, Farnham. pp. 101–114.
Gaugler, R., I. Brown, D. Shapiro-Ilan, and A. Atwa, 2002. Automated technology for in vivo mass production of entomopathogenic nematodes. *Biol. Control* 24: 199–206.
Gaugler, R. and R. Han, 2002. Production technology. In: R. Gaugler (ed), *Entomopathogenic nematology*. CAB International, Wallingford. pp. 289–310.

Georgis, R., 2002. The biosys experiment: an insider's perspective. In: R. Gaugler (ed), *Entomopathogenic nematology*. CAB International, Wallingford. pp. 357–372.

Georgis, R. and H.K. Kaya, 1998. Formulation of entomopathogenic nematodes. In: H.D. Burges (ed), *Formulation of microbial biopesticides: beneficial microorganisms, nematodes and seed treatments*. Kluwer, Dordrecht. pp. 289–308.

Glare, T.R. and M. O'Callaghan, 2000. *Bacillus thuringiensis: biology, ecology and safety*. Wiley, Chichester.

Goettel, M.S. and D.W. Roberts, 1992. Mass production, formulation and field application of entomopathogenic fungi. In: C.J. Lomer, C. Prior (eds), *Biological Control of locusts and grasshoppers*. CAB International, Wallingford. pp. 230–244.

Goodman, C.L., A.H. McIntosh, G.N. El Sayed, J.J. Grasela, and B. Stiles, 2001. Production of selected baculoviruses in newly established lepidopteran cell lines. *In Vitro Cell Dev. Biol. – Anim.* 37: 374–379.

Granados, R.R. and B.A. Federici, 1986. *The biology of baculoviruses. Vol. II. Practical application for insect control*. CRC Press, Boca Raton, FL.

Green, S., S.M. Stewart-Wade, G.J. Boland, M.P. Teshler, and S.H. Liu, 1998. Formulating microorganisms for biological control of weeds. In: G.J. Boland, L.D. Kuykendall (eds), *Plant-microbe interactions and biological control*. Marcel Dekker, New York, NY. pp. 249–281.

Grewal, P.S., 2002. Formulation and application technology. In: R. Gaugler (ed), *Entomopathogenic nematology*. CAB International, Wallingford. pp. 265–287.

Grewal, P. and R. Georgis, 1994. Fundamental research on entomopathogenic nematodes: an industrial perspective. *Proceedings of the VIth International Colloquium on Invertebrate Pathology and Microbial Control*, August 28–September 2, 1994. Montpellier, France. pp. 126–130.

Grimm, C., 2001. Economic feasibility of a small-scale production plant for entomopathogenic fungi in Nicaragua. *Crop Prot.* 20: 623–630.

Gröner, A., 1987. Massenproduktion von baculoviren (in vivo und in vitro) und aspekte der kommerzialisierung von viruspräparaten. *Med. Fac. Landbouww. Rijksuniv. Gent.* 52(2a): 139–145.

Grzywacz, D., K.A. Jones, G. Moawas, and A. Cherry, 1998. The in vivo production of *Spodoptera littoralis* nuclear polyhedrosis virus. *J. Virol. Methods* 71: 115–122.

Guillon, M., 1993. Formulation of biopesticides: an industrial view of biological control agents. In: P. Lepoivre (ed), *The Proceedings of a Workshop in the EC Programme "Competitiveness of Agriculture and Management of Agricultural Resources" (CAMAR) on Biological Control of Fruit and Foliar Disease*, September 16–17, 1993, Gembloux. EC D.-G. Agriculture, Brussels. pp. 99–105.

Guillon, M., 1995. Industrial production of insect viruses for biological control: technical aspects and economical interest. *Proceedings of the Microbial Control Agents in Sustainable Agriculture,* October 18–19, 1995. Italy. pp. 65–72.

Guillon, M., 1997. Production of biopesticides: scale up and quality assurance. In: H.F. Evans (ed), *Microbial Insecticides: Novelty or Necessity? BCPC Symposium Proceedings* No. 68, Coventry, April 16–18, 1997. BCPC, Farnham. pp. 151–162.

Gwynn, R.L., 2006. Progress, issues and constraints in Africa for the development of entomopathogenic nematodes. *Paper presented at the Summit Workshop COST 850 "Biocontrol symbiosis"*, June 02–05, 2006, Salzau Castle, Kiel, Germany. www.cost850.ch. Cited January 22, 2008.

Hall, R.A., 1980. Effect of relative humidity on survival of washed and unwashed conidiospores of *Verticillium lecanii. Acta Oecologia* 1(3): 265–274.

Hall, R.A., 1981. The fungus Verticillium lecanii as a microbial insecticide against aphids and scales. In: H.D. Burges (ed), *Microbial control of pests and plant diseases 1970–1980*. Academic Press, London. pp. 483–498.

Hallsworth, J.E. and N. Magan, 1994. Improved biological control by changing polyols/trehalose in conidia of entomopathogens. *Brighton Crop Protection Conference* 1994 – *pest and diseases*. BCPC Publications, Farnham. pp. 1091–1096.

Harman, G.E., 2006. Overview of mechanisms and uses of *Trichoderma* spp. *Phytopathology* 96(2): 190–194.

Harris, J.G., 1997. Microbial insecticides – an industry perspective. In: H.F. Evans (ed), *Microbial Insecticides: Novelty or Necessity? BCPC Symposium Proceedings* No. 68, Coventry, April 16–18, 1997. BCPC, Farnham. pp. 41–50.

Hazzard, R.V., B.B. Schult, E. Groden, E.D. Ngollo, and E. Seidlecki, 2003. Evaluation of oils and microbial pathogens for control of lepidopteran pests of sweet corn in New England. *J. Econ. Entomol.* 96(6): 1653–1661.

Hegedus, D.D., M.J. Bidochka, and G. Khachatourians, 1990. *Beauveria bassiana* submerged conidia production in a defined medium containing chitin, two hexosamines or glucose. *Appl. Microbiol. Biotechnol.* 33: 641–647.

Hirano, S.S. and C.D. Upper, 2000. Bacteria in the leaf ecosystem with emphasis on *Pseudomonas syringae* – a pathogen, ice nucleus and epiphyte. *Microbiol. Mol. Biol. Rev.* 64(3): 624–653.

Hofstein, R. and B. Fridlender, 1994. Development of production, formulation and delivery systems. *Brighton Crop Protection Conference* 1994 – *pests and diseases*. BCPC Publications, Farnham. pp. 1273–1280.

Huber, J., 2005. Viruspräparate. In: H. Schmütterer, J. Huber (eds), *Natürliche Schädlingsbekämpfungsmittel*. Ulmer, Stuttgart. pp. 15–28.

Huber, J. and H.G. Miltenburger, 1986. Production of pathogens. In: J.M. Franz (ed), *Biological plant and health protection*. G. Fisher Verlag, Stuttgart. pp. 168–181.

Hunter-Fujita, F.R., P.F. Entwhistle, H.F. Evans, and N.E. Crook, 1998. *Insect viruses and pest management*. Wiley, Chichester.

Hynes, R.K. and S.M. Boyetchko, 2006. Research initiatives in the art and science of biopesticide formulations. *Soil Biol. Biochem.* 38: 845–849.

Ignoffo, C.M., 1973. Development of a viral insecticide: concept to commercialization. *Exp. Parasitol.* 33: 380–460.

Ignoffo, C.M. and T.L. Couch, 1981. The nucleopolyhedrosis virus of Heliothis species as a microbial insecticide. In: H.D. Burges (ed), *Microbial control of pests and plant diseases 1970–1980*. Academic Press, London. pp. 330–362.

Ikonomou, L., Y.-J. Schneider, and S.N. Agathos, 2003. Insect cell culture for industrial production of recombinant proteins. *Appl. Microbiol. Biotechnol.* 62: 1–20.

Inceoglu, A.B., S.G. Kamita, A.C. Hinton, Q. Huang, T.F. Severson, K. Kang, and B.D. Hammock, 2001. Recombinant baculoviruses for insect control. *Pest Manag. Sci.* 57: 981–987.

Inglis, G.D., M.S. Goettel, T.M. Butt, and H. Strasser, 2001. Use of hyphomycetous fungi for managing insect pests. In: T.M. Butt, C. Jackson, N. Magan (eds), *Fungi as biocontrol agents: progress, problems and potential*. CAB International, Wallingford. pp. 23–69.

Inglis, G.D., S.T. Jaronski, and S.P. Wraight, 2002. Use of spray oils with entomopathogens. In: G.A.C. Beattie, D.M. Watson, M.L. Stevens, D.J. Race, R.N. Spooner-Hart (eds), *Spray Oils Beyond 2000 – Sustainable Pest and Disease Management. Proceedings of the Conference* October 25–29, 1999, Sydney, Australia. University of Western Sydney Press, Sydney. pp. 302–312.

Jackson, M.A., S. Cliquet, and L.B. Iten, 2003. Media and fermentation process for the rapid production of high concentrations of stable blastospores of the bioinsecticidal fungus *Paecilomyces fumosoroseus*. *Biocontl. Sci. Technol.* 13(1): 23–33.

Jackson, M.A., C.A. Dunlop, and S.T. Jaronski, 2010. Ecological considerations in producing and formulating fungal entomopathogens for use in insect biocontrol. *BioControl* 55: 129–145.

Jackson, M.A., S. Erhan, and T.J. Poprawski, 2006. Influence of formulation additives on the desiccation tolerance and storage stability of blastospores of the entomopathogenic fungus *Paecilomyces fumosoroseus* (Deuteromycotina: Hyphomycetes). *Biocontl. Sci. Technol.* 16(1): 61–75.

Jackson, M.A., M.R. McGuire, L.A. Lacey, and S.P. Wraight, 1997. Liquid culture production of desiccation tolerant blastospores of the bioinsecticidal fungus *Paecilomyces fumosoroseus*. *Mycol. Res.* 101(1): 35–41.

Jackson, T.A., J.F. Pearson, M. O'Callaghan, H.K. Mahanty, and M.J. Willocks, 1992. Pathogen to product-development of Serratia entomophila (Enterobacteriaceae) as a commercial biological control agent for the New Zealand grass grub (Costelyta zealandica). In: T.A. Jackson, T.R. Glare (eds), *Use of pathogens in scarab pest management*. Intercept, Andover. pp. 191–198.

Jaronski, S.T., 1986. Commercial development of Deuteromycetous fungi of arthropods: a critical appraisal. In: R.A. Samson, J.M. Vlak, R. Peters (eds), *Fundamental and applied aspects of invertebrate pathology*. Foundation of the Fourth International Colloquium of Invertebrate Pathology, Wageningen, Netherlands. pp. 653–656.

Jenkins, N.E. and M.S. Goettel, 1997. Methods for mass-production of microbial control agents of grasshoppers and locusts. *Memoirs Entomol. Soc. Canada* 171: 37–48.

Jenkins, N.E., G. Heviefo, J. Langewald, A.J. Cherry, and C.J. Lomer, 1998. Development of mass production technology for aerial conidia for use as a mycoinsecticide. *Biocont. News Inform.* 19(1): 21–31 N.

Johnson, V.W., J.F. Pearson, and T.A. Jackson, 2001. Formulation of *Serratia entomophila* for biological control of grass grubs. *N. Z. Plant Prot.* 54: 125–127.

Jones, K.A. and D. Burges, 1997. Product stability: from experimental preparation to commercial reality. In: H.F. Evans (ed), *Microbial Insecticides: Novelty or Necessity? BCPC Symposium Proceedings* No. 68, Coventry, April 16–18, 1997. BCPC, Farnham. pp. 163–171.

Jones, K.A. and H.D. Burges, 1998. Technology of formulation and application. In: H.D. Burges (ed), *Formulation of microbial biopesticides: beneficial microorganisms, nematodes and seed treatments*. Kluwer, Dordrecht. pp. 7–30.

Jones, K.A., A.J. Cherry, D. Grzywacz, and D.H. Burges, 1997. Formulation: is it an excusive for poor application? In: H.F. Evans (ed), *Microbial Insecticides: Novelty or Necessity? BCPC Symposium Proceedings* No. 68, Coventry, April 16–18, 1997. BCPC, Farnham. pp. 173–180.

Keller, S., 2000. Use of *Beauveria brongniartii* in Switzerland and its acceptance by farmers. *IOBC/WPRS Bull.* 23(8): 67–71.

Kiewnick, S., 2001. Advanced fermentation and formulation technologies for fungal antagonists. *IOBC/WPRS Bull.* 24(1): 77–79.

Kiss, L., J.C. Russel, O. Szentiványi, X. Xu, and P. Jeffries, 2004. Biology and biocontrol potential of *Ampelomyces* mycopararasites, natural antagonists of powdery mildew fungi. *Biocontl. Sci. Technol.* 14(7): 635–651.

Kleespies, R.G. and G. Zimmermann, 1992. Production of blastospores by three strains of *Metarhizium anisopliae* (Metch) Sorokin in submerged culture. *Biocontl. Sci. Technol.* 2: 127–135.

Kleespies, R.G. and G. Zimmermann, 1994. Viability and virulence of blastospores of *Metarhizium anisopliae* (Metch) Sorokin after storage in various liquids at different temperatures. *Biocontl. Sci. Technol.* 4: 309–319.

Kleespies, R.G. and G. Zimmermann, 1998. Effect of additives on the production, viability and virulence of blastospores of *Metarhizium anisopliae*. *Biocontl. Sci. Technol.* 8: 207–214.

Knowles, A., 2006. *Adjuvants and additives: 2006 edition*. Agrow Reports, London.

Koike, M., T. Higashio, A. Komori, K. Akiyama, N. Kishimoto, E. Masuda, M. Sasaki, S. Yoshida, M. Tani, K. Kuramoti, M. Sugimoto, and H. Nagao, 2004. *Verticillium lecanii* (*Lecanicillium* spp.) as epiphyte and its application to biological control of arthropod pests and diseases. *IOBC/WPRS Bull.* 27(8): 41–44.

Koppenhöfer, A.M., 2007. Nematodes. In: L.A. Lacey, H.K. Kaya (eds), *Field manual of techniques in invertebrate pathology*, 2nd edition. Springer, Dordrecht. pp. 249–264.

Koppenhöfer, A.M. and E.M. Fuzy, 2003. *Steinernema scarabaei* for the control of white grubs. *Biol. Control* 28: 47–59.

Lacey, L.A. and H.K. Kaya, 2007. *Field manual of techniques in invertebrate pathology*, 2nd edition. Springer, Dordrecht.

Lasa, R., C. Ruiz-Portero, M.D. Alcázar, J.E. Belda, P. Caballero, and T. Williams, 2007. Efficacy of optical brightener formulations of *Spodoptera exigua* multiple nucleopolyhedrovirus (SeMNPV) as a biological insecticide in greenhouses in southern Spain. *Biol. Control* 40: 89–96.
Lello, E.R., M.N. Patel, G.A. Matthews, and D.J. Wright, 1996. Application technology for entomopathogenic nematodes against foliar pests. *Crop Prot.* 15(6): 567–574.
Lewis, F.B., 1981. Control of the gypsy moth by a baculovirus. In: H.D. Burges (ed), *Microbial control of pests and plant diseases 1970–1980*. Academic Press, London. pp. 363–378.
Lisansky, S.G., 1985. Production and commercialization of pathogens. In: N.W. Hussey, N. Scopes (eds), *Biological pest control. The glasshouse experience*. Blanford Press, Poole. pp. 210–218.
Lisansky, S.G., 1993. Production: the key technology for biopesticides. *IMPACT AgBioIndustry*, February 1993: 7–13.
Lisansky, S.G., R. Quinlan, and G. Tassoni, 1993. *The Bacillus thuringiensis production handbook: laboratory methods, manufacturing, formulation, quality control, registration*. CPL Press, Newbury.
Magan, N., 2001. Physiological approaches to improving the ecological fitness of fungal biocontrol agents. In: T.M. Butt, C. Jackson, N. Magan (eds), *Fungi as biocontrol agents: progress, problems and potential*. CAB International, Wallingford. pp. 239–251.
Mayoral, F., M. Benuzzi, and E. Ladurner, 2006. Efficacy of the *Beauveria bassiana* strain ATCC 74040 (Naturalis) against whiteflies on protected crops. *IOBC/WPRS Bull.* 29(4): 83–88.
McCoy, C.W., 1981. Pest control by the fungus Hirsutella thompsonii. In: H.D. Burges (ed), *Microbial control of pests and plant diseases 1970–1980*. Academic Press, London. pp. 499–512.
McCoy, C.W., 1996. Pathogens of eriophyoid mites. In: E.E. Lindquist, M.W. Sabelis, J. Bruin (eds), *World crop pests. Eriophyoid mites. Their biology, natural enemies and control*, Vol. 6. Elsevier, Amsterdam. pp. 481–490.
McCoy, C.W., A.J. Hill, and R.F. Kanavel, 1975. Large-scale production of the fungal pathogen *Hirsutella thompsonii* in submerged culture and its formulation for application in the field. *Entomophaga* 20(3): 229–240.
Mendonça, A.F., 1992. Mass production, application and formulation of Metarhizium anisopliae for control of sugarcane froghopper, Mahanarva posticata, in Brasil. In: C.J. Lomer, C. Prior (eds), *Biological control of locusts and grasshoppers*. CAB International, Wallingford. pp. 239–244.
Milner, R.J. and D.M. Hunter, 2001. Recent developments in the use of fungi as biopesticides against locusts and grasshoppers in Australia. *J. Orthoptera Res.* 10(2): 271–276.
Montesinos, E., 2003. Development, registration and commercialization of microbial pesticides for plant protection. *Int. Microbiol.* 6: 245–252.
Moscardi, F., 1999. Assessment of the application of baculoviruses for control of Lepidoptera. *Ann. Rev. Entomol.* 44: 257–289.
Nahar, P.B., S.A. Kulkarni, M.S. Kulye, S.B. Chavan, G. Kulkarni, A. Rajendran, P.D. Yadav, Y. Shouche, and M.V. Deshpande, 2008. Effect of repeated in vitro sub-culturing on the virulence of *Metarhizium anisopliae* against *Helicoverpa armigera* (Lepidoptera: Noctuidae). *Biocontl. Sci. Technol.* 18(4): 337–355.
Nicholls, C.I., N. Pérez, L. Vasques, and M.A. Altieri, 2002. The development and status of biologically based integrated pest management in Cuba. *Integr. Pest Manag. Rev.* 7: 1–16.
Patel, A., B. Slaats, J. Hallmann, R. Tilcher, W. Beitzen-Heineke, and K.-D. Vorlop, 2005. Verkapselung von bakteriellen Antagonisten und eines nematophagen Pilzes. *Gesunde Pflanzen* 57: 30–33.
Patel, M.N., M. Stolinski, and D.J. Wright, 1997. Neutral lipids and the assessment of infectivity in entomopathogenic nematodes: observations of four *Steinernema* species. *Parasitology* 114: 489–496.
Pell, J.K., J. Eilenberg, A.E. Hajek, and D.C. Steinkraus, 2002. Biology, ecology and pest management potential of Entomophthorales. In: T.M. Butt, C. Jackson, N. Magan (eds), *Fungi*

as biocontrol agents: progress, problems and potential. CAB International, Wallingford. pp. 71–153.

Perrin, B., 2000. Improving insecticides through encapsulation. *Pesticide Outlook*, April 2000: 68–71.

Piggott, S.J., D.J. Wright, and G.A. Matthews, 2000. Polymeric formulation for the application of entomopathogenic nematodes against foliar pests. *Proc. BCPC Conference-Pest and Diseases 2000* 1: 1063–1068.

Pijlman, G.P., E. van den Born, D.E. Martens, and J.M. Vlak, 2001. *Autographa californica* baculoviruses with large genomic deletions are rapidly generated in infected insect cells. *Virology* 283: 132–138.

Quinlan, R.J., 1988. Use of fungi to control insects in glasshouses. In: M.N. Burge (ed), *Fungi in biological control systems*. Manchester University Press, Manchester. pp. 19–36.

Quinlan, R.J. and S.G. Lisansky, 1983. Microbial insecticides. In: H.-J. Rehm, G. Reed (eds), *Biotechnology*, Vol. 3. Verlag Chemie, Weinheim. pp. 234–254.

Reinecke, P., W. Andersch, K. Stenzel and J. Hartwig, 1990. Bio 1020, a new microbial insecticide for use in horticultural crops. *Brighton Crop Protection Conference* 1990 – *Pests and diseases*. BCPC Publications, Farnham. pp. 49–54.

Rhodes, D.J., 1990. Formulation requirements for biological control agents. *Aspects App. Biol.* 24: 145–153.

Rhodes, D.J., 1996. Economics of baculovirus – insect cell production systems. *Cytotechnology* 20: 291–297.

Roditakis, E., I.D. Couzin, K. Balrow, N.R. Franks, and A.K. Charnley, 2000. Improving secondary pick up of insect fungal pathogen conidia by manipulating host behaviour. *Ann. Appl. Biol.* 137: 329–335.

Rombach, M.C., R.M. Aguda, and D.W. Roberts, 1988. Production of *Beauveria bassiana* (Deuteromycotina: Hyphomycetes) in different liquid media and subsequent conidiation of dry mycelium. *Entomophaga* 33(2): 315–324.

Roussos, S., B.K. Lonsane, M. Raimbault, and G. Viniegra-Gonzalez, 1997. Advances in solid state fermentation. *Proceedings of the 2nd International Symposium on Solid State Fermentation FMS-95*, Montpellier, France. Kluwer, Dordrecht.

Rowe, G.E. and A. Margaritis, 2004. Bioprocess design and economic analysis for the commercial production of environmentally friendly bioinsecticides from *Bacilllus thuringiensis* HD-1 *kurstaki*. *Biotechnol. Bioeng.* 86(4): 377–388.

Schroer, S. and R.-U. Ehlers, 2005. Foliar application of the entomopathogenic nematode *Steinernema carpocapsae* for biological control of diamondback moth larvae (*Plutella xylostella*). *Biol. Control* 33: 81–86.

Schroer, S., D. Sulistyanto, and R.-U. Ehlers, 2005b. Control of *Plutella xylostella* using polymer-formulated *Steinernema carpocapsae* and *Bacillus thuringiensis* in cabbage fields. *J. Appl. Entomol.* 129(4): 198–204.

Schroer, S., X. Yi, and R.-U. Ehlers, 2005a. Evaluation of adjuvants for foliar application of *Steinernema carpocapsae* against larvae of the diamondback moth (*Plutella xylostella*). *Nematology* 7(1): 37–44.

Shah, F.A., M.A. Ansari, M. Prasad, and T.M. Butt, 2007. Evaluation of black vine weevil (*Otiorhynchus sulcatus*) control strategies using *Metarhizium anisopliae* with sub-lethal doses of insecticides in disparate horticultural growing media. *Biol. Control* 40(2): 246–252.

Shah, F.A., C.S. Wang, and T.M. Butt, 2005. Nutrition influences growth and virulence of the insect-pathogenic fungus *Metarhizium anisopliae*. *FEMS Microbiol. Lett.* 251: 259–266.

Shapiro, M., 1986. In vivo production of baculoviruses. In: R.R. Granados, B.A. Federici (eds), *The biology of baculoviruses. Vol. II. Practical application for insect control*. CRC Press, Boca Raton, FL. pp. 31–61.

Shapiro-Ilan, D.I. and R. Gaugler, 2002. Production technology for entomopathogenic nematodes and their bacterial symbionts. *J. Ind. Microbiol. Biotechnol.* 28: 137–146.

Shapiro-Ilan, D.I., D.H. Gouge, S.J. Piggott, and J. Patterson Fife, 2006. Application technology and environmental considerations for use of entomopathogenic nematodes in biological control. *Biol. Control* 38: 124–133.

Shieh, T.R., 1989. Industrial production of viral pesticides. *Adv. Virus Res.* 36: 315–343.

Shuler, M.L., H.A. Wood, R.R. Granados, and D.A. Hammer, 1995. *Baculovirus expression systems and biopesticides*. Wiley-Liss, New York, NY.

Skovmand, O., I. Thiéry, and G. Benzon, 2000. Is Bacillus thuringiensis standardisation still possible? In: J.-F. Charles, A. Delécluse, C. Nielsen-LeRoux (eds), *Entomopathogenic bacteria: from laboratory to field application*. Kluwer, Dordrecht. pp. 275–296.

Smits, P.H., 1987. Nuclear polyhedrosis virus as biological control agent of *Spodoptera exigua*. PhD Thesis, Laboratory of Virology, University Wageningen, Wageningen. p. 127.

Stock, D., 1997. Do we need adjuvants? Mechanistic studies and implications for future developments. *Proceedings of the 50th N.Z. Plant Protection Conference*. New Zealand Plant Protection Society, Incorporated, Hastings, New Zealand. pp. 185–190.

Stockdale, H. and R.A.J. Priston, 1981. Production of insect viruses in cell culture. In: H.D. Burges (ed), *Microbial control of pests and plant diseases 1970–1980*. Academic Press, London. pp. 314–328.

Strauch, O., I. Niemann, A. Neumann, A.J. Schmidt, A. Peters, and R.-U. Ehlers, 2000. Storage and formulation of the entomopathogenic nematodes *Heterorhabditis indica* and *H. bacteriophora*. *BioControl* 45: 483–500.

Summers, M.D., 2006. Milestones leading to the genetic engineering of baculoviruses as expression vector systems and viral pesticides. *Adv. Virus Res.* 68: 3–71.

Sun, X., X. Sun, B. Bai, W. van der Werf, J.M. Vlak, and Z. Hu, 2005. Production of polyhedral inclusion bodies from *Helicoverpa armigera* larvae infected with wild-type and recombinant HaSNPV. *Biocontl. Sci. Technol.* 15(4): 353–366.

Swanson, D., 1997. Economic feasibility of two technologies for production of a mycopesticide in Madagascar. *Memoirs Entomol. Soc. Canada* 171: 101–113.

Szewczyk, B., L. Hoyos-Carvajal, M. Paluszek, I. Skrzecz, and M. Lobo de Souza, 2006. Baculoviruses – re-emerging biopesticides. *Biotech. Adv.* 24: 143–160.

Ugine, T.A., S.P. Wraight, and J.P. Sanderson, 2004. Acquisition of lethal doses of *Beauveria bassiana* conidia by western flower thrips, *Frankliniella occidentalis*, exposed to foliar spray residues of formulated and unformulated conidia. *J. Invertebr. Pathol.* 90: 10–23.

Van Beek, N. and D.C. Davis, 2007. Baculovirus insecticide production in insect larvae. In: D.W. Murhammer (ed), *Baculovirus and insect cell expression protocols*. Humana Press, Totowa, NJ. pp. 367–378.

VanderGheynst, J., H. Scher, H.-Y. Guo, and D. Schulz, 2007. Water-in-oil emulsions that improve the storage and delivery of the biolarvicide *Lagenidium giganteum*. *BioControl* 52: 207–229.

Van der Pas, R.K., W.J. Ravensberg, and E. Cryer, 1998. Insect pathogenic fungi for environmentally-friendly pest control in the glasshouse: investigating oil formulations. *IOBC/WPRS Bull.* 21(4): 129–132.

Van Oers, M.M., G.J. Messelink, S. Peters, and J.M. Vlak, 2005. Een nieuw baculovirus van en voor de Turkse mot. *Gewasbescherming* 36(6): 268–269.

Verhaar, M.A., T. Hijwegen, and J.C. Zadoks, 1999. Improvement of the efficacy of *Verticillium lecanii* used in biocontrol of *Sphaerotheca fuliginea* by addition of oil formulations. *BioControl* 44: 73–89.

Vimala Devi, P.S., T. Ravinder, and C. Jaidev, 2005. Cost-effective production of *Bacillus thuringiensis* by solid state fermentation. *J. Invertebr. Pathol.* 88: 163–168.

Vlak, J.M., C.D. de Gooijer, J. Tramper, and H.G. Miltenburger, 1996. *Insect cell cultures. Fundamental and applied aspects*. Kluwer, Dordrecht.

Weiss, S.A., B.F. Dunlop, R. Georgis, D.W. Thomas, P.V. Vail, D.F. Hoffmann, and J.S. Manning, 1994. Production of baculoviruses on industrial scale. *Proceedings of the XXVIIth Annual Meeting Society Invertebrate Pathology*. Montpellier, France. pp. 440–446.

Weiss, S.A. and J.L. Vaughn, 1986. Cell culture methods for large-scale propagation of baculoviruses. In: R.R. Granados, B.A. Federici (eds), *The biology of baculoviruses. Vol. II. Practical application for insect control*. CRC Press, Boca Raton, FL. pp. 63–88.

Wilson, M. and E. Ivanova, 2004. Neutral density liquid formulations for nematode-base biopesticides. *Biotechnol. Lett.* 26: 1167–1171.

Wood, H.A. and P.R. Hughes, 1996. Recombinant viral insecticides: delivery of environmental safe and cost-effective products. *Entomophaga* 41: 361–373.

Wraight, S.P. and R.I. Carruthers, 1999. Production, delivery, and use of mycoinsecticides for control of insects pests on field crops. In: F.R. Hall, J.J. Menn (eds), *Biopesticides: use and delivery*. Humana Press, Totowa, NJ. pp. 233–269.

Wraight, S.P., M.A. Jackson, and S.L. de Kock, 2001. Production, stabilization and formulation of fungal biocontrol agents. In: T.M. Butt, C. Jackson, N. Magan (eds), *Fungi as biocontrol agents: progress, problems and potential*. CAB International, Wallingford. pp. 253–287.

Young, S.Y. and W.C. Yearian, 1986. Formulation and application of baculoviruses. In: R.R. Granados, B.F. Federici (eds), *The biology of baculoviruses. Vol. II. Practical application for insect control*. CRC Press, Boca Raton, FL. pp. 157–180.

Ypsilos, I.K. and N. Magan, 2004. Impact of water-stress and washing treatments on production, synthesis and retention of endogenous sugar alcohols and germinability of *Metarhizium anisopliae* blastospores. *Mycol. Res.* 208(11): 1337–1345.

Zimmermann, G., 2005. Pilzpräparate. In: H. Schmütterer, J. Huber (eds), *Natürliche Schädlingsbekämpfungsmittel*. Ulmer, Stuttgart. pp. 87–109.

Chapter 4
Quality Control

Abstract Quality control can de divided in production control, process control, and product control. The first two refer to internal quality control of the production of a microbial pest control product, and ensure a stable production process with a minimum of failures. Product control refers to the quality of the final product that leaves the factory and which needs to perform according to registration and customer satisfaction requirements. Products must meet product specifications, which are set by the manufacturer, until the end of the claimed shelf-life. Registration requirements in terms of quality control are reviewed. Complete quality control procedures and data for validation must be established, although there are no officially recognized criteria. Practical challenges in quality control procedures are reviewed per type of pathogen. Natural variation makes efficacy testing via bio-assays difficult, and setting an internal standard is required. Recommendations for standardization and criteria will be provided. Research needs are identified that may facilitate quality control in the future. Quality control must ensure that end-users receive high quality products. Total quality control covers all aspects of quality, including the field use of a biopesticides. Biocontrol companies should ensure that the whole chain is well aware of quality issues and that those involved act accordingly. A list of benefits of quality control is provided which illustrates that both the biocontrol industry and its customers benefit from proper quality control.

1 Introduction

Biopesticides are often criticized for their variable performance and lack of reliability. Quality control (QC) is therefore of paramount importance in order to ensure that products are delivered that comply with pre-determined specifications and deliver the efficacy within the prescribed conditions for use. Quality control does not only refer to the final end-use product, but also to the production and the production processes. In general, quality control objectives are to ensure that:

W.J. Ravensberg, *A Roadmap to the Successful Development and Commercialization of Microbial Pest Control Products for Control of Arthropods*, Progress in Biological Control 10,
DOI 10.1007/978-94-007-0437-4_4,

(1) properties of incoming raw materials comply with the manufacturer's specifications;
(2) there is consistency between production runs and products;
(3) end-use products meet criteria set by registration authorities; and that
(4) product performance meets the end-user's perception of quality in relation to price, and leads to repeat purchases of the product.

Quality control is performed in all manufacturing fields, but interpreted differently. Thus, each industry must define quality control precisely. In biocontrol, quality control has received much attention in the field of sterile insect technology and in arthropod natural enemies. For an overview of quality control matters regarding mass-reared sterile insects I refer the reader to Boller and Chambers (1977), and for arthropod natural enemies to Van Lenteren (2003a). A brief historical overview on the development of quality control for beneficial insects is provided by Leppla (2008). In quality control of natural enemies, Leppla (2003) distinguished three functions: production, process and product control. In the field of microbial pesticides, however, clear definitions are lacking. Few authors extensively treat quality control in its entirety. Most papers dealing with quality control focus their comments to end-use product control, while aspects of process control are frequently included. Definitions of quality control used for microbials will preferably be similar to the ones used for natural enemies and other beneficial biocontrol agents, for the sake of convenience. I will use the separation of functions as given by Leppla (2003) and use the following slightly modified definitions:

(1) production control is a procedure intended to ensure that incoming raw materials comply with defined specifications, and that production equipment be properly maintained, and that all inputs needed to run the production (materials, labour, energy, etc.) be available at the right time;
(2) process control is a procedure intended to ensure that a manufacturing process runs according to previously determined process parameter profiles so that expected yields and formulated products be delivered;
(3) product quality control is a procedure intended to ensure that a manufactured end-use product adheres to the product specifications, a previously defined set of quality standards, in order to meet the efficacy norms within described conditions for use, and to meet criteria set by registration authorities.

To determine the process parameters and the specifications of incoming raw production materials and formulation ingredients, and the outgoing end-use products, research is needed to establish the appropriate standards and to develop methods to measure these standards. A company can decide itself on specific standards, although for the final product some of these are requirements of registration authorities. As a result, standard operating procedures (SOPs) should be developed. This research is imperative and is a considerable amount of work. It should be conducted within the developmental phase of the production and of the final product.

The results are the basis for a consistent manufacturing process once products are routinely made and quality control is implemented.

The term quality assurance is often used in this context; it covers all activities from designing and developing quality control and the implementation of it during all phases, including reacting to deviations and improving processes and procedures to ensure good quality of the end-use product. It can be described as the continuing process of ensuring good quality. International standards have been developed for manufacturing activities and models are described and certified under ISO 9000 series. Implementation of such a quality management system can ensure consistent production and high quality products. Furthermore, a producer may choose to comply with Good Manufacturing Practices (GMP), although this has been developed for the pharmaceutical industry for which it is more appropriate. ISO and GMP are, however, rarely implemented in the biocontrol industry. The biopesticide companies Andermatt Biocontrol, Futureco and Prophyta are ISO 9001-2000 certified, Valent BioSciences is ISO 9000-2008 certified. The Bio-Fly plant, a subsidiary of Bio-Bee, Israel, operates in compliance with the ISO 9001-2000 quality standards in the production of sterile flies of the Mediterranean fruit fly. Bio-Fly is the only company in the field of mass production of arthropods which works with this certified quality system. Generally, companies find these systems too complicated and too expensive.

The production of MPCAs is a complex system that involves living organisms. This calls for the monitoring of crucial aspects in the whole line of processes, from production to the use of the end-use products. Quality control and feedback from deviations should be an integral part of the development and the use of biocontrol products within all biocontrol companies. This overall process is called Total Quality Control (TQC). An elaborate view on TQC for natural enemies is given by Leppla (2003); the relevant aspects of TQC apply to microbial control agents as well. Quality control systems for MPCPs are much more comprehensive and specialized than those used for chemical pesticides, where relatively simple analytical techniques can be used. MPCPs contain living organisms, which are inherently variable in performance, and as a result the products will never be as stable as technical products. This variability demands a rigorous quality control system in order to ensure that biological products are delivered within specifications. Manufacturers are required to develop their own systems and regulatory procedures are not harmonized, nor are any standards available. As a result, quality control in biopesticides is neither transparent nor standardized. I will discuss the critical aspects of quality control in the commercial development of biopesticides, focussing on product quality control, and provide recommendations for a more standardized quality control.

2 History of Quality Control in Microbial Pest Control Products

The market of MPCPs is characterized by the frequent appearance and disappearance of products as well as companies (Lisansky 1997). For this, many different explanations have been put forward, including poor performance of products. The lack of effective quality control is often cited as a possible reason for the failure of a product. Some examples are given in the literature for baculovirus products produced in Egypt respectively India (Grzywacz et al., 1997; Battu et al., 2002). Likewise, quality control of EPN products by independent scientists has revealed inconsistent quality. Gaugler et al. (2000) investigated mail-ordered *H. bacteriophora* and *S. carpocapsae* products from cottage industry in the USA and found only 60% of the number of nematodes stated on the label, on average. Also, reduced pathogenicity was often found. Few resources are dedicated to quality control, whose importance is often underestimated in the biocontrol industry. Small companies may not have the knowledge or the facilities for proper quality control checks. To be fair, proper testing methods were lacking in the past.

Registration does require well-described quality control procedures and quality control data, but quality control is seldom enforced following approval. Some countries do not require registrations, or products are sold without formal approvals, and quality control may well be lacking in such situations. I have seen many unregistered MPCPs on the market which did not comply with the label claims. On most of the occasions this concerned the number of infective propagules, which was too low, and sometimes, even infective propagules were lacking. More complicated features such as performance and purity were not checked, but were bound to deliver inferior results. These products damage the image of biocontrol. This underscores the importance of a good quality control programme.

In the field of natural enemies, similar situations have occurred, leading to failures in biocontrol and the disappearance of some companies (Van Lenteren 2003a). The biocontrol industry together with scientists did develop quality control criteria and test guidelines for most natural enemies (Van Lenteren et al. 2003b), and now most of the larger companies do conduct quality control according to these tests and criteria (Bolckmans 2003).

In the field of microorganisms, an attempt was made in 2001 to initiate a similar collaboration between industry and scientists. Scientists of CABI and the International Biocontrol Manufacturers Association (IBMA) initiated a working group on quality control and standards of MPCPs. A joint IBMA – COST 830 workshop was held in 2001 to try to set appropriate standards for microbial products (Migheli and Ruiz Sainz 2003). As a result of a lack of funding, this laudable initiative did not produce many tangible results and the IBMA Working Group is no longer active. Today, there is no mutually agreed set of testing methods and criteria for MPCPs, unlike for natural enemies. Registration procedures for MPCPs generally require quality control methods and data, but standardized guidelines and clear criteria are lacking, even within the EU Directive 91/414 (EC 1991) on registration of plant protection products. Every company is allowed to develop its own methods and to set its own standards with which products should comply. Jenkins and Grzywacz (2003) recognized the lack of standardization and they have suggested a minimum set of parameters for quality control for fungi and baculoviruses.

In the field of EPNs, quality control was first discussed between industry and scientists at the eighth workshop of the IOBC Working Group on "Quality control of mass-reared arthropods" in 1995. All companies agreed that QC criteria needed to be developed for nematodes as with beneficial arthropods (Mason et al., 2002). Ultimately, successful collaboration between scientists and the industry in the EU COST Action 819 led to the development and publication of QC criteria and testing protocols (Grunder 2005). Clear standards, as determined and accepted for natural enemies, however, are still lacking for most quality control parameters for EPNs.

3 Production Quality Control and Process Quality Control

Quality control of production concerns internal procedures for checking whether inputs (purchased raw materials: medium ingredients, formulation compounds and packaging materials) comply with their specifications, whether standard operating procedures (SOPs) are well set-up and followed, and whether production and downstream equipment is maintained properly and is operating as expected. Warrior (2000) emphasized that the quality of raw materials contributes to a large extent to the quality of the final products and that specifications of raw materials must be checked before release for production purposes. He referred mainly to the large scale production of *Bacillus thuringiensis* (Bt).

Quality control of production processes relates to monitoring vital process parameters, and whether they run according to expected profiles. Conducting these procedures should result in minimum production failures, and in expected yields of good quality. Continuous feedback and improvements should also lead to production stability with minimal extra costs, and to products that match planning. As production systems and processes vary greatly for MPCAs and between companies, details cannot be given here. Nevertheless, quality control is important and should be an integral part of the mass production. Process parameters can be relatively simple such as room temperature and percentage relative humidity in a rearing room for insects in in vivo production of, for example, a baculovirus. On the other hand, it can be a complicated set of parameters that influence each other in liquid production of EPNs. In this case, variables such as pO2, pCO2, pH, temperature, foaming and hydrostatic pressure need to be monitored and adjusted where necessary. Some variables can be adjusted in various ways, as for example oxygen pressure, which can be adjusted by aeration, stirrer speed, temperature and by vessel pressure. This illustrates the complexity of such production processes and computer steered programmes are needed to regulate the process. Process control also includes downstream and formulation processes, packaging and storage. At various steps in the processes, parameters must be monitored and checked. Data should be gathered during all processes and should be used to improve the production on a continuous basis. Production control and process control are often combined in SOPs and performed by production personnel. Production and process control have the greatest effects on the quality of the resulting product and therefore should be taken very seriously within a company.

Quality control on inoculum stability is crucial. A procedure for storage and use of inoculum material for the production is described in Chap. 3 where the chief goal is maintaining inoculum stability. The inoculum needs to be monitored for phenotypic and genotypic changes and for contaminants. When genetic stability is an issue, specialists may need to be consulted to make sure the inoculum quality is still as desired. Examples of quality problems are loss of plasmids in bacteria, and mutations in baculoviruses. If the original inoculum is no longer of good quality, the strain needs to be recovered anew from a culture collection or from a stored product. Starting with a new isolate is a delicate decision and requires very detailed checking of the strain and its properties before it should replace the older strain and be taken into production. Companies prefer to avoid this as it costs time and money. Long term storage of stable inoculum is the preferred option; if this is not possible new inoculum must be made and tested rigorously. In the case of the production of Green Muscle (*Metarhizium anisopliae* var. *acridum*), the standard isolate is passaged through the desert locust every 6 months to avoid loss of virulence (Cherry et al., 1999). Inoculum storage should be taken extremely seriously and measures should be taken to avoid any problems with the stored material as much as possible. For instance, it is wise to keep inoculum in two different places to spread risks and to have an alarm system on the storage facility or deep freezer.

4 Product Quality Control of Microbial Pest Control Products

Quality control generally refers to product control, i.e. to the final formulated product. In manufacturing, however, there may be several in-between products, like the "technical grade active ingredient" (TGAI), that need to be checked. Material from succeeding production batches may be stored to be formulated together. Checks on propagule numbers, microbial contaminants are needed and some batches may be discarded or handled differently, stored longer or shorter. During formulation, propagule numbers may be set to the specification of the final product, thus propagule counts are needed before final formulation. The final formulated product is subject to quality control and the product needs to conform to previously determined product specifications ("specs"). These specifications are usually determined with regard to product performance in the field in a broad sense and to registration requirements. Product performance relates to efficacy, obviously, but also to applications characteristics (particle size, emulsion separation, sedimentation in the package) and to shelf-life parameters (moisture content, microbial purity).

The "external" objective of quality control is to deliver a product to the end-user that complies with the specifications at the moment of use. It should also deliver the expected results. This objective focuses at satisfying customers and on complying with registration criteria. Each batch has to be tested prior to sale in order to prevent release of products that do not meet the "specs". The "internal" objective of quality control is to highlight any problem areas which should then be subject to corrective action. This should lead to improved production of products with fewer and fewer discarded batches, and to stability of the production and the products.

A number of papers refer to the overall issue of quality control of MPCAs. Burges (1981) recognized two main aspects of quality control: identity of the pathogen and level of contaminants. In addition he discussed activity, physical features, and safety to vertebrates in an example treating baculoviruses. Jenkins and Grzywacz (2000) reviewed the need for quality control procedures in fungal and viral biocontrol agents in the production as well as for the final products. Product specifications should be determined with regard to physical properties, microbial contaminants, efficacy and storage. They illustrated this by providing recommendations for specifications for fungal and viral products. Lisansky (1985; CPL 2006b) listed five aspects that should be subject to QC: efficacy, microbiological purity, absence of mammalian toxicity, physical characteristics and shelf-life. Quality control methods and specifications are discussed and given in detail for Green Muscle (Jenkins et al. 1998). The need for quality assurance in the industrial production of biopesticides is briefly discussed for bacterial, fungal and viral products by Guillon (1997a). One company, Verdera, Finland, has disclosed some information on their chosen quality criteria for biofungicides. Specifically, they refer to QC during production and to end-use product criteria. Assessed product criteria are viability and efficacy tests, related to storage stability (Palin-Holmberg et al. 2003). In the literature, product quality control generally comprises the features named below.

4.1 Identity of the Microbial Pest Control Agent

The organism needs to be properly identified at strain level using the most appropriate scientific techniques. Molecular techniques are needed to do this. The area of taxonomic identification is still developing and species and strain concepts are changing with the development of new technologies. Identification is a difficult topic, but very important with regard to a proper risk assessment. Therefore, the Biopesticide Steering Group (BPSG) of the OECD is developing a guidance document that should help regulators and the industry in assessing the taxonomic identification. Most companies lack the expertise to do this in-house and external laboratories are asked to conduct the identification. The OECD recommends that two laboratories identify the strain to ascertain the identification (OECD 2008). Typically, this identification is done at the beginning of a product development and not as part of the quality control procedure of every batch. Once a master stock culture has been made of the properly identified inoculum, production is started with a specimen of that stock, and batch-wise identification is superfluous. When a new master stock is made, re-identification of the organism should be conducted. Nevertheless, cultures should always be checked visually for deformations or other deviations (genetic or phenotypic changes) to make sure the organism still shows its usual features. At the slightest suspicion, re-identification should be conducted. I consider this aspect to be part of the process control rather than of product control. It cannot be done for every batch, and furthermore, there is no need for it. I recommend re-identifying the inoculum every 3–4 years by a specialist. Then, a recent identity confirmation is available for registration purposes and liability.

4.2 Number of Infective Propagules

A product contains a certain number of infective or virulent propagules. This needs to be specified on the label. This number of infective propagules, when applied according to the recommended methods of application and under the right conditions of use, should achieve the claimed level of control of the pest. Counting of propagules can be done, but is often difficult in technical grade products as well as in formulated products in which remnants of the production medium and/or carrier and particles of formulation ingredients may be difficult to distinguish from spores or virus particles. The number of propagules, however, does not necessarily reflect virulence directly. Accordingly, the actual activity needs to be determined by germination tests or by establishing the number of colony forming units (cfu). This method is used in the case of bacteria and fungi. The active ingredients in Bt's are spores and endotoxins (crystallized proteins), the latter can be measured by analytical techniques. But virulence of the product needs to be assessed in bio-assays. Baculoviruses cannot be counted, usually, because they are hard to distinguish from formulation particles or insect remnants. Even if they can be counted, viable and non-viable particles cannot be distinguished. Since they are obligate parasites that can only replicate in insect cells, a bio-assay is needed to assess the product's virulence. EPNs can usually be counted and living and dead dauer juveniles (DJs) can be distinguished under the microscope. Since yields may vary per batch, counts of infective units needs to be carried out to be able to adjust the number during the formulation process to reach the total number in the end-use product as specified. Next, packaging samples need to be counted again to confirm the "specs". Checking the product on the number of infective propagules is a basic element of product control.

4.3 Microbial Purity

The total number or percentage of microbial contaminants in a MPCP should be limited. These contaminants may negatively influence product quality in terms of shelf-life, efficacy and even physical characteristics. Furthermore, contaminants may pose a risk to the applicator and the consumer. Registration criteria therefore only accept low numbers of contaminants and require absence or near-absence of human pathogens. The total number of contaminating organisms should not exceed a certain number, or, it should not exceed a certain percentage of the number of the active ingredient. The standard depends on what the company decides or on what the registration requirements demand. The standard is often set at 0.1% of the number of infective propagules and is just a pragmatic amount chosen by many companies. There are, however, no obligatory standards within the EU, EPA, OECD and national regulations for MPCPs. For more details, see below at registration requirements. Determination of the level of contaminating organisms, usually bacteria and fungi, is done by cfu testing on microbe-specific media and at certain temperatures.

A total count of mesophiles (fungi, bacteria and yeasts) is often conducted and counts of bacteria, specifically human and mammalian pathogens, by selective tests. Some countries demand animal toxicity tests for Bt products and baculoviruses with every batch to exclude human pathogens. Determination of the level of contaminants needs to be part of process control as well as of product control.

4.4 Presence of Toxins

Presence of toxic metabolites is a concern of regulatory bodies and information needs to be presented to the authorities for each specific MPCA where this is considered relevant. The issue of toxins is not well regulated and a discussion between the industry and regulators is ongoing. Clear and appropriate requirements still need to be defined. This topic is dealt with in Chapter 5 in more detail. Standard criteria are not defined and a case by case approach is taken. Harmful effects should be minimal. In the production of some fungi and bacteria, metabolites may be formed and this may depend on the conditions of the production process. Deviations in the process may lead to a higher production of metabolites, and metabolite levels may need to be checked. Toxins are difficult to monitor and biochemical analytical techniques have to be developed to identify them and to monitor them. Routinely monitoring of toxins in the production is only required in some cases following risk assessments by the evaluating authority. An example is the fungus *Isaria fumosorosea (=Paecilomyces fumosoroseus)* where absence of secondary metabolites must be checked in each fermentation batch. This has subsequently been decided and documented by the EC (2002). When toxins are found, the batch has to be discarded.

4.5 Physical, Chemical and Technical Characteristics

Registration requirements demand product stability with regard to physical and chemical stability, and information on technical properties. The formulation needs to be physically stable during the shelf-life period. This means no clump-forming in wettable powders should occur, nor should irreversible sedimentation of propagules in suspensions or separation of carriers in the case of emulsions take place. Chemical stability (like pH) also needs to be considered and checked according to registration requirements, but in reality this is of lesser importance. Each type of formulation has its particular characteristics that need to be checked. These are referred to as technical properties. Examples are wettability, suspensibility, and particle size distribution in wettable powders and water dispersible granulates. For emulsions it is emulsifiability and stability, for suspensions pourability, etc. These properties have to be tested according to recognized CIPAC methods, which have been developed for chemical pesticides, and comply with standard criteria.

4.6 Efficacy

Efficacy is the most significant parameter in quality control. The product's field performance is the most valuable aspect; not only for registration, but also for the company's revenues from the product. If batches of the product do not perform up to expectations, repeated sales will decline. Therefore, every batch should be tested prior to release. This can only be done quickly and in a cost-effective way by using bio-assays. These should reflect the product's performance in the field as much as possible. The relationship between bio-assay results and field results needs to be established from the outset. This study should include batches with poor quality in order to see whether the bio-assay really distinguishes good batches from poor batches. This is often not done and then the test is useless. Bio-assays must be developed in such a way that they give a reliable prediction of how a product will work in the field. Obviously, field studies cannot be conducted for each batch because of time and costs issues.

Bio-assays differ per pathogen and per target. The same bio-assay that was developed for strain selection can also be used in QC tests. There are numerous tests available, see Chapter 2 for more information. Registration requirements demand a description of a QC efficacy test, but specific guidance is not available. Every company is allowed to use its own method as long as it gives reliable data. Bio-assays must be standardized and must be done correctly, otherwise they are meaningless. A major problem in using bio-assays in quality control testing is accuracy and reproducibility. Variations due to insect-rearing fluctuations, assay conditions and insect stage can influence the insect's response to the pathogen. As a result, it is desirable to include a standard product as a reference in the test and to compare the results of the sample with the results of the standard. This is possible in bacteria and baculoviruses where material can be stored for years without a change of properties. In fungi this is more difficult. In EPNs this is almost impossible due to difficulties in storing material for along time. Testing for efficacy is clearly the most important aspect of product control.

4.7 Quality Control and Shelf-Life

A product has to comply with its set QC criteria up to the end of its shelf-life period. Usage of a product at the end of its shelf-life should give similar results as does usage of a fresh product. As a consequence, the length of the shelf-life period is determined by the QC criteria and product specifications which it still has to meet at the end of the shelf-life period. For example, if spore germination rate in a fungal product has a minimum set criterion of 80% and after 6 months the germination rate drops under this level, the shelf-life period is determined at 6 months, even when other criteria are still above their minimum criteria. If one parameter does not comply to "specs" anymore, the batch has to be discarded.

There are no clear registration requirements for storage stability for biopesticides in the EU. The company determines the product's shelf-life period in combination with the storage temperature. For instance, a mycoinsecticide has a shelf-life of 6 months when stored between 5 and 10°C. Evidence of product stability should be presented in the registration dossier and then this will be accepted by the authorities. For registration tests, QC parameters need to be tested at initiation and at termination of the storage period. Tolerance levels are, however, only specified for chemical pesticides, and those are not appropriate for biopesticides. Currently, the OECD BPSG (Biopesticide Product Steering Group) and the IBMA are in discussion regarding the establishment of specific testing procedures and tolerance limits for storage stability of MPCPs (OECD 2008). I will discuss this more in detail in Chapter 5. Product quality parameters need to be checked at regular intervals during the shelf-life period to investigate whether product characteristics are stable. This will ensure confidence in the determined specifications. Where deviations are found, corrective measurers need to be taken in the preceding processes. If, in the end, improvements are not possible, the length of the shelf-life period must be reconsidered.

4.8 Quality Control Parameters for End-Use Products

Product quality control needs to be carried out on the formulated product. It may also be done on bulk material just before packaging if this process is known not to influence quality. Testing should require as little time as possible so that no valuable storage time is lost. Parameters subject to testing at this point in time are:

(1) number of infective propagules;
(2) physical-chemical properties;
(3) microbial contaminants;
(4) efficacy.

Once results are conform "specs", the batch can be released for sale. If one of the parameters does not comply with its specification, the batch has to be discarded or send back to production in order to be improved, if possible. Batches where human pathogens or harmful toxins have been detected are to be discarded early in the production process. An identity check is not part of product control.

5 Critical Aspects of Product Quality Control with the Various Types of Pathogens

Product quality control is an integral part of manufacturing biopesticides to ensure sale of high quality products. I believe that today this concept is well recognized in the biocontrol industry and that QC has become a routine procedure. In addition, registration requires detailed description of QC procedures, accompanied by real data to illustrate that products do meet previously determined specifications. Entomopathogenic fungi and bacteria, and baculoviruses are subject to registration, EPNs generally not. Registration requirements with regard to product control of biopesticides are discussed below. In many papers, quality control of MPCPs is discussed. Researchers acknowledge the importance of quality control, but few papers treat this subject systematically and in-depth. I will review the specific aspects of each type of pathogen related to product control. Each type of pathogen and each type of product requires monitoring of specific characteristics related to the number of infective propagules, the formulated product, and its field of application. On the other hand, all products require monitoring of similar parameters that are relevant to all of them such as microbial purity and technical properties. The most determinative parameters are the number of infective propagules, the biological activity of the product, microbial purity, and technical properties. Counting propagules is difficult in bacterial and viral products and generally it does not give reliable information on efficacy. A bio-assay must be conducted to obtain useful information on efficacy. In fungal products, the germination rate and bio-assays produce reliable information on virulence. In nematode-based products counts of dauer juveniles and bio-assays need to be conducted to measure quality.

In bio-assays, standardization of the methods is crucial, as is comparison with a reference standard. This reference ideally is a stored reference standard of the product; if this is not available, then the reference can be a chemical insecticide, or a specific mortality rate. A standard should be developed in the developmental phase of the product and should have a predictable relationship with field efficacy. This must have been established by concurrent testing of the product in the field and in bio-assays. In bacteria and baculoviruses, a reference product can be stored for many years. This is not the case in fungi and nematodes; there a set mortality rate can be the "standard".

Standardization has been investigated and discussed within each group of pathogens. This is, however, still ongoing and appears to be a complex matter. As of today, little standardization has been agreed upon and it is hardly implemented in the industry except for Bt's. For fungal and viral products, standardization of quality control has been proposed by Jenkins and Grzywacz (2003). The authors recommended minimum quality control parameters and methods for each type of product. Research has delivered a good understanding of the critical quality aspects in each type of pathogen. Further investigations are still needed to improve quality control testing methods and to make testing easier and cheaper.

5.1 Entomopathogenic Bacteria

In bacterial insecticides, the essential parameters subject to quality control are the same as for fungal products: the number of active propagules, the biological activity, and the microbial purity. Most of the literature discussing QC in bacterial products refers to Bt products. Warrior (2000) listed five parameters for QC as used by a major producer (Valent BioSciences, formerly Abbott) of Bt products: strain identity, metabolite profile, spore count, biological stability, and physical properties. Potency and physical specifications are the main QC parameters according to Couch (2000). The insecticidal activity of Bt is based on spores and toxins. There are many strains and types of toxins available, each with its specific biological activity. Pathogenicity of Bt's is often primarily due to δ-endotoxins (crystallized proteins). Accordingly, spore counts do not correlate well with the biological activity, and hence a bio-assay is a prerequisite for quality product control. Standardization of the bio-assay, the use of a reference standard Bt, and which test insect to use have been the focus of many studies and debates. Detailed bio-assay protocols are needed. Otherwise, reproducibility is poor as was demonstrated by Skovmand et al. (1998) when a *B. thuringiensis* var. *israelensis* (Bti) preparation was tested in different laboratories. Factors such as age, stage and strain of larvae used, amount and type of food used and rearing conditions influence bio-assay results. Also the method and duration of sample homogenization may strongly influence results, as well as the method of administration. Analytical biochemical techniques (HPLC and electrophoresis) can be used to determine the concentration of the δ-endotoxins and this could make QC easier (Bernhard 1992). Within Valent BioSciences this is used in combination with bio-assays (D. Avé, personal communication). If the analytical method is reliable and shows a strong correlation with efficacy, bio-assays may no longer be necessary. This would give quicker and cheaper QC results, but today bio-assays are still required.

Another problem is the lack of standardization when expressing the active ingredients. This can be done in spores/g, but in Bt's this is meaningless in most cases. Endotoxin levels can be expressed in dry weight, but this often includes fermentation products as it is difficult to measure only the crystal proteins, and because other metabolites may also be involved in the pathogenicity process. As a result, standardization in Bt has been a subject of discussion for a long period of time. For two reasons: first, industrial standardization which refers to the need of a producer when checking his product's quality, and second, there is a need for an international standardization to be able to compare products with different strains, and from different producers (Burgerjon and Dulmage 1977). The latter is desirable for scientists, regulators and users. The discussion has resulted in the acceptance of an international reference standard for Bt's where the biological activity is measured in a bio-assay on a single insect species and compared to a reference Bt strain. The biological activity of products, often called potency in Bt's, is expressed in International Units (IU)/mg based on a comparative bio-assay with the product and the standard, assessed against a particular insect. Generally, *Trichoplusia ni* or *Ephestia kühniella* are used as reference insects. Sometimes diamondback moth and *Spodoptera* units are used. This way of expressing potency of products in order to make them comparable is unique to Bt's within the field of biopesticides. Still, some problems occur with standardization such as the lack of available standard Bt, and the need for more detailed bio-assay and sample preparation protocols (Skovmand et al. 2000).

The increasing number of discovered Bt toxins as well as the number of products warrant suggestions for improvements for bio-assays and internationally accepted standard reference material as given by Skovmand et al. (2000). They suggested the use of three test insect species in the bio-assays because of the differing susceptibilities of lepidopteran species to various endotoxins, and the establishment of a standard based on three Bt strains containing a mix of toxins. Further on, they suggested the combination of bio-assay testing with quantitative protein analysis. Finally, they called for an international podium to further develop and acknowledge the standardization, as was carried out earlier by the USDA and the WHO. A historical overview on the progress in this area over the last 50 years and on the still existing problems is given by Asano (2006).

The above paragraphs are relevant for lepidopteran Bt's. Bti products for mosquito control all contain the same endotoxin which makes comparisons easier. For *Bt tenebrionis*, for control of Coleoptera, international standards have not been developed (Skovmand et al., 2000). For Bti products, the FAO and WHO have jointly developed specifications for the active ingredient (identity and biological activity), relevant impurities, bacterial contaminants, physical properties and storage stability (FAO/WHO 2006). The document includes detailed physical and biological testing methods. These specifications could be used as a reference for all Bt formulations.

Standardization as discussed above refers to a QC parameter for product efficacy. It does not include other QC parameters such as microbial purity, physical properties and safety.

Registration requires monitoring of these parameters also. Concerning microbial purity, Couch (2000) gave a list with acceptable levels of various contaminants based on internationally recognized levels for animal feed. Generally, this is similar for all entomopathogen-based products that need to follow registration requirements. Below these requirements will be discussed in detail. In a Danish post-registration survey on specifications of biopesticides, Bt products were found to be free of contaminants (Winding 2005). For safety reasons, each batch should be tested on toxicity to mice according to Couch (2000). When Bt products were first commercialized, there was concern that possible contamination with *B. anthracis* would not be detected with conventional testing for contaminants. Therefore, a mouse toxicity test has been required by the EPA in the USA for each batch and companies still perform this test on a regular basis (CPL 2006a).

Specifications for physical properties for Bt products should be developed by a producer and checked for in every batch. The properties depend on the type of formulation. Protocols and specifications as for chemical pesticides should be used as a reference.

Quality control in Bt's has been intensely investigated by many researchers and companies and quality control is well developed. Many organizations and regulatory authorities agree on the methods and reference standards. As a result, Bt products are of high quality and are the most reliable products in the MPCP market.

In a non-Bt product, Invade, based on *Serratia entomophila*, the quality control system QC has been described by Pearson and Jackson (1995). QC parameters are cell density, purity, virulence and longevity. Purity is defined as confirmation of the production strain by cell growth characteristics on a specific agar plate and the expected phages pattern. Microbial contaminants are not separately tested. Virulence is checked by visualization of plasmids and by a bio-assay confirming pathogenicity on the target insect, i.e. grass grubs. In the longevity test, the decline of cell viability is periodically checked. The authors do not give product criteria or tolerance limits. This demonstrates that QC in bacterial products varies per product and per species and depends on the active ingredient, the mode of action and the target insect(s).

5.2 Entomopathogenic Fungi

In general, biopesticides based on entomopathogenic fungi need to be registered and a detailed description of quality procedures as well as real data from tests are required to illustrate that the product complies with pre-determined specifications. In fungal products, the essential parameters subject to quality control are the number of active propagules, the biological activity and the microbial purity. Fungal biocontrol products cover a range of species and applications such as antagonists, mycoinsecticides, plant growth enhancers, and weed control products. Production of these fungi generally is in vitro and it is well-known that several passages on artificial medium can influence their virulence. For all these products, there is a need to check QC parameters to guarantee an effective and safe product in the market. Jenkins and Grzywacz (2003) gave recommendations for minimum quality control for fungal products, including aspects of process control as well as product control. To test the product quality the following should be checked: the level of microbial contaminants, viability, virulence, moisture content and particle-size distribution. Specifications and methods were given for microbial purity, viability and particle-size distribution. The difficulty of setting up a system to check virulence was discussed.

Counting the number of active ingredients is relatively easy in fungal preparations. The infective propagules can be conidiospores, blastospores or mycelial particles, which can be plated out on artificial medium and counted to assess the percentage viability. There is no standard accepted viability rate. Jenkins and Grzywacz (2000) originally recommended 85% for fungal products, up to the advised expiry date when kept at recommended temperature. But they later reduced it to 80% (Jenkins and Grzywacz 2003) without explanation. Speed of germination is a quality factor too and can be assessed under the microscope in water or on an artificial medium.

In entomopathogenic fungi, virulence is a complex process in which many factors such as cuticle degrading enzyme production, spore binding factors to the cuticle, and metabolite production play a role (Pernfuss et al. 2004). The loss of any of these factors, a process called attenuation (Butt et al. 2006), may have a detrimental effect on the product's performance. This phenomenon has been observed in many fungi where it affected virulence, but it also can lead to phenotypic changes such as alterations in growth and reduced spore production. Therefore a QC check on biological activity in bio-assays is essential. Bio-assays can be conducted in many ways, depending on the product and on the target insect. Guidance on bio-assays for entomopathogenic fungi can be found in Lacey (1997) and in many other papers. The vital factor in bio-assays is that they be standardized in terms of target insect instar, insect-rearing method, and inoculum. Ideally, batches should be compared with a reference product, a standard, but since stability in long time storage of fungal material is uncertain, this cannot be done. The alternative is to test the recommended label dose and to set a certain percentage mortality as a specification. This needs to correlate with the desired efficacy in the field. This correlation has to be studied first before a specification can be determined.

Every producer develops its own methodology and sets its own specifications, depending on the product and the registration requirements. Because of the lack of general accepted protocols, a standardized laboratory bio-assay method was developed by Landa et al. (1994) for entomogenous fungi against whitefly. They described a method using fourth instar nymphs of whitefly and placing them in a droplet of a conidial suspension on a microscope glass slide. Assessments were made on fungal growth rate and mortality. This bio-assay does not need any plant material and can easily be standardized. Several entomopathogenic whitefly fungi were compared in this test. Many variables can be tested in this way such as dosage, temperature, relative humidity, effects of adjuvants and chemical pesticides. The authors believe that they have developed a sensitive and rapid assay for fungal insecticides that could be used as QC tool. The assay certainly seems quick and cheap and easier to standardize than a bio-assay involving plant material and populations of insects. Whether there is a good correlation with field efficacy, however, has not been studied and this could be a weak point. This is essential for product QC. Thus far this method has not been implemented within the industry or accepted as a testing method by regulatory authorities.

Bio-assays are time-consuming and laborious and they may give variable results. Loesch et al. (2007) have made an attempt to develop a quick screening technique. They tried to correlate strain attenuation with carbon utilization profiles in *Beauveria brongniartii*. This was studied in a microtiter plate test-system with 128 different media and assessed for germination rate and speed. Correlation with bio-assays still needs to be done. This could prove to be an easier and cheaper virulence testing method once tested and shown to be reliable in several fungi.

Microbial contamination in fungal products needs to be examined as part of a QC procedure. It involves the assessment of the total level of contaminants and checking for the presence of human pathogens and of other opportunistic bacteria. The total level should not exceed a certain number, and human and mammalian pathogens should not be present at all or at very low levels. Jenkins and Grzywacz (2003) stated that the acceptable level of contaminants will be product-specific and that higher levels are acceptable depending on the nature of the contaminants. This may be true when only technical aspects of the product are taken into account, such as storability and efficacy. When safety and product liability are taken into account, however, the total level of contaminants is not just product-specific. This depends on general standards, and registration requirements that should apply regardless of the product concerned. The risk that these contaminants pose depends on their nature and their number. Assessing the total number of mesophiles is rather easy, but identification of pathogens is more difficult. Various methods can be used by plating samples on selective and differential media and at different temperatures, including 37°C. Most of these are standard methods for assessing bacterial contamination in food; they can be done by the producer itself or it can be out-contracted to a specialized laboratory. These standard tests to monitor microbial purity can be used for all pathogen products. Safety testing can be required by the regulatory authority when metabolites are produced by the fungus. An example is described above concerning *I. fumosorosea (P. fumosoroseus).*

Physical properties like moisture content and particle size distribution are important and need to be checked too. Moisture may affect storage stability, while particle size may influence the proper application of the product. The properties that need to be checked depend on the formulation type. The above-mentioned refer to wettable powders, which is the most common formulation with mycoinsecticides.

Standardization for mycoinsecticides has been proposed by Jenkins and Grzywacz (2003) for products with similar activity. They recommended that product performance be compared with that of a standard and that tolerance limits be set. Establishing such a set of agreed criteria for should be done between manufacturers and researchers and regulators to cover the needs of all stakeholders. This initiative, however, has not been followed up yet.

5.3 Baculoviruses

In baculovirus products, the essential parameters in the quality control procedure are the assessment of the number of propagules, the biological activity and the microbial purity (Shapiro 1986; Smits 1987; Shieh 1989; Guillon 1997b). Counting of occlusion bodies (OBs) or polyhedral inclusion bodies (PIBs) is difficult in formulated products due to their small sizes and due to insect remnants from the production as well as particles of formulation ingredients. PIBs of nucleopolyhedroviruses (NPV) are bigger than OBs of granuloviruses (GV) and can be counted under a light microscope; GVs are more difficult to count because they are smaller than 0.5 μm. Analysis of the protein concentration in a suspension is another method of assessing the concentration of GVs. Details of various methods is given by Hunter-Fujita et al. (1998). The actual number of OBs or PIBs does not necessarily reflect the product's biological activity since active and non-active viral particles can not be distinguished. Hence a bio-assay is needed. Bio-assays can be conducted in many ways depending on the product and the target insect. Guidance on bio-assays for baculoviruses can be found in Smits (1987), Lacey (1997), Hunter-Fujita et al. (1998) and Jones (2000). The significant factor in bio-assays is that they be standardized in terms of target insect instar, insect rearing method and inoculum. If this is not the case, the results are meaningless. A virus reference standard sample can be tested simultaneously with the produced batch and results can be compared. This is possible because a standard stock of virus can be stored for a long period without any loss of activity.

Microbial purity is a critical parameter in baculovirus products due to the production in vivo. Details for monitoring contaminants in baculovirus products are given by Hunter-Fujita et al. (1998) for determination of various groups of bacteria. Human pathogens are normally not present in baculovirus samples. Smits and Vlak (1988) did not find any in produced batches of SeMNPV; neither did Lasa (2007). Smits and Vlak (1988) did find a relatively high rate of contaminants, about 1–10 bacteria per 100 polyhedra. A formulated viral product typically contains 10^{12}–10^{13} PIBs or OBs per litre which would mean > 10^7–10^9 contaminants/ml. Lasa (2007) found similar ratios. The upper part of this range of contaminants is generally considered to be too high. In that case purification is needed or other methods to control the level of microbes, such as the use of antibiotics or other antimicrobial products. These results, however, originate from research projects in which production was relatively small scale. There are no recent reports publicly available from samples of commercial mass production and end-use products. The above indicates that microbial purity in baculovirus products is a critical quality control aspect that needs considerable attention, not only in the final product, but surely also during production where preventative measures should reduce the level of contamination as much as possible.

There are no quality standards for baculovirus products that are internationally accepted, either by the industry or by regulatory bodies. The lack of standards was recognized by researchers in Eastern Europe and an attempt was made within the IOBC/EPS region in the late 1980s to establish the main criteria with regard to quality control of viral insecticides (Ciuhrii 1996). A draft proposal was set up in which it is recommended to determine the following QC parameters: the pathogen concentration indicated by the normal amount of nucleocapsides, identification of the strain based on the number of virions, nucleocapsides and envelopes, and the microbial contaminants. The methods described are too academic and too difficult for industry since an electron microscope is needed. This draft proposal was more about methods then about setting standards. It was never broadly adopted and further details have not been published on this issue.

Registration of baculoviruses requires product specifications, but again there are no standards and no tolerance limits. The producer determines the product specifications. The number of propagules, as decided by the producer, must be mentioned on the label, but as discussed above, this is only of limited value. Other product specifications are not mentioned on the label, it is also generally not required to do so. Biological activity should be measured in a standard way against a company's internal standard and it should be consistent for all batches. The specification is usually set by the manufacturer. Jenkins and Grzywacz (2003) recommend testing product virulence "in a standardized bio-assay on final product, using defined instars of a laboratory reared strain of the target pest. Results should confirm that the potency ratio is within defined limits specified for the product and determined during product development or > 0.5 of a standard virus preparation". The maximum level of microbial contaminants can be set by the company itself, provided that the number is acceptable in terms of safety and in terms of the product's shelf-life and performance. This level is then documented, including the assessment methods, in the registration dossier and the authority may accept that level. Some regulatory authorities, though, have set an acceptable limit for the total number of contaminants. According to Guillon (1997a) the legal maximum is 10^6 cfu/ml for baculoviruses, but it is not specified in which country. Jenkins and Grzywacz (2003) recommend that "formulated products should not exceed a maximum of non-pathogen contaminants of 1×10^8 cfu/ml in for liquid formulations with an activity of 1×10^9 OBs or PIBs/ml and 5×10^8 cfu/gr for dry powder formulations". The rationale here is that bacteria that develop in in vivo production on insect cadavers are usually associated with dying insects and do not pose a risk to humans.

According to Guillon (1997a) *B. cereus* is the most common contaminant in insect guts and it represents 99% of the contaminants. This is a ubiquitous bacterium in many environments and not considered very hazardous. In the REBECA project, a maximum level of 10^7/ml product is recommended as the acceptable level for this species (REBECA 2008). Nevertheless, human pathogens should be absent or at very low levels. Safety to mammals is also mentioned as a QC parameter (Burges 1981; Shapiro 1986; Smits 1987). The EPA demands a mouse intraperitoneal (IP) test for harmful effects with the TGAI before release of a production batch for formulation and selling (OECD 2009). The rationale in this case is the monitoring for human pathogens. I believe this is not needed as a routine QC check on every batch when other parameters are within specifications and human pathogens are absent. This can be tested on microbe-specific media. Only in specific cases, animal testing should be included as a standard requirement. It is costly, time-consuming, and requires unnecessary use of testing animals.

Physical and chemical parameters are required for registration by the EPA (Shieh 1989), but are of minor concern and hardly relevant as a routine check of product QC.

5.4 Entomopathogenic Nematodes

In most countries, EPNs are not considered a biopesticide in terms of registration and strict registration criteria, including quality control criteria, do not apply to EPN products. In that respect they are similar to natural enemies, but unfortunately, EPNs were not considered in the IOBC/EC project with beneficial arthropods in which testing methods and standard criteria were developed (Van Lenteren 2003a). Consequently, quality criteria are set by the producers themselves. With the use of EPNs, field performance can be quite variable and as a result the quality of the product frequently is a point of discussion. Many authors, from research institutes as well as from industry, have recognized the importance of QC and a range of testing methods have been developed to assess various aspects of quality. In EPN based products, quality control parameters are viability, *i.e.* the percentage of viable nematodes and the minimum total number of viable nematodes per unit of product, virulence, age (shelf-life), morphological and physiological characteristics, and field performance (Miller 2002). Grewal (2000) considered maintenance of high viability and virulence as the backbone of an effective quality control strategy. Testing virulence in bio-assays, however, can be set up in many different ways. Grewal proposed a sand-well bio-assay with parameters for seven nematode species and strains as a standard tool for assessing virulence. The parameters are the *Galleria* larvae-nematode ratio to be tested and the temperature, and he determined the expected mortality after 72 h as a specification. This was a worthy attempt to standardize testing virulence in nematodes. Peters (2000) proposed another type of bio-assay for three nematode species as a standard test, although without determining specifications. These standard tests, however, are not widely accepted and have not been adopted by the industry. Later, Grewal and Peters (2005) defined nematode product quality broader, including packaging, instructions for use, ease of transport, etc. In my opinion these are not product QC parameters strictly speaking, rather aspects of total quality management. This indicates the importance of defining quality control and product quality control as mentioned above.

The easiest parameter to assess is viability, the ratio of live to dead nematodes, and the total number of viable DJs per unit of product. Although in some cases it may be difficult to see whether a DJ is alive or not, especially for a non-trained end-user or distributor, this can be overcome by providing the right technique to assess viability. The number of living DJs should meet the number as stated on the label, up to the end of the shelf-life period. Still unresolved topics for stakeholders are the percentage of viable DJs that is the minimum acceptable amount and whether over-packing compensates for low viability. This depends on the origin of the death of the DJs. If it is caused by mechanical factors, there is little concern. If it is caused by a slow decrease in quality, for instance through a long storage period, the surviving specimens are most likely also of lower quality and then there is a reason for concern. However, an end-user is not likely to accept a high percentage of dead larvae, even when the living DJs are still of good quality. In this case quality determination becomes highly subjective. No more than 10% dead DJs in a product seems an acceptable limit in the market. This is also dependent on product quality of similar products in the market. When one product out-performs the others in this respect, users will demand a higher quality.

Viability assessment should go along with the testing of virulence or biological activity. Many different types of bio-assays have been developed for testing biological efficacy. A number of variables can be tested in these assays. These can be abiotic factors such as temperature, soil type, moisture content, time period, depth of target in a soil column, as well as biotic factors like the nematode species or strain, the number of DJs, age of DJs, the target species and its developmental stage. These tests are often used to screen species and strains for suitability against a certain insect, and for quality control purposes; numerous papers describe such testing. Tests should be done under sub-optimal conditions and conditions should mimic field conditions as much as possible. Positive results in laboratory bio-assays do not automatically correlate with good field efficacy, as was demonstrated with *H. bacteriophora* products from different production systems (Gaugler and Georgis 1991). Testing under optimal conditions is likely to mask lower quality and is therefore meaningless.

Despite more than 30 years of research and production of EPNs it is still not clear which parameters can be best used to monitor QC in relation to field performance. Many different intrinsic QC parameters have been studied in *Heterorhabditis* species by Jung (1999), such as mortality, body contents, presence of secondary cuticle, wave-like movements per time unit, body length, lipid and glycogen contents, number of bacteria carried within a DJ, movement in sand or agar, penetration capacity in wax moth larvae and infectivity. These factors were studied in relation to storage temperature, storage period, formulation type (with differences in osmotic value, pH, and carrier material). A semi-field test with strawberry plants infested with vine weevil larvae was performed to correlate the parameters with efficacy. Jung found that energy reserves and migration ability correlated best with efficacy, although it varied between species and isolates. She recommended that each laboratory test for the determination of quality be validated by successful control in the field. This example shows that energy reserves and migration ability may be used as QC parameters, but also that they should be tested and validated for each strain. The validation of bio-assay data with field data is highly desirable, but this is extremely difficult since efficacy in the field depends on many factors and is therefore highly variable.

Energy reserves in relation to quality have been studied by many scientists (e.g. Patel and Wright 1997; Patel et al. 1997) and it is generally concluded that this is a useful laboratory tool for quality assessments. The determination of lipid content was a standard procedure in Biosys to monitor the quality of its products (Georgis 2002). Grewal and Georgis (1999) also stated that differences in lipid content between species do not always reflect proportionate differences in storage stability and field efficacy. This indicates that quality control is a complex matter where various factors interact, and that species behave differently. More research is needed to find out which parameters need to be assessed in relation to QC. It is apparent that there is no single parameter that can be used as a QC parameter, and certainly not for all different taxonomic groups (Jung 1999). If quality assessments of commercial products are done by independent organizations, the lack of accepted protocols and criteria becomes apparent. Quality may then be judged on a limited set of criteria and, although all QC parameters should meet their set criteria, concerns may be raised based on too limited testing. An example is Caamano et al. (2008) where only viability in commercial products was assessed and biological activity was not.

Scientists, extension workers and the biocontrol industry collaborated for many years developing standardization of quality control tests for EPNs and defining standards. This work was done within the EU COST Actions 819 and COST 850. It resulted in the publication of a handbook on quality control of EPNs (Grunder 2005) describing protocols for testing biochemical, physical and behavioural characteristics related to quality, and protocols for bio-assays and field testing. The book presents commonly accepted protocols for quality testing in relation to field efficacy, but only a few standards have been given and were accepted among all involved. It further recognized the need for a common database for some commonly used nematodes so that companies and scientists can share their experiences and results with the accepted protocols. This would be a first step in developing and agreeing on standard criteria and to a widely acceptable self-regulated product quality control with EPNs. The remaining difficulty is that there is no reference stock that can be tested simultaneously with the tested batch. This is due to the difficulty of having a stable stock as a comparison (Peters 2000). Furthermore, there is no culture collection of EPNs which could store such a reference stock, as in the case of Bt where the Pasteur Institute keeps the reference material.

Still, a simple and quick test should be developed for and be accepted by the industry. I suggest to take the work of Grewal (2000) as a starting point. He measured virulence in a sand-well bio-assay with a pre-determined host: nematode ratio and compared it with an expected mortality rate as a standard. According to the research of Grewal this method is sensitive to "impaired" nematodes, and therefore capable of distinguishing good quality batches from poor quality batches.

5.5 Product Quality Parameters

In general, the quality parameters for MPCPs of the four types of pathogens are similar. Particular aspects of quality control, however, need more attention then others due to the production system of the pathogen and due to the nature of the product. In Table 4.1 an overview summarizes this point. In all products the number of propagules and the viability are essential parameters. Virulence is the most critical factor for obvious reasons. Microbial contaminants in EPN products are not a serious problem when production takes place in sterile fermentors; it does need a lot of attention when nematodes are produced on animal offal. For bacteria and fungi produced under sterile conditions, the presence of contaminants is less critical than

Table 4.1 Importance of product quality control parameters per type of pathogen

Parameter/pathogen	Bacteria	Fungi	Baculo-viruses	EPN
Number of propagules	++++	++++	++++	++++
Viability	+++	+++	+++	++++
Virulence	++++	++++	++++	++++
Microbial contaminants	+++	+++	++++	++
Safety	++	++	+	–
Physical stability	+	+	+	+
Chemical stability	+	+	+	–
Shelf-life	++	+++	++	++++

+ to ++++: Increasing amount of attention needed during quality control.
–: Not relevant.

for baculoviruses produced in vivo. Safety is connected with possible production of metabolites and this plays a role in bacteria and fungi. Technical properties are a minor concern for all products. Shelf-life, on the other hand, is challenging in EPNs while bacterial spores and baculoviruses are the most stable products stored when under appropriate conditions. These observations are a result of the production system and the biology of a pathogen. This is also the basis for the registration requirements where a case by case approach is taken.

6 Registration and Quality Control

6.1 Registration Requirements Related to Quality Control

Registration requires description of both the production process and the product, and mandates quality control procedures. The EU Directive 91/414 sets out the general requirements for active substances for placing plant protection products on the EU market. EU Directive 2001/36 Part B (EC 2001) lays down detailed requirements for those active substances consisting of microorganisms, and plant protection products based on microorganisms. Quality control requirements regarding production of the active substance are given in a descriptive way only: "quality control assurance criteria for the production should be submitted. The techniques to ensure a uniform product and the assay methods for its standardization and purity of the microorganism must be described". For the plant protection product as well "quality criteria for the product should be submitted." Further, analytical methods must be provided for the analysis of the product. This information is required for post-registration control.

In EU 2005/25/EC (EC 2005a) Uniform Principles are laid down to ensure an equivalent evaluation by EU member states. The Uniform Principles stipulate that an authorization can only be granted if full information is provided on the continuous quality control of the production method, production process and the plant protection product. There should be emphasis in the evaluation on occurrence of changes in characteristics of the microorganism and on contaminating organisms. This directive, however, only states that "proposed quality criteria must be evaluated." The three directives, thus, do not stipulate obligatory standards with which products should comply. Quality criteria are left open for evaluation by the competent authority to which a dossier is submitted. This still leads to various interpretations of the requirements and to differences in acceptance of data by EU Member States.

Producers need to set their own criteria and develop their own methods to demonstrate that products comply with these criteria. Only with regard to contaminants does the Directive state that "it is desirable to have a plant protection product without contaminants, if possible. The acceptable amount should be judged by the competent authority." OECD Guidance documents also do not provide standard criteria (OECD 2003; 2004a, b). The FAO and WHO have developed specifications for pesticides in order to ". . . .provide an international point of reference against which the quality of products can be judged, either for regulatory purposes or in commercial dealings . . ." (FAO/WHO 2006). A chapter on "Specification guidelines for microbial pesticides" is included. The chapter content lists bacterial, viral and fungal pesticides, but details are only given for bacterial larvicides (Bt's) for mosquito control. The document is clearly not yet finished. Testing methods are described in detail regarding bio-assays for potency determination and physical and chemical properties. Some criteria are given for these properties. Maximum acceptable levels of microbial contaminants have not yet been determined.

For biopesticides, the lack of official criteria refers particularly to the number of viable propagules and its tolerance range, and to the level of contaminants in products. For a producer it is hard to estimate which data are acceptable. My experience is that acceptance of the total level of microbial contaminants varies per country. This is further complicated since acceptance has shown changes over time. An example of the USA illustrates my point. When the USDA Forest Service registered the gypsy moth virus in the 1970s, they set the maximum level of aerobic bacteria at 10^9/g. Later, when Elcar (*Heliothis* NPV) was registered, the EPA had set the level at 10^7/g (Battu et al. 2002). Currently the EPA has not defined maximum levels, at least not in their guidelines for registrants. Only when a dossier is being evaluated, does an applicant learn whether his proposed levels are accepted.

Only for chemical and physical properties such as acidity, viscosity and technical properties of a formulation such as wettability, dry sieve test, emulsifiability, etc., testing must be done according to recognized CIPAC or OECD methods designed for chemicals. These do contain set criteria to which products must comply.

Regarding human pathogens, the Uniform Principles state that an authorization for a microorganism is not granted if it appears that it is pathogenic to humans under the conditions of use. This is not necessarily the same as a complete absence of human pathogens in the product. The EPA (Environmental Protection Agency) in the USA has similar requirements: "human and non-target animal pathogens must not be present at hazardous levels in the product before formulation". They require a toxicity test on mice with an intraperitoneal injection of $> 10^6$ cfu before formulation for bacterial and viral products. Canada's Pest Management Regulatory Agency (PMRA) asks for an assessment of potential microbial hazards, using criteria and methods that are consistent with international standards for food. The language used by authorities attempts to be clear, nevertheless it is ambiguous. Presence of human pathogens is accepted so long as they do not reach hazardous levels. For a producer this is to difficult to work with, so I recommend that absence of human pathogens must be the goal for the producer. A proposal has been developed within the OECD to clarify this issue on acceptable maximum levels of contaminants, and particularly of human pathogens (OECD, unpublished). This is still under discussion within the authorities (see below for more details).

Industry, regulators and scientists have discussed these omissions in the regulations within the REBECA project, an EU policy support action with the goal to propose alternative, less bureaucratic and more efficient regulation procedures maintaining the same level of safety for human health and the environment while at the same time accelerating market access and lowering registration costs. Results of this project will be discussed in detail in Chap. 5.

6.2 Regulation Concerning Label Information

Label information related to product quality control parameters, as required by registration in the EU, only concerns the content of active ingredients such as number of spores, cfu's or viral particles, per unit weight or volume of product. Neither infectivity nor biological activity is required on the label, except for Bt's where potency (units/mg) is indicated on the label. Label information of chemical pesticides only indicates the content of active substances. Some authors plea for standardization so that activity quantification can be accomplished and put on the label. This will offer the ability to compare different formulations of the same pathogen or different pathogens on biological activity. Guillon (1997b) reported that the IBMA strongly expressed the need for harmonization and that the organization wanted to develop this together with organizations like the IOBC and the European Committee for Standardization (ECS). This has not been done, however. Would this be a useful exercise? Chemicals can not directly be compared with each other on the basis of activity quantification on the label.

An activity indication would refer to a certain target insect, but this does not mean that activity towards another target can be deduced from that. It may be very different. This is seen in Bt products where potency is determined on a standard insect. This allows comparison between products, but does not give information on the efficacy on other insects. Skovmand et al. (2000) recommend the use of three different reference insects, but does this really improve the situation? Even after 30 years of experience with Bt's there is lack of consensus on standardization for Bt's. For other biopesticides standard methods for determining potency have not yet been developed. As a result there is no label information required.

6.3 Post-registration Product Inspection by Authorities

Plant protection products are subject to post-registration checks on compliance with the regulations in a broad sense. For MPCPs, aspects eligible for control are the identity of the microorganism, the number of active propagules, the microbial purity and physical-chemical properties, and compliance with obligatory label information. Assessed data will then be compared to what has initially been provided by the registrant in the registration dossier. Authorities seldom perform these checks and in more than 25 years working with biopesticides I have only once experienced such a control. This was conducted by several Danish institutes under the auspices of the Danish EPA in 2004 on all microbial products that were on the market in Denmark (Winding 2005). Thirteen products were checked, three Bt products, the others mycoinsecticides and mycofungicides. The objectives were the identity of the microbial agent, the number of propagules and the number of contaminants, and to compare this with information provided earlier by the producer. Deviations in quantity as well as identity were found for some MPCPs. In some the occurrence of contaminating microorganisms was higher than specified by the producers. Earlier investigations in Denmark on seven MPCPs revealed similar results (Løschenkohl et al. 2003).

The report also mentioned that a query among nine EU countries was performed on checks on MPCPs and the quantity of microorganisms in them. It did not give information on tolerance levels and apparently these checks were not being performed in any country as a routine inspection. In a postscript in the report, the Danish EPA comments on the maximum level of deviation of content of active ingredients. In chemical pesticides these are fixed in the EU Directives, but this is not the case for microbial pesticides. For MPCPs the EU directive asks for information on the nominal content and the maximum and minimum, but criteria are lacking. The Danish EPA suggests a margin of maximum five-fold be acceptable. Further, the amount of acceptable contaminants is also not regulated and the Danish EPA suggests initiating discussions on these two topics within the EU to establish agreed levels of deviations in MPCPs. This topic has been further taken up within the REBECA project; see for more details on this project Chapter 5. The most interesting topic in this report is that authorities seem to accept a five-fold deviation in the level of propagules in products on the market until the expiry date. This is nowhere to be found in any guideline and offers producers insight into what level is acceptable to authorities. Acceptance between countries, however, is likely to vary considerably until levels of acceptable deviations are officially established within the EU. Whether such a deviation is acceptable for the product's efficacy is a matter that a producer has to decide himself.

6.4 Other Regulations that Have an Impact on Product Quality

Food crops are increasingly produced taking into account HACCP (hazard analysis and critical control points; EU Regulation 2073/2005/EC (EC 2005b)) protocols. This is to prevent introduction of microorganisms in quantities that present an unacceptable risk for human health. Supermarkets impose this on growers of food crops because of growing concerns of public health. It is obvious that MPCPs should not have any contaminants that may be a potential food pathogen. The EU data requirements even refer to HACCP in respect to microbial purity of a plant protection product. For a biopesticide producer, the use of HACCP protocols is another good reason to strive for absence of human pathogens and other opportunistic bacteria in MPCPs.

Residue levels on food crops are regulated and maximum residue levels (MRL) are generally set for plant protection products. Microorganisms are exempt of a MRL unless secondary metabolites pose a health concern. Both HACCP protocols and MRL regulation relate to food safety, a growing concern in our society.

Obviously, product accountability is a concern of every manufacturer. The producer or the importer of a product into the EU is responsible for the quality of a product and any harmful events or damage that may arise from the use of the product. Liability claims may be filed against a producer if gross negligence and irresponsible behaviour with the product can be shown.

7 Practical Aspects of Product Quality Control

7.1 Protocols, Sampling and Testing

Methods and protocols have to be developed in order to monitor the quality parameters of products. This needs research and validation of the methods on their predictive value related to the performance of the product in the field. Specifications must be established. Methods may be available from a screening programme or from the literature. Developing this quality control programme takes a lot of research and should not be underestimated. Facilities, insect rearing and production of product are needed. It will take time and money and contributes to the costs of the product. Such a testing programme needs to be established during the developmental phase of a product. It also needs to generate methods and real data for registration purposes.

Sampling is an important part of a quality control programme and many questions have to be answered. Some examples are: how many samples need to be tested, what sample size is required, what is the number of replicates per sample, what tolerance is acceptable for the various parameters, how frequently are samples tested during the shelf-period and even afterwards. Unformulated as well as formulated material needs to be tested. If a batch has to be discarded, preferably this is determined before formulation and/or before packaging. Production batches are not necessarily the same as product batches. Several production batches may be formulated together. TGAI may be stored first before formulation is conducted, and formulated product may be stored before packaging. Where is a need to monitor quality? Some of these procedures are process control. Finally, product control must be conducted. The line between process control and product control is not always that clear. Batch numbers are given to material because of traceability needs. Material must be treated in a uniform way in order to be called a batch, otherwise material must receive different batch numbers and they must be tested separately. Each batch that is to be tested must be sampled separately.

A guideline for sampling for quality control of bulk technical grade active ingredients (TGAI) and formulated products is given by the FAO/WHO for pesticides (FAO/WHO 2006). The FAO/WHO guideline can be used for biopesticides too. Other regulatory authorities do not provide any guidance on this topic. Companies determine themselves sampling methods, frequency and number of samples tested for quality control purposes. Statistical methods should be followed in order to have reliable results that will be acceptable for official regulatory purposes.

Other aspects of quality control are monitoring properties of packages. Packages must be filled with the correct volume of material and within the tolerance limits, or each package must contain a minimum of material. Packages must be sealed and labelled correctly.

Still another aspect is who will be conducting the tests. It is usually done within a company, but parts of the quality control can be outsourced to independent laboratories. Testing is preferably done by a quality control or an R & D department rather than by the production people themselves. This guarantees independent results and independent interpretation of the results. In small companies this may be a problem. It is difficult to discard a batch on a minor deviation from the standards when people have worked hard to produce that batch. If possible, testing methods should be tested by an independent third party laboratory to confirm their reproducibility. A critical part of QC is determination of microbial contaminants, particularly the presence of human pathogens. One may consider having this conducted by an independent laboratory.

In the case of Invade, the producer outsources QC to the research laboratory AgResearch. Once, a batch was sold that contained non-pathogenic bacteria. This may have been a reason to have the QC carried out by the research laboratory, but this was not explained in the paper (Pearson and Jackson 1995). Perhaps some methods cannot be done by the producer due to complexity of the tests or availability of certain expensive equipment. Using an independent organization for QC may give more accurate data. At Koppert, we also use an independent accredited laboratory for an extra check of Mycotal and Vertalec on microbial contaminants. The reason here is not the difficulty of the tests as we do test ourselves for contaminants. Rather, we want an independent confirmation of the quality of the product with regard to contaminants and human pathogens. Since our products are used in food crops up to the day of harvest, this item ranks as the most critical risk with regard to any harmful effects and serious complaints.

7.2 Determination of the Standards and Tolerance Ranges

Each manufacturer has to establish product specifications and tolerance ranges. There is hardly any guidance to this from regulatory bodies, except in the case of Bt's. For Bt's, standards have been developed as is discussed above. A producer needs to know what will be acceptable for the authorities, however. This is not an easy exercise, but relevant. Communication may be sought with the authority in order the obtain clarity on this subject. Decisions on specifications and on tolerance levels should take into account biological and ecological factors in the field, and even in the bio-assay. Biological variability may overshadow set criteria and therefore broad tolerance levels may be needed. The tough question is what a realistic and relevant tolerance range is. Each situation demands its own research. Generally, there are normal batch to batch variations, as well as variations within the testing methods. A company has to tackle this problem in a pragmatic way and decide by what limits they still feel confident in selling the product, simultaneously taking registration requirements into account.

In Bti bio-assays with strict protocols for measuring bio-potency results may vary to $\pm 25\%$ from the average (FAO/WHO 2006). Still, FAO/WHO recommends the following minimum standard to be met: no more than 10% loss in bio-potency below the labelled potency value when stored at 5°C for 2 years. This example illustrates the complexity of determining specifications and tolerance limits. In bio-assay testing of Mycotal we have seen a similar variation in mortality and germination levels between batches. These two parameters show a significant correlation (linear

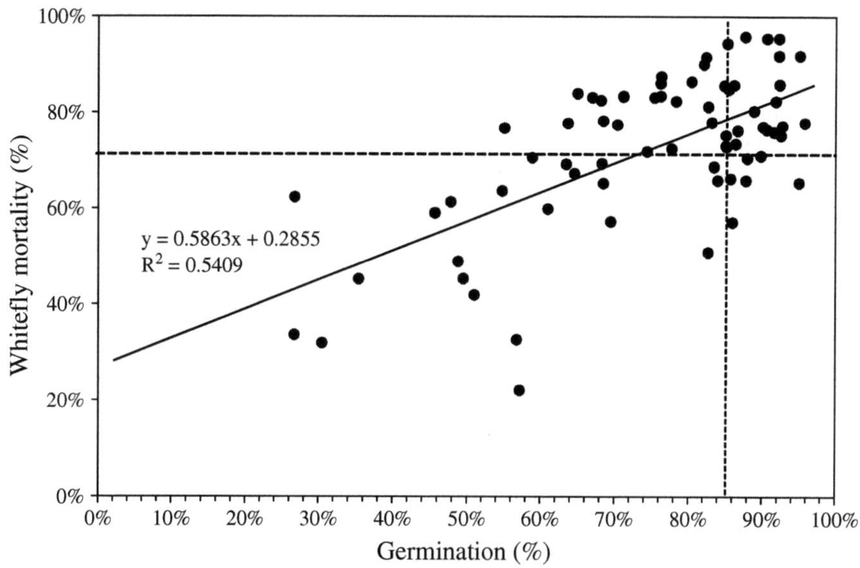

Fig. 4.1 Mycotal: relationship between germination and whitefly mortality in bio-assays (-------: standard for germination rate; – – – – – –: standard for mortality rate)

regression analysis, $P < 0.001$) (Fig. 4.1; unpublished data, Koppert). Nevertheless, batches that comply with the germination rate do not necessarily comply with the specification for mortality (R-Sq is only 54.1%), so the germination rate is not a reliable predictor for the mortality. Both parameters need to be tested independently, unfortunately. At Koppert, we have set the standard for germination > 85% and for virulence at > 70% mortality at a critical relative humidity (see Box 4.1). Jenkins and Grzywacz (2003) recommended for baculoviruses that potency be >0.5 of a standard virus preparation. For fungal products they did not recommend a specification due to difficulties in establishing it.

These examples illustrate that each case needs it own research and considerations before a specification for virulence can be established. For parameters such as microbial contaminants, safety tests, and technical properties, generally minimum or maximum standards are established which makes decisions easier to take. Generally, a batch can only be released for sale once all the QC checks have been performed and the outcome approved. When one of the "specs" is not met, a batch should be discarded. If possible, a batch could be returned to production to improve it.

7.3 Virulence Testing and Determination of the Specifications

Efficacy of the product is the all-important quality trait. Bio-assays are needed to test this, but they are time-consuming and costly; they may not be timely and may not allow deadlines to be met. Scientists often suggest checking LD50 values as a product QC parameter (Jenkins and Grzywacz 2003; FAO/WHO 2006; CPL 2006b). This implies testing various dosages above and below the expected LD50 dose in bio-assays which makes testing laborious and expensive. I recommend testing just the label dosage in a bio-assay. This keeps the test easier and less costly. Batches should be run against a reference standard preparation, if available. This makes testing more reliable. If a reference standard is not available which is usually the case in nematodes and fungi, the outcome can be compared with the untreated. The corrected activity should then be compared with the specification that has been set for the bio-assay test in previously conducted research. Preferably, in bioinsecticides, the label dose gives 80–90% mortality in the bio-assay and this can be used as the specification for bio-assay activity for each batch. Correlation with good field efficacy must be studied in order to know what the bio-assay specification means under field conditions.

Few authors discuss setting a specification for virulence. In Elcar, the baculovirus product for control of *Heliothis* spp., potency was determined by bio-assays developed for Bt products. As a reference, material was used that had shown good results in the field in earlier years (Shieh 1989). In product control of Green Muscle, Jenkins et al. (1998) used a standard stock culture of the fungus for comparison. Virulence expressed as average survival time should be within 0.5 days of that obtained with the reference in a bio-assay with the desert locust. The FAO/WHO has set a minimum standard of 90% mortality for Bti in the standardized bio-assay (FAO/WHO 2006).

Bio-assays must be highly standardized in order to get accurate and reproducible results year-round. Insects used in bio-assays should be reared under standard conditions so that testing results are comparable over a period of time. Stressed insects are more susceptible to disease, and this may give higher mortalities and unreliable outcomes of bio-assays.

Untreated should be tested in all cases. The result indicates whether the standardized bio-assay performed as expected, and whether the test has been conducted properly. Only a certain percentage of mortality in the control treatment should be allowed. If this is too high, the bio-assay should be assessed as invalid. Seasonal effects are known to influence natural mortality as well as bio-assay outcomes, even in well-standardized testing methods.

Robertson et al. (1995) investigated the effect of natural variation when performing bio-assays with one chemical and two microbial insecticides. They analysed the variations in bio-assay responses of three insect species. They concluded that when results of bio-assays are used to determine resistance, treatment efficacy or product quality, the range of natural variations should have been assessed first by collecting a large data set. With this information other phenomena can be separated from natural variations. For quality control purposes they recommended, first, to production managers to be aware that LC values are estimates, not constants, and second, to establish whether test results of a new batch fall within the 95% confidence limits of the existing large data set. Only then significant differences can be determined. This study shows the complexity of making decisions on quality control standards and tolerance ranges, not only for manufacturers, but also for regulators. Still, bio-assays are the only pragmatic tool for quality control checks. They must be performed with great care, and although they take up considerable resources, they are indispensable for product quality control.

7.4 Examples of Product Specifications

Details of product control are generally kept confidential or are just not published. An exception is the mycoinsecticide Green Muscle where details of product control are presented by Jenkins et al. (1998). The authors gave the specifications to which the product must comply and briefly described the testing methods. Tolerance limits were not given. Specifications for a baculovirus product, Elcar, were given by Shieh (1989). Besides bio-assay potency and a polyhedral count, biochemical methods were given which were used to characterize the product and to compare it with stored standard material. The company Verdera disclosed some information on their chosen quality criteria for biofungicides (Palin-Holmberg et al. 2003). Specifications for Mycotal used at Koppert BV are presented in Box 4.1. Product specifications can also be found in publicly available reviews of registration dossiers by regulatory authorities. Examples are the Dutch Ctgb (Board for Authorisation of Plant Protection Products and Biocides) which publishes evaluation reports on their database on internet, and the EC which discloses review reports of new active substances including specifications and methods on http://ec.europa.eu/food/plant/protection/evaluation/newactive.

7.5 Post-shipment Quality Control

Taking into account the vulnerability of biological products during shipping, sampling should also regularly be carried out along the distribution chain in order to check quality. Sampling may be carried out at any point in the distribution chain, from the producer to the end-user. Usually it is carried out at the factory only. If quality shows a decrease in the distribution chain, corrective measurers have to be taken to improve shipping and storing conditions. Post-shipment quality control has shown that quality loss during this process is occurring and that it needs appropriate attention, particularly in nematodes (Gaugler et al. 2000; Caamano et al. 2008). Enema and Koppert have developed protocols for end-users to check aspects of quality of EPN products (Ricci and van der Pas 2005; Peters et al. 2003). Post-shipment QC is part of total quality control and essential to guarantee delivery of high-quality products to customers.

7.6 Costs of Quality Control

Bio-assay testing calls for laboratory facilities, insect rearing, host plants, and labour. Facilities may need to be separate from the production location because of the risk of contamination and different staff members should be involved. Lisansky

(CPL 2006b) estimated that costs are about $60,000 annually per insect species and recommends planning budgets for QC. Guillon (1995) considered QC as the main cost factor in the production of baculoviruses. I have not found any other reference to this topic. At Koppert, it is hard to estimate costs since we rear most of the pest insects anyway for production of natural enemies. Labour costs involved in the QC of Mycotal are approximately €5,000 on an annual base. On top of this, overhead costs should be calculated. From this figure and the estimated insect rearing costs, QC costs are likely to be between €10,000 and €20,000 for a bioinsecticide depending on the pest insect, the product and the activities of a company. QC is a considerable expense that must be attributed to the full product cost price.

Box 4.1 Product Quality Control for Mycotal: Specifications and Methods Used at Koppert Biological Systems

Every batch is tested before sale, after 3 months and at the end of the shelf-life period of 6 months. Microbial purity, moisture content and solubility are only checked before sale. The product must comply with all standards.

Number of propagules: Standard: $1 \pm 0.15 \times 10^{10}$ spores per gram of product	*Method*: a diluted spore solution is prepared (spreading agent and anti-foam added) by stirring for 60 min. Counting spores is done with a Bürker-Türk counting chamber under the microscope. Two samples of 50 g are taken from one end-use product package. From each sample four sub-samples are counted. The average number is compared to the standard.
Germination rate and speed of germination: Standard: $\geq$ 85% within 16 h at 21°C	*Method*: prepare spore solution as above. Samples are incubated on a slide with SDA medium. The slides are stained after 16 h with lactophenol, followed by counting germinated and non-germinated spores at 1,000 × under the light microscope. Germinated is defined as having a germination peg at least as long as the diameter of the spore. One sample of 50 g is taken from one end-use product package. From this sample three sub-samples are counted. The average number is compared to the standard.
Efficacy: Standard: $\geq$ 70% mortality 8 days after treatment and incubated at RH 75% and 21°C	*Method*: one sample of 5 g from a product package is tested. Bio-assay leaf discs containing young instars (L2 and L3) of *Trialeurodes vaporariorum*, embedded on water-agar, are sprayed by a Potter-tower with the label rate. Surviving and dead larvae are counted under the binocular microscope. Five leaf discs are counted and about 100 larvae per disc are assessed. Average mortality is compared to the standard.

Microbial contaminants: Standard: < 50,000 cfu/g of product at 30°C and at 37°C; Standard for external testing by accredited laboratory: – Total aerobic plate count at 37°C: < 50,000 cfu/g – Yeasts and moulds at 37°C: < 10/g – Coliforms: < 50/g – Staphyloc. aureus: < 50/g – Fecal streptococci: < 50/g – Salmonella: absent in 25 g	*Method for internal testing*: one sample of 50 g is taken from one end-use product package and prepared as in germination test. Four sub-samples in two dilutions are plated on two specific agar media: eight CLED plates for bacteria and eight MEA plates for moulds and yeasts. Incubation takes place for 2 days at 30 and 37°C for bacteria, for 3 days at 30°C for moulds and yeasts. *Method for external testing*: one sample of 250 g from an end-use product package is tested according to standard testing methods.
Technical properties: (1) *moisture content*: Standard: between 5 and 9%	*Method*: one sample of 1 g is dried at 100°C in an IR moisture analyser. The weight difference before and after drying is loss of water.
(2) *wet sieve test/solubility*: Standard: < 2% residue on 75 μm sieve	*Method*: the solubility of the product is tested by the wet sieve test. A sample of 5 g is diluted in 5 l (=label rate) and stirred for 30 min. The solution is poured over a 75 μm sieve; the remnants are dried and weighed. The percentage left on the sieve is determined and compared to the standard.

8 Conclusions and Recommendations

8.1 The Need for Quality Control

Every batch should be subject to all quality control tests. Essential parameters that need to be checked are listed in Table 4.1. Too many poor quality products have been put on the market in the past and this still occurs today. Poor quality leads to inadequate efficacy, and could even be a risk to human health and the environment. This is unfavourable for the producer and the users; ultimately it is bad for biocontrol in general. Product control should only be needed to confirm that the end-use product is of good quality. If production and formulation are done properly they should deliver products of high and consistent quality. This is an ideal situation. If product control finds deviations from the specifications, batches should be discarded. Improvement of those batches is rarely possible. Feedback from product control must be used to improve processes at an earlier point in the production line. This is one pivotal aspect of product control for the producer. QC is also a process that ensures that remedial efforts, if necessary, have produced satisfactory results and that detects recurrences or new instances of quality problems. Ultimately, quality control must ensure that end-users consistently receive high quality products.

8.2 *Standardization of Biological Activity*

Standardization for measuring efficacy has been developed for Bt's, but not for other biopesticides. Standardization of biological activity of Bt's allows scientists and regulators to compare products, although there are still difficulties (Skovmand et al. 1998, 2000). In addition to standardization in Bt's, a number of authors have expressed a desire for commonly accepted standard methods and criteria in other biopesticides as well (Burges 1981; Guillon 1997b; Glazer and Lewis 1998; Jenkins and Grzywacz 2003). I presume standardization of Bt's was initiated by researchers because demonstration of efficacy is not required for product approval in the USA. In the EU efficacy requirements are strict and in order to get approval a considerable amount of efficacy data has to be generated. Users can be confident that label claims of an approved product have been evaluated by the authorities. In this respect, biopesticides are treated similarly as chemical pesticides. Therefore, I believe that standardization of biological activity is not necessary and that efforts should not be spent on this activity as was carried out with Bt's over the last five decades. This would require a considerable amount of research in co-operative programmes between industry and academia and would need considerable resources.

For EPNs, on the other hand, it is desirable to establish a standardization of activity since they are generally not registered and efficacy data are not required. I recommend following the same approach as with beneficial arthropods. Here a pragmatic approach was chosen whereby it was determined whether mass-reared organisms were of an acceptable quality rather than of an optimal quality (Van Lenteren 2003a). Quality was considered in terms of whether a natural enemy is in a condition to properly control the pest instead of quality in strictly scientific terms. This was tested under relevant laboratory conditions. As I suggested above, the research of Grewal (2000) can be used as a starting point. In my opinion this effort should be made by the industry itself, with the help of scientists, and a very pragmatic approach should be taken.

8.3 *Recommendation for Tolerance Range of the Number of Active Ingredients*

Registration requirements do not give any guidance on acceptable tolerance levels of active propagules in products during their shelf-life period. The Danish EPA suggests accepting a five-fold deviation in the level of propagules in products on the market until the expiry date. This means that on the lower side only 20% of the set specification is acceptable and at the upper side 500% of the amount over the specification. Both the upper and lower ranges seems unrealistic to me. A product will generally not achieve its field performance with only 20% of its active ingredients. This renders dose-mortality research and determination of the most efficacious dose meaningless. A producer will not over-formulate its product by a factor of five, simply due to economics. The suggested range is nowhere to be found in any guideline and offers producers insight into what level is acceptable to authorities. Acceptance between countries, however, is likely to vary considerably until levels of acceptable deviations are officially established within the EU.

When comparing acceptable ranges of active ingredients (a.i.) in chemical pesticides the largest accepted deviation (tolerance ranges depend on the amount of a.i. per kg product) is ±15% of the declared content in homogeneous formulations when the a.i. is $\leq$ 25 g/kg. For heterogeneous formulations the accepted range is ±25% (FAO/WHO 2006). Looking at dose-mortality relationship and economics, this seems an acceptable tolerance range for all biopesticides to me. At Koppert, specifications of Mycotal and Vertalec were set with a maximum deviation of 15%, and this has been workable and achievable for many years without jeopardizing the products' performance. There is a strong need to determine the permissible tolerance range for active propagules in biopesticides with which they must comply during the shelf-life period. The competent authorities should make clear requirements for the industry and harmonize regulation on this point. I recommend allowing a tolerance range of ±25%. This seems to be a workable range for most biopesticides and this range is accepted in chemical pesticides for heterogeneous formulations. I consider biopesticides similar to heterogeneous formulations since they contain particles. In case it is desirable for a certain formulation to deviate from this percentage, a reasoned case should be acceptable.

8.4 Recommendation for an Acceptable Level of Microbial Contaminants

The amount of acceptable contaminants is also not regulated. The Danish EPA suggests initiating discussions on this as well as on the above topic within the EU to establish agreed levels of deviations in MPCPs (Winding 2005). This topic has been further taken up within the OECD BPSG (OECD 2009) and within the REBECA project. The reader is referred to Chapter 5 for more details on this project. The OECD BPSG wrote an issue paper "Microbial contaminants for microbial pest control products" which was discussed with the biopesticide industry during a workshop in July 2009 (OECD, unpublished). The approach of the regulators was based on accepted levels of contaminants in food and drinking water. The industry argued that biopesticides cannot be compared to food products. Biopesticides are different products that are generally applied in highly diluted solutions (100–1000×) onto crops. Moreover, microorganisms generally decrease quickly in numbers in the environment. The risks from contaminated biopesticides are therefore much lower than from contaminated food or feed products in which contaminants may reproduce rapidly. IBMA proposed limits for products based on baculoviruses and for bacterial and fungal products. Tests should be performed for a limited number of indicator species and higher levels should be accepted than those for food and drinking water. Further, the IBMA would like to see the general contaminant level (total aerobic count) set 10 times higher as proposed by the OECD document, to 10^8 cfu/g of product. In specific cases levels may be adjusted if there are good reasons for it. An example is *B. cereus* in baculovirus products (REBECA 2008). For human pathogens, the levels should be strict and clear, industry agrees with the regulators on this aspect of microbial purity. The toxicity test on mice, however, should only be required by the authorities as part of the five batch analysis in the phase of the dossier generation, and not as a routine batch test. The OECD BPSG is revising the issue paper and an improved version is due in 2010 for further discussion with regulators and the biopesticide industry.

8.5 Research Needs for Alternatives to Virulence Testing

The determination of virulence is only practical by performing laboratory bio-assays. This still requires considerable resources and time and may delay the release of product for sale. Alternatives to bio-assays have been looked for. Examples are the assessment in Bt's by analysis of endotoxins (CPL 2006b; Oestergaard et al. 2007), or simple bio-assays without plant material (Landa et al. 1994) or virulence testing by measuring enzyme activity (Pernfuss et al. 2004; Shah et al. 2007).

Alternatives to bio-assays are desired that would be quicker and cheaper and would still give reliable results. Biochemical analyses on material could be quick and relatively simple. Some attempts have been made with fungi and baculoviruses. The correlation with field efficacy must be determined and the results of the analysis should be able to give a consistent and reliable predictive value. There is a need to further develop these kinds of methods that would make quality control testing easier. It then can be done earlier and more frequently in the production line, and in products. This would greatly improve testing for virulence and it would provide an easier way to guarantee good quality products.

8.6 Total Quality Assurance

Quality plays a role at producer, distributor, and extension and users level. All have a responsibility with regard to good storage conditions and shipment conditions. Instructions on labels and leaflets with directions for use should be carefully read and followed. Producers should indicate instructions to all links in the supply chain in a proper way. For instance, in Koppert, we have developed "corporate guidelines for temperature-controlled shipments for natural enemies and microbials" for our subsidiaries and distributors all over the world in order to ensure delivery of consistent high quality products. Every year these are updated with incorporation of lessons learned and trainings are given to all people involved.

Post-shipment product control is pivotal for the users of natural enemies which are often shipped in active stages. These stages are vulnerable to harmful shipping conditions. This is not the case in entomopathogens which are generally shipped as dormant phases. EPNs, however, are more vulnerable than bacteria, baculoviruses and fungi, and do need more careful handling and shipping. Guidelines for checking quality have been produced for EPNs where it is relevant and tests can be performed within the chain. Training could be provided for proper quality checks. For other biopesticides this appears to be too complicated without relative expensive facilities and expertise. At the same time, producers have to deliver products and instructions for handling in such a way that quality checks are superfluous and that quality is guaranteed by the method of handling products.

8.7 The Benefits of Quality Control

Quality control is an essential part of a commercial production process and it is vital with regard to customer satisfaction and from a product liability viewpoint (CPL 2006b). It is imperative because of the variable biological nature of the products and constraints concerning their storage life (Guillon 1997a). In addition, registration authorities require solid quality control procedures and relevant standards with which products must comply. The responsibility for the quality of a product ultimately lies at the producer. Quality control starts with the production and continues until the product leaves the factory. Even after this point, total quality control should take the application of the product into account to guarantee customer satisfaction. A producer should instruct his distributors and customers properly so they know how to maintain a good quality once the products are in their hands.

Sixty years after the first biopesticides were launched on the market, quality control must be standard in the industry and only tested batches that meet the predetermined criteria should be released for selling. Manufacturers that are member of the IBMA undersign the Charter of Principles which implies a sound quality control programme. Members of the Biopesticide Industry Alliance (BPIA) of the USA subscribe to similar principles. In this way, the biocontrol industry attempts to self-regulate its activities. Obviously, subscribing to a code of conduct is not by definition a guarantee for high quality products, and many small companies are not member of an industry association. Therefore, these associations should try to broaden their membership and promote supply of high quality products only.

Manufacturers should promote quality awareness within their companies, not only within the management, but on the work floor where the day to day production is executed. Similarly, it should penetrate to the sales people and feedback should come from all persons involved so that a process of continuous improvement becomes an automatism within the whole company. Irrespective of official

Table 4.2 Benefits of quality control

- Keeps production personnel alert and aware of quality demands, and rewards them for improvements in quality of produced agents
- Detects problems and deviations in the production and in the product
- Ensures proper production and production processes
- Checks results of corrective efforts and recurrences of problems
- Ensures consistent delivery of high quality products
- Gives confidence in products to sales personnel and distributors
- Gives satisfaction to users and guarantees repeated sales
- Guarantees safe products that comply with registration requirements
- Benefits all stakeholders in biocontrol

regulations, quality control must become a natural striving for improvement of processes and products. Quality control is not a burden, but a benefit for all stakeholders (Table 4.2). Product quality control should be used to assure users that they are buying and using high quality products. Where there is transparency in product specifications and in what a product should achieve, confidence in the product is improved. Proper quality control will be a benefit both to the biocontrol industry and its customers.

References

Asano, S., 2006. Standard bio-assay method using insects for quality control of *Bacillus thuringiensis* products – the history and problems (in Japanese). *Jpn. J. Appl. Entomol. Zool.* 50: 101–114.

Battu, G.S., A. Ramesh, and G.S. Dhaliwal, 2002. Prospects of baculoviruses in integrated pest management. In: O. Koul, G.S. Dhaliwal (eds), *Microbial biopesticides*. Tailor and Francis, London. pp. 215–238.

Bernhard, K., 1992. Quantitative determination of δ-endotoxin contents in spray-dried preparations of *Bacillus thuringiensis* strain GC-91. *World J. Microbiol. Biotechnol.* 8: 24–29.

Bolckmans, K., 2003. State of affairs and future directions of product quality assurance in Europe. In: J.C. van Lenteren (ed), *Quality control and production of biological control agents: theory and testing procedures*. CAB International, Wallingford. pp. 215–224.

Boller, E.F., D.L. Chambers (eds), 1977. Quality control: an idea book for fruit fly workers. *IOBC/WPRS Bull.* 5, 162 pp.

Burgerjon, A. and H. Dulmage, 1977. Industrial and international standardisation of microbial pesticides. 1. Bacillus thuringiensis. *Bacillus thuringiensis Entomophaga* 22(2): 121–129.

Burges, H.D., 1981. Safety, safety testing and quality control of microbial pesticides. In: H.D. Burges (ed), *Microbial control of pests and plant diseases 1970–1980*. Academic Press, London. pp. 737–767.

Butt, T.M., C. Wang, F.A. Shah, and R. Hall, 2006. Degeneration of entomogenous fungi. In: J. Eilenberg, H.M.T. Hokkanen (eds), *An ecological and societal approach to biological control*. Springer, Dordrecht. pp. 213–226.

Caamano, E.X., R.A. Cloyd, L.F. Solter, and D.J. Fallon, 2008. Quality assessment of two commercially available species of entomopathogenic nematodes, *Steinernema feltiae* and *Heterorhabditis indica*. *Hort. Technol.* 18(1): 84–89.

Cherry, A.J., N.E. Jenkins, G. Heviefo, R. Bateman, and C.J. Lomer, 1999. Operational and economic analysis of a West African pilot-scale production plant for aerial conidia of *Metarhizium* spp. for use as a mycoinsecticide against locusts and grasshoppers. *Biocont. Sci. Technol.* 9: 35–51.

Ciuhrii, M., 1996. Main criteria for standardisation of quality assessment of viral products based on baculoviruses. *IOBC/WPRS Bull.* 19(9): 257–260.

Couch, T.L., 2000. Industrial fermentation and formulation of entomopathogenic bacteria. In: J.-F. Charles, A. Delécluse, C. Nielsen-LeRoux (eds), *Entomopathogenic bacteria: from laboratory to field application*. Kluwer, Dordrecht. pp. 297–316.

CPL, 2006a. *Biopesticides 2007*. CPL Business Consultants, Wallingford.

CPL, 2006b. *Biopesticides 2007, Vol. 5. A how to do it guide to biopesticides*. CPL Business Consultants, Wallingford, UK.

EC, 1991. Council Directive 91/414/EEC of 15 July 1991 concerning the placing of plant protection products on the market. *Official Journal of the European Union L 230*, 19.08.1991: 1–32.

EC, 2001. Commission Directive 2001/36/EC of 16 May 2001 amending Council Directive 91/414/EEC concerning the placing of plant protection products on the market (Data requirements microorganisms). *Official Journal of the European Union L 164*, 20.6.2001: 1–38.

EC, 2002. Review report for the active substance *Paecilomyces fumosoroseus* (Apopka strain 97, PFR 97 or CG 170, ATCC20874) 4203/VI/98-final, 24 January 2002. http://ec.europa.eu/food/plant/ protection/evaluation/new_subs_rep_en.htm. Cited January 30, 2008.

EC, 2005a. Council Directive 2005/25/EC of 14 March 2005 amending Annex VI to Directive 91/414/EEC as regards plant protection products containing microorganisms. *Official Journal of the European Union L 90*, 8/4/2005: 1–34.

EC, 2005b. Commission Regulation (EC) No 2073/2005 of 15 November 2005 on microbiological criteria for foodstuffs. *Official Journal of the European Union L 338*, 22/12/2005: 1–26.

FAO/WHO, 2006. *Manual on development and use of FAO and WHO specifications for pesticides*. March 2006 revision of the first edition. Only available on internet: http://www.fao.org/ag/AGP/AGPP/Pesticid/Specs/Pdf/Manual_update%202006.pdf. Cited February 2, 2008.

Gaugler, R. and R. Georgis, 1991. Culture method and efficacy of entomopathogenic nematodes (Rhabditidae: Steinernematidae and Heterorhabditidae). *Biol. Control* 1: 269–274.

Gaugler, R., P. Grewal, H.K. Kaya, and D. Smith-Fiola, 2000. Quality assessment of commercially produced entomopathogenic nematodes. *Biol. Control* 17: 100–109.

Georgis, R., 2002. The Biosys experiment: an insider's perspective. In: R. Gaugler (ed), *Entomopathogenic nematology*. CAB International, Wallingford. pp. 357–372.

Glazer, I. and E. Lewis, 1998. Virulence and environment: from petri-dish to the field. In: N. Simões, N. Boemare, R.-U. Ehlers (eds), *Proceedings of the COST 819 workshop on entomopathogenic nematodes. Pathogenicity of entomopathogenic nematodes versus insect defence mechanisms: impact on selection of virulent strains*. EUR 17776-COST 819, European Community Press, Luxembourg. pp. 203–215.

Grewal, P.S., 2000. Formulation and application technology. In: R. Gaugler (ed), *Entomopathogenic nematology*. CAB International, Wallingford. pp. 265–287.

Grewal, P. and R. Georgis, 1999. Entomopathogenic nematodes. In: F.R. Hall, J.J. Menn (eds), *Biopesticides: use and delivery*. Humana Press, Totowa. pp. 271–299.

Grewal, P. and A. Peters, 2005. Formulation and quality. In: P.S. Grewal, R.-U. Ehlers, D.I. Shapiro-Ilan (eds), *Nematodes as biocontrol agents*. CAB International, Wallingford. pp. 79–90.

J.M. Grunder (ed), 2005. Quality control of entomopathogenic nematodes. *COST action 819*. Agroscope FAW, Wädenswil, Switzerland. Berti Druck AG, Rapperswil.

Grzywacz, D., D. McKinley, K.A. Jones, and G. Moawad, 1997. Microbial contamination in *Spodoptera littoralis* nuclear polyhedrosis virus produced in insects in Egypt. *J. Invertebr. Path.* 69: 151–156.

Guillon, M., 1995. Industrial production of insect viruses for biological control: technical aspects and economical interest. *Proceedings of the microbial control agents in sustainable agriculture*, October 18–19, 1995, Italy. pp. 65–72.

Guillon, M., 1997a. Production of biopesticides: scale up and quality assurance. In: H.F. Evans (ed), *Microbial insecticides: novelty or necessity? BCPC symposium proceedings* No. 68, Coventry, April 16–18, 1997. BCPC, Farnham. pp. 151–162.

Guillon, M., 1997b. Quantification of biopesticide activity – a rapid survey of methods and standardization problems. *EPPO Bull.* 27: 123–125.

Hunter-Fujita, F.R., P.F. Entwhistle, H.F. Evans, and N.E. Crook, 1998. *Insect viruses and pest management.* Wiley & Sons, Chichester.

Jenkins, N.E. and D. Grzywacz, 2000. Quality control of fungal and viral biocontrol agents – assurance of product performance. *Biocont. Sci. Technol.* 10: 753–777.

Jenkins, N.E. and D. Grzywacz, 2003. Towards the standardisation of quality control of fungal and viral biocontrol agents. In: J.C. van Lenteren (ed), *Quality control and production of biological control agents: theory and testing procedures.* CAB International, Wallingford. pp. 247–263.

Jenkins, N.E., G. Heviefo, J. Langewald, A.J. Cherry, and C.J. Lomer, 1998. Development of mass production technology for aerial conidia for use as mycopesticides. *Biocont. News Inform* 19 (1): 21 N–31 N.

Jones, K.A., 2000. Bioassays of entomopathogenic viruses. In: A. Navon, K.R.S. Ascher (eds), *Bio-assays of entomopathogenic microbes and nematodes.* CAB International, Wallingford. pp. 95–139.

Jung, K., 1999. Qualität insektenpathogener Nematoden (Heterorhabditidae: Nematoda). Identifikation und Quantifizierung qualitätsbestimmender Eigenschaften sowie deren Beeinflussung durch ausgewählte abiotische Faktoren während der Lagerung. Ph.D. Thesis, Technische Universität, Darmstadt. 172 pp.

Lacey, L.A., 1997. *Manual of techniques in insect pathology, biological techniques.* Academic Press, San Diego.

Landa, Z., L. Osborne, F. Lopez, and J. Eyal, 1994. A bioassay for determining pathogenicity of entomogenous fungi on whiteflies. *Biol. Control* 4: 341–350.

Lasa, R., 2007. Formulation and efficacy of *Spodoptera exigua* nucleopolyhedrosisvirus as a biological insecticide for beet armyworm control in the greenhouses of southern Spain. Ph.D. Thesis, Laboratoria de Entomologia Agricola y de Patologia de Insectos, University of Navarra. p. 136.

Leppla, N.C., 2003. Aspects of total quality control for the production of natural enemies. In: J.C. van Lenteren (ed), *Quality control and production of biological control agents: theory and testing procedures.* CAB International, Wallingford. pp. 19–24.

Leppla, N.C., 2008. Roots and traditions of the *IOBC* Global Working Group on arthropod mass rearing and quality control. www.users.ugent.be/~padclerc/AMRQC/intro.html. Cited January 31, 2008.

Lisansky, S.G., 1985. Production and commercialization of pathogens. In: N.W. Hussey, N. Scopes (eds), *Biological pest control. The glasshouse experience*, Blanford Press, Poole. pp. 210–218.

Lisansky, S.G., 1997. Microbial biopesticides. In: H.F. Evans (ed), *Microbial insecticides: novelty or necessity? BCPC symposium proceedings* No. 68, Coventry, April 16–18, 1997. BCPC, Farnham. pp. 3–10.

Loesch, A., S. Hutwimmer, B. Pernfuss, and H. Strasser, 2007. Assessment of virulence test-systems for quality assurance using sub-cultivated *Beauveria brongniartii* conidia. *IOBC/WPRS Bull.* 30(7): 149–154.

Løschenkohl, B., K. Thygesen, and S.L. Nielsen, 2003. Måling af bioaerosoler under udbringning af mikrobiologiske bekæmpelsesmidler og ved efterfølgende arbejdsprocesser i potteplanter (in Danish). Danish EPA Bekæmpelsesmiddelforskning, nr. 79.

Mason, K., R.L. Ridgeway, and L.A. Merrill, 2002. Entomopathogenic nematode products and quality evaluations. In: N.C. Leppla, K.A. Bloem, R.F. Luck (eds), *Proceeding of the eighth and ninth workshops of the IOBC Working Group on "Quality control of mass-reared arthropods".* pp. 58–62.

Migheli, Q. and J.E. Ruiz Sainz, 2003. Quality control and efficacy assessment of microbial inoculants: need for standard evaluation protocols. *Proceedings of the joint meeting held in Seville (Spain) 25–28 October 2001 of the Cost 830 Working Group 4, IBMA Quality and Standards*

Working Group. EUR 20444. Office for Official Publications of the European Communities, Luxembourg.

Miller, R., 2002. Entomopathogenic nematode total quality management. In: N.C. Leppla, K.A. Bloem, R.F. Luck (eds), *Proceeding of the eighth and ninth workshops of the IOBC working group on "Quality control of mass-reared arthropods"*. pp. 51–57.

OECD, 2003. Guidance for registration requirements for microbial pesticides. *Series on Pesticides No. 18*. OECD, Paris (available at www.oecd.org).

OECD, 2004a. Guidance for country data review reports on microbial pest control products and their pest control agents, February 2004. *Series on Pesticides* No. 22. OECD, Paris (available at www.oecd.org).

OECD, 2004b. Guidance for industry data submissions for microbial pest control products and their pest control agents, February 2004. *Series on Pesticides* No. 23. OECD, Paris (available at www.oecd.org).

OECD, 2008. Working document on the evaluation of microbials for pest control. OECD Environment, Health and Safety Publications, *Series on Pesticides* No. 43. ENV/JM/MONO (2008)36. OECD, Paris (available at www.oecd.org).

OECD, 2009. Report of workshop on the regulation of biopesticides: registration and communication issues, 15–17 April 2008, EPA, Arlington, USA. *Series on Pesticides* No. 44. ENV/JM/MONO (2009)19. OECD, Paris (available at www.oecd.org).

Oestergaard, J., S. Voss, H. Lange, H. Lemke, O. Strauch, and R.-U. Ehlers, 2007. Quality control of *Bacillus thuringiensis* ssp. *israelensis* products based on toxin quantification with monoclonal antibodies. *Biocont. Sci. Technol.* 17(3): 295–302.

Palin-Holmberg, G., O. Mohammadi, P. Seiskari, and P. Pulkkanen, 2003. Quality control protocol for biofungicides. In: Q. Migheli, J.E. Ruiz Sainz (eds), *Proceedings of the joint meeting held in Seville (Spain) 25–28 October 2001 of the Cost 830 Working Group 4, IBMA Quality and Standards Working Group "Quality control and efficacy assessment of microbial inoculants: need for standard evaluation protocols"*. EUR 20444. Office for Official Publications of the European Communities, Luxembourg. p. 73.

Patel, M.N. and D.J. Wright, 1997. Glycogen: its importance in the infectivity in aged juveniles of *Steinernema carpocapsae*. *Parasitology* 114: 591–596.

Patel, M.N., M. Stolinski, and D.J. Wright, 1997. Neutral lipids and the assessment of infectivity in entomopathogenic nematodes: observations on four *Steinernema* species. *Parasitology* 114: 489–496.

Pearson, J.F. and T.A. Jackson, 1995. Quality control management of the grass grub microbial control product, Invade. *Proc. Agron. Soc. NZ.* 25: 51–53.

Pernfuss, B., D. Abendstein, and H. Strasser, 2004. Virulence of entomopathogenic fungi. *Laimburg J.* 1(2): 242–247.

Peters, A., 2000. Insect based assays for entomopathogenic nematode infectiousness: definitions, guidelines, problems. *IOBC/WPRS Bull.* 23(2): 109–114.

Peters, A., F. Stepper, K. Iwahn, and U. Kölzer, 2003. Qualitätssicherung von entomopathogenen Nematoden. *DGaaE-Nachr.* 17(1): 16.

REBECA, 2008. *REBECA – Regulation of biological control agents, Specific support action project no. SSP E-CT-2005-022709*. Project Coordinator, Ralf-Udo Ehlers, Christian-Albrechts-University of Kiel, Germany. http://www.rebeca-net.de. Cited April 10, 2008.

Ricci, M. and R. van der Pas, 2005. Practical recommendations for end-users and extension services. In: J.M. Grunder (ed), *Quality control of entomopathogenic nematodes. COST action 819*, Agroscope FAW, Wädenswil, Switzerland. Berti Druck AG, Rapperswil. pp. 97–103.

Robertson, J.L., H.K. Preisler, S.S. Ng, L.A. Hickle, and W.D. Gelernter, 1995. Natural variation: a complicating factor in bioassays with chemical and microbial pesticides. *J. Econ. Entomol.* 88(1): 1–10.

Shah, F.A., N. Allen, C.J. Wright, and T.M. Butt, 2007. Repeated in vitro subculturing alters spore surface properties and virulence of *Metarhizium anisopliae*. *FEMS Microbiol. Lett.* 276: 60–66.

Shapiro, M., 1986. In vivo production of baculoviruses. In: R.R. Granados, B.A. Federici (eds), *The biology of baculoviruses. Vol. II. Practical application for insect control*. CRC Press, Boca Raton. pp. 31–61.
Shieh, T.R., 1989. Industrial production of viral pesticides. *Adv. Virus Res.* 36: 315–343.
Skovmand, O., I. Thiery, G.L. Benzon, G. Sinègre, N. Monteny, and N. Becker, 1998. Potency of products based on *Bacillus thuringiensis* var. *israelensis*: interlaboratory variations. *J. Am. Mosq. Control Assoc.* 14(3): 298–304.
Skovmand, O., I. Thiéry, and G. Benzon, 2000. Is Bacillus thuringiensis standardisation still possible? In: J.-F. Charles, A. Delécluse, C. Nielsen-LeRoux (eds), *Entomopathogenic bacteria: from laboratory to field application*. Kluwer, Dordrecht. pp. 275–295.
Smits, P.H., 1987. Nuclear polyhedrosis virus as biological control agent of *Spodoptera exigua*. Ph.D. Thesis, Laboratory of Virology, Wageningen University, Wageningen. 127pp.
Smits, P.H. and J.M. Vlak, 1988. Quantitative and qualitative aspects in the production of a nuclear polyhedrosis virus in *Spodoptera exigua* larvae. *Ann. Appl. Biol.* 112: 249–257.
Van Lenteren, J.C., 2003a. Need for quality control of mass-produced biological control agents. In: J.C. van Lenteren (ed), *Quality control and production of biological control agents: theory and testing procedures*. CAB International, Wallingford. pp. 1–18.
Van Lenteren, J.C., A. Hale, J.N. Klapwijk, J. van Schelt, and S. Steinberg, 2003b. Guidelines for quality control of commercially produced natural enemies. In: J.C. van Lenteren (ed), *Quality control and production of biological control agents: theory and testing procedures*. CAB International, Wallingford. pp. 265–303.
Warrior, P., 2000. Living systems as natural crop-protection agents. *Pest Manag. Sci.* 56: 681–687.
Winding, A., 2005. Quantification and identification of active microorganisms in microbial plant protection products. *Environmental project No. 982*. Danish Ministry of the Environment. Only available on the internet: http://www2.mst.dk/common/Udgivramme/Frame.asp?pg= http://www2.mst.dk/udgiv/Publications/2005/87-7614-521-2/html/default_eng.htm. Cited January 23, 2009.

Chapter 5
Registration of Microbial Pest Control Agents and Products and Other Related Regulations

Abstract Several regulations apply to the handling and the use of microbial pest control agents. Microorganisms used for control of pests are subject to registration as a "plant protection product". The history of the development of data requirements for entomopathogens is presented. Current registration requirements are reviewed, particularly for the European Union. For several topics, data requirements are unclear or even lacking; procedures are both long and expensive. Initiatives to ameliorate these impediments are reviewed. The EU project REBECA provided recommendations for improvement of procedures as well for data requirements. The OECD BioPesticides Steering Group continues this work and is developing guidance for several data requirements issues. A general estimate of costs for the generation of a registration dossier is provided.

Entomopathogenic nematodes are not covered by the legislation of plant protection products in the EU. Several countries, however, have a registration for macroorganisms in place. Regulations for import and for export, and for release of macroorganisms have become complex and an administrative burden for biocontrol companies. The current situation in the Netherlands is highlighted as an example.

Various other regulations, such as laws on biodiversity, importation and release of exotic organisms, and safe handling apply to the use of microorganisms, and are briefly treated. Intellectual property rights are important for a company. The possibility to obtain a patent for a biopesticide and the value thereof will be discussed. Regulations form a major obstacle in the commercialization of biopesticides. Recommendations for improvements are presented for data requirements and for regulatory procedures. The role that the biocontrol industry should have in this political field is outlined.

1 Introduction

Various regulations apply to microbial pest control agents (MPCAs) and products (MPCPs). The person responsible for handling these organisms in any way and particularly who releases them into the environment is accountable for possible

W.J. Ravensberg, *A Roadmap to the Successful Development and Commercialization of Microbial Pest Control Products for Control of Arthropods*, Progress in Biological Control 10,
DOI 10.1007/978-94-007-0437-4_5, © Springer Science+Business Media B.V. 2011

adverse effects. Compliance with the many different legislative requirements is integral to sustainable use of those organisms and to sustainable business activities of a biopesticide manufacturer. In order to bring a MPCP on the market it must be registered as a plant protection product. Entomopathogenic nematodes are often treated separately from microorganisms and in most cases other regulations apply. Regulations for both groups of organisms will be discussed in this chapter. Legislation is established in directives and guidelines and a large data package has to be generated to fulfil the data requirements. A registration dossier is submitted to the competent authority for a risk assessment. After the evaluation, they decide whether a product can be released to the market and what precautions are necessary for safe use. This procedure was designed for chemical plant protection products and has been in place for many decades. For MPCAs/MPCPs this procedure was more or less copied, and gradually adapted to the specific characteristics of this type of products. Still, many data requirements are not appropriate to living organisms and this raises considerable problems. Registration is reported as the greatest challenge in the development of MPCPs. In this chapter I will review the development of the data requirements for MPCAs and MPCPs and the procedures of registration with a strong emphasis on the situation in Europe. For reasons of comparison, reference will be made to regulations and procedures in the USA and other developed countries in cases where this illustrates differences and gives directions for improvements.

Further, I will give recommendations for the generation of a registration dossier and the information that can be used to complete the data package. Attention will be given to those data requirements that are ambiguous and difficult to answer. Where possible I will give recommendations for solutions. Often though, these matters are very complex and experts need to work together to make the requirements more appropriate to these biocontrol agents. Several initiatives have been established to improve the current situation and some examples will be presented, as well as recommendations from the lessons learned in these projects. The costs of registration are often an impediment for companies, particularly due to the uncertainty to what amount they may accumulate. I will attempt to give an estimate of the registration costs for each type of entomopathogen.

Registration for entomopathogenic nematodes is quite different from bacteria, fungi and baculoviruses. The situation will be reviewed, including other regulations for import and export. The complexity of legislation to biocontrol companies is reviewed for invertebrate biocontrol agents, and ideas for improvements and harmonization are presented.

I will pay attention to the use of exotic organisms, but I will not treat regulations for importation and release of exotic entomopathogens for classical biological control. This topic is reviewed by Bigler et al. (2005), Loomans (2007), Hajek et al. (2007) and Hunt et al. (2008). Regulation on transgenic organisms is also out of the scope of this book and they are not used in crop protection in Europe anyway.

Registration procedures are complicated and expensive and they do not really encourage the development of new products. In contrast, governments advocate the development of non-chemical products, but they lack appropriate regulatory procedures. The main obstacles in legislation for use of entomopathogens are identified, and recommendations for improvements will be presented, particularly from the viewpoint of the biocontrol industry, and based on more than 25 years of personal experience with registration of entomopathogens in Europe. New regulations may offer new possibilities for easier registration of MPCPs and these will be mentioned.

Intellectual property rights are a valuable asset. The possibility of patenting a MPCA and/or a MPCP is an important consideration for a company in order to protect their investment. Aspects to consider are presented, as well as some examples where patents on entomopathogens have been exploited by biocontrol companies. The data package generated for registration is considered protection as well and I will briefly discuss this topic.

2 Registration of Microbial Pest Control Agents and Products

Microbial pest control agents and microbial pest control products, also called plant protection products (PPPs), are subject to registration, just like chemical substances and PPPs. The registration process evaluates the risks of the active substance and of the products that contain these microorganisms to human and animal health and to the environment in relation to the intended use of the products. The regulating authority performs a risk assessment based on the product dossier submitted by the applicant. After the evaluation of the risk assessment an approval may be issued including legal label recommendations and standard phrases for special risks and safety precautions. The fulfilment of the data requirements in a dossier forms a substantial part of the investment in the development of a microbial pesticide.

Registration is undoubtedly the largest barrier for commercialization of biopesticides, particularly in Europe. Ever since the development of biopesticides, registration has been the subject of an ongoing discussion concerning data requirements, risk assessment, administrative procedures and costs (Summers et al. 1975; Burges 1981; Lisansky 1986; Quinlan 1990; Tooby 1997; Van Lenteren 1996; Waage 1997; Neale 2000; Ravensberg and Elad 2002; Franceschini and Jondini 2004; Zimmermann et al. 2004; Mensink and Scheepmaker 2007; Waage 2007; Chandler et al. 2008). Here are the main reasons:

- data requirements are not well-defined and often adapted from requirements for chemicals which are generally not appropriate for microorganisms;
- test methods and guidance documents are set up for chemical substances and do not allow proper testing of microbials;
- end-points of risk assessments are not clearly established which allows for differences in interpretation and often leads to more data being required;
- regulators and risk assessors lack expertise and consequently they tend to overask information;

- procedures are lengthy, non-transparent and costly: both the EU procedure for the active substance and the national procedures for product authorizations;
- the onset of sales is unpredictable.

Noteworthy is that Marrone (2007) does not mention registration as a barrier for biopesticides in the USA, but she refers to the long and expensive registration process in the EU as a reason for the few available products on the market. Chandler et al. (2008) reviewed the registration process for biopesticides in the EU and its member states and they concluded that "the regulatory failure arises from the application of an inappropriate synthetic pesticides model and lack of regulatory innovation". In consequence, MPCP developers and manufacturers face an uncertain procedure without a clear estimation of costs and timeframe before they can even start generating a return on the investment. Therefore, many potential MPCAs are not being developed, or, companies which have tried commercializing products have failed due to high investments and no revenues for a long period.

2.1 The Development of Data Requirements

The first microbial pesticide was registered in 1948 in the USA (Schneider 2006). This was *Bacillus popilliae* (= *Paenibacillus popilliae*) for the control of Japanese beetle. Since the early developments of biopesticides, registration has been a difficult issue. Appropriate data requirements did not exist and had to be developed. Initially, applicants for the registration of a biopesticide followed chemical guidelines. Through contacts with regulators, data requirements were gradually adapted and they became more appropriate for biopesticides.

Interest in baculoviruses as biological insecticides goes back more then 50 years, particularly in the USA. The first baculovirus product was registered by the USA Environmental Protection Agency (EPA) in 1975 (Ignoffo and Couch 1981). This was preceded by much research in the public sector in the 1960s and the development of data requirements by the EPA. In this process, many safety studies on vertebrates were conducted which demonstrated the safety of these viruses (Burges et al. 1980a). In a USDA-EPA symposium, new test methods and criteria were jointly developed with regard to safety aspects of viral insecticides (Summers et al. 1975). In 1983, four viral pesticides were registered and in the course of these evaluations the EPA also established guidelines and data requirements for microbial pesticides. A detailed overview of the history of registration of viral products in the USA is given by Betz (1986). The discussion about data requirements started somewhat later in Europe, mainly in Germany and the UK (Burges et al. 1980b; Miltenburger 1980a, b). The first viral insecticide (Madex) was approved in 1987 in Switzerland. The first virus for use in greenhouse crops (Spod-X) was approved in 1993 in the Netherlands. Many of the data requirements were answered by means of published studies, and specific product data were generated via experimental studies (Smits and Vlak 1994).

The first bacterial product (Thuricide) was registered in 1961 in the USA. The development of guidelines and data requirements was initiated by members of the American Society for Invertebrate Pathology (SIP). A working Group on "Safety of Microbial Control Agents" was founded at the annual meeting in Montpellier in 1971. This group mainly worked on Bt and viruses, but also on fungi, protozoa and EPNs (Laird 1981). In Europe, guidelines were drafted by an IOBC study group in 1980 for bacterial pesticides (Burges et al. 1982). The first product (Thuricide) was approved in Europe in 1964 in Germany. In the Netherlands, the first Bt products (Dipel and Thuricide) were approved in 1971 for use in greenhouses.

The first proposal for data requirements for entomopathogenic fungi was provided by Hall et al. (1982). The first registration for an entomopathogenic fungus in Europe was obtained in the UK in 1981 by Tate & Lyle for Vertalec, a product based on an aphid-specific strain of *Verticillium lecanii* (= *Lecanicillium longisporum*). Quinlan (1990) reported on the data requirements in the UK at the time. He also described the difficulties with requirements and procedures in some other European countries. Gradually, other products appeared on the market e.g. Mycotal (*V. lecanii* (= *L. muscarium*)), Bio1020 (*M. anisopliae*), Preferal (*Paecilomyces fumosoroseus* (= *Isaria fumosorosea*)), and Botanigard and Naturalis (*B. bassiana*). In the USA, Mycar (*Hirsutella thompsonii*) was the first registered entomopathogenic fungus for control of citrus rust mites, and approval was granted to Abbott Laboratories in 1981. However, the product was never commercially successful and was withdrawn in 1985 (McCoy 1996).

In the Netherlands, the author was personally involved with the registration of Mycotal, which was the first fungus to be submitted for registration. A special application form for microbial products had been developed following more than 10 years of discussions, workshops, meetings and proposals. Data requirements were similar to those for chemical products, and only during pre- and post-submission meetings was it decided which data was needed to be generated and how. The first application (1982) was not approved in 1985 due to insufficient efficacy. With an improved formulation, approval was received in 1992. Attempts to register Mycotal in France and Germany in the mid 1980s were abandoned because data requirements and procedure costs were unclear and the authorities over-required data in the eyes of Tate & Lyle.

Despite the many proposals for data requirements for entomopathogens by scientists and the biocontrol industry, most countries used their own set of data requirements or used the ones for chemicals with some modifications. Only 8 years after the implementation of Directive 91/414/EEC (EC 1991a), in 2001 when 2001/36/EC (EC 2001a) came into force, a standardized set of data requirements became available for all EU countries. These were established within the EU without consultation with the industry and academic experts.

It is beyond the scope of this book to review the historical development of these requirements in detail. For applicants who wish to register a product, it is more useful to understand the current requirements. Below, I will give recommendations on how to build a dossier and which sources, in addition to the guidelines, can be used to better understand what exactly the authorities require.

2.2 The Role of Researchers and Industry in the Development of Data Requirements

The process of establishing and improving data requirements has taken place independently in many countries over the last five decades. Scientists have documented this process and have often indicated the need for appropriate guidelines both for the industry and for the regulators. They also have conducted much research to demonstrate that entomopathogens (and other biocontrol agents) are safe to the environment and to human health. Methods have been developed to study the effects, and based on this research new and appropriate guidelines have been formulated and proposed. Many papers have been presented on this subject during conferences, followed by intensive discussions. Specific workshops (IOBC, EPPO, OECD, and others) have been held on this topic since the 1970s. The following list of publications is a selection that presents an elaborate illustration of this debate over a long period until today: Hall et al. (1982), Laird et al. (1990), Garibaldi (1995), Cook et al. (1996), EPPO/CABI (1997), OECD (1998), Strasser (2000), Zimmermann et al. (2004), Jaronski et al. (2003), Kiewnick (2007) and Chandler et al. (2008). These debates at scientific conferences were often held in the absence of regulators and this was very unfortunate since without their understanding and commitment to improve requirements and guidelines, these discussions had little impact.

The industry view on registration of bioinsecticides, bioherbicides and biofungicides has been given by a number of authors who all asked for specific data requirements and harmonization of procedures (Lisansky 1986, 1994; Cross and Polonenko 1996; Neale 1997, 2000; Panetta 1999; Ravensberg and Elad 2002; Franceschini and Jondini 2004). Novel products faced this challenge of registration and many companies failed to get approval, or only received approval after a long time and with high costs. As a result, many companies stopped their activities with biopesticides, or put them on hold. Even today, applicants face difficulties in answering the questions. Still, parts of the requirements are "modified chemical requirements" and are not appropriate to the specific registration issues of microorganisms. At the same time, they make the evaluation for risk assessors difficult and allow for differences in interpretation and in decisions. Change in regulation is a slow and difficult process with many stakeholders involved. Harmonization is strongly desired, but it is only slowly finding its way into these regulations and procedures. Industry and scientists are still in discussion with regulators on improving the requirements and regulations. This is an ongoing process by companies when they apply for a biopesticide registration, and by activities of the International Biocontrol Manufacturer Association (IBMA). This association plays an important role as a spokesman of the industry and participates in many activities with authorities and other organizations.

The EU has realized that the registration hurdle is keeping products from the market and that procedures need to be improved. In some countries, specific initiatives have been set up to facilitate product registration on a national level. Examples will be given below. Legislation is intermittently revised and societal and political influences play a more prominent role then ever. Unfortunately, the focus is still on the chemical pesticides and the biopesticides are only a minor aspect, so to achieve improvements in the regulation of biopesticides will be an ongoing challenge for the biocontrol industry.

2.3 Regulatory Framework for Registration of Microbial Pesticides

In the EU, the placing on the market of a PPP is regulated by the Council Directive 91/414/EEC (EC 1991a) which came into force on 26 July 1993. The EU registers the active substance (term also used for microorganisms) whereas any final formulated product needs to be registered in each individual Member State (MS). Before 1993, any active substance and PPP needed to be registered on a national level. The EU Directive was an attempt to harmonize registration within the EU for all involved parties. Biopesticides needed to be registered according to this legislation, but until 2001 there were no specific data requirements. Only when the Commission Directive 2001/36/EC (EC 2001a) was published in May 2001 as Annex IIB and IIIB of 91/414/EEC were data requirements differentiated from chemical substances, and specified for microorganisms as active substances (Annex IIB), as well as for the preparations based on them (Annex IIIB). The definition of a microorganism applies to bacteria, fungi, protozoa, viruses and viroids. The publication of the specific data requirements was a major improvement. However, both the industry and the regulators found that while working with these requirements several questions were ambiguous and needed clarification.

Once a microorganism is approved, it is included in Annex I, the list of all approved active substances in the EU. Annex III lists the data requirements for the PPPs. Annex IV and V of 91/414/EEC (Commission Directive 2003/82/EC (EC 2003)) describe the standard phrases for special risks and safety precautions (labelling). Only in 2005 were the Uniform Principles for microorganisms laid down in Council Directive 2005/25/EC (EC 2005) as Annex VI part II of 91/414/EEC, while the Uniform Principles for chemicals originated from 1997. Uniform Principles ensure that evaluations and decisions with regard to authorization of plant protection products are conducted according to these principles by Member States. It was only decided in 2005 by the EU that inclusion on Annex I of microbial pest control agents has to be done at the strain level (SANCO/10754/2005 rev. 5). Further, in 2008 it was decided that baculoviruses will be included at the species level, although new isolates must be notified to the EU (SANCO/0253/2008 rev. 2). An overview of the registration process and the data requirements are provided by Neale and Newton (1999).

The EU has revised this Directive and the new version has become Regulation (EC) No 1107/2009 (EC 2009a) which makes it directly law in all MSs. This Regulation has been published on the 24th November 2009 and will apply 18 months after publication, thus from 14 June 2011. A few changes may be helpful in getting more biopesticides in a wider market. These will be discussed in the conclusions of this chapter.

In the USA, the requirements for biopesticides were recently updated as the "Pesticides; Data Requirements for Biochemical and Microbial Pesticides, Federal Register Notice: October 26, 2007 (Volume 72, Number 207), EPA 40 CFR Part 158" (EPA 2007a). The definition was changed from "microbial agent" to strictly "microorganism". Other changes include the requirement for more mutagenicity studies with metabolites, immunotoxicity studies and an additional avian study. The intra-cerebral study was removed. Details on the registration process and the requirements are provided by McClintock (1999). In Canada, data requirements are outlined in Regulatory Directive DIR 2001-02. Data requirements in the USA and Canada are largely similar in practice, although they differ somewhat on paper (O. Messerschmidt, personal communication). The only substantial difference is that PMRA requires a complete dossier on efficacy and phytotoxicity data, and the EPA does not. A comparison on legislation in Europe, Canada, the USA and the regulatory processes is made by Hauschild and Speiser (2007). Formal data requirements are similar, the registration processes, however, are different whereby the process in the EU is the most complex and lengthy.

Further, the OECD has published three guidance documents (OECD 2003a, 2004a, b) on data requirements and the evaluation thereof. These documents are a great improvement in terms of harmonization. A dossier should be set up according to the OECD format and this format facilitates the process for all stakeholders in many countries. Below I will discuss the actions that the OECD BioPesticides Steering Group (BPSG) is currently undertaking to improve registration of biopesticides.

2.4 The Generation of a Dossier

In the EU an Annex II dossier is required for registration of the active substance at the EU level, and an Annex III dossier for the product for national authorizations. An Annex II dossier provides all the required information for the microbial agent per se including at least one representative use. This is a label recommendation such as, for instance, the spraying of a tomato crop against whitefly in a greenhouse with a certain dose rate, frequency and intervals. This use is needed to establish exposure routes and to perform risk assessments for that particular use. The chapters in such a dossier consist of data on the microorganism: identity and biological properties; analytical methods; human toxicology; residue on food and feed, fate and behaviour in the environment, and ecotoxicology.

An Annex III dossier contains information on the product, with all the desired label uses. The chapters present data on the formulated product: physical, chemical and technical properties; application methods and mode of action; analytical methods; efficacy; human toxicology; residue; fate and behaviour; and ecotoxicology. The difference between an Annex II and III dossier is small when the active substance and the formulation are very similar, or when the formulants are considered safe.

In the USA and in Canada, a registration must be obtained for the active ingredient as a manufacturing use product (MUP) as well as for the separate end-use product (EP). The registration process is handled by one authority in each country. Both Canada and the USA as well as the EU accept a dossier in the OECD format which facilitates international exchange of information.

Most of the regulations allow for answering questions in one out of three ways: (1) via a data waiver; (2) via a statement based on scientific literature; or (3) via an experimental study. This can be decided on a case by case approach and an applicant is free to choose how to fulfil the requirements.

2.4.1 Data Waivers

The request for data waivers, e.g. justifications for non-submission of data, is an important possibility in the building of a dossier. The acceptance of waivers reduces the need to conduct expensive experimental studies. Non-submission of data can be justified when exposure does not occur or is limited given the method of application of the MPCP. A data waiver can also be argued for from published literature from which it can be demonstrated that adverse effects are unlikely to occur. Justification for data waivers should be written using sound scientific arguments. It is recommended to discuss the possibility of waivers during a pre-submission meeting with the regulatory authority.

The use of waivers is under discussion in the OECD BPSG (OECD 2009) in order to harmonize decision taking on this subject. A guidance document on waivers is in development in which the focus lies on ecotoxicology, and on infectivity tests and clearance studies (elimination of the microorganism from the body of the test animal). Acceptance of waivers will depend on the type of microorganism and its level of familiarity. It will be a case by case approach. Data waivers are largely accepted for the group of baculoviruses. There are more areas where data waivers are applicable such as on genotoxicity, residue, ecotoxicology, environmental fate and behaviour. Another area is efficacy regarding questions on effects on yield, adjacent crops, succeeding crops and non-target plants.

2.4.2 Statements

Questions can be answered by statements based on scientific literature. There is a wealth of information on microorganisms and safety. Many studies demonstrate that microbial biocontrol agents do not cause any harmful effects. These studies often addressed non-target organisms and environmental fate and behaviour. Some animal toxicity, residue, and metabolite studies have been conducted. I will give an overview of useful information about bacteria, fungi and baculoviruses.

When using generic literature it is necessary to address the topic in relation to the strain that is subject to registration. When it can be demonstrated that the information also refers to the strain used, regulators are more likely to accept the statement. When the strain used differs from the information in the literature, then the differences should be addressed and it should be made clear to what extent the data do apply to the strain used. Bridging data may still be needed to accept statements which are based on information of similar strains or species. The more convincing evidence can be given in the statement, on a solid scientific level, the more valuable the statement is for the risk assessment. Statements may form a considerable part of a dossier when the organism is well-studied. For a newly-discovered species this is not possible. It is recommended to use a consultant with experience in writing statements or to ask a specialized scientist to address the particular topic. Sometimes it is easier and cheaper to have a test conducted than to have a statement written by an expert or consultant.

2.4.3 Experimental Studies

For certain requirements, experimental studies have to be performed according to official Good Laboratory Practice (GLP) guidelines such as OECD guidelines. Often, toxicological animal studies need to be performed in this way, as do environmental studies that investigate non-target organisms and fate in the environment, and physical-chemical property studies. These study protocols have been developed for chemical substances and often adaptations are needed to make them appropriate for testing microbials agents. These need careful and accurate modifications in the protocol which need to be discussed in detail with the contracting research organization (CRO). The product developer needs to take the lead since these CROs often lack experience with microorganism. If protocols deviate considerably from the standard protocol, it is recommended to ask the evaluating regulatory body for their approval of the adapted protocol to prevent difficulties later in the process. If communication with the regulators is difficult on this point, assure that solid scientific argumentation is provided to justify the protocol adaptations.

Efficacy studies need to be conducted by CROs that are certified for conducting trials under Good Experimental Practice (GEP) conditions, and according to EPPO guidelines. These are developed for chemical pesticides and protocols often need to be modified in order to properly assess the activity of a biopesticide. Consultation with the authority is strongly recommended again.

2.4.4 Useful Information on Bacteria

The safety of entomopathogenic spore-forming bacteria has been documented in numerous publications and reviews. The majority refers to Bt species, subspecies and strains and, in general, these organisms pose little risk to human and animal health (Glare and O'Callaghan 2000; Lacey and Siegel 2000; Libman and MacIntosh 2000; Siegel 2001; Federici and Siegel 2007) and to the environment (Meadows 1993; Klier 2000; Glare and O'Callaghan 2003; Jackson 2003; Lacey and Merritt 2003). Moreover, there have not been any reports of adverse effects on humans and the environment arising from the widespread use of Bt's in the last five decades (Otvos et al. 2005). The approval of transgenic Bt crops in many countries and the wide-scale planting of these crops and the use of products from these crops in food and feed further suggest the safety of Bt to human and animal health and to the environment (Federici and Siegel 2007). A review of Bale et al. (2008) indicated that Bt crops do not lead to any direct adverse effects on parasitoids and predators. Several OECD guidance documents have been published that provide valuable information for environmental risk assessments (OECD 1997), and methods with regard to identification and detection of bacteria (OECD 2003b, 2004c).

A concern for regulators is the close relationship between *B. thuringiensis*, *B. cereus* and *B. anthracis*. The last two are serious human pathogens. Species differentiation is difficult and the genetic relationship between these three bacteria is not fully known. The same holds for the exchange of plasmids between these three species. Nor is their role in pathogenesis and toxin production understood (Vilas-Bôas et al. 2007). Moreover, the toxins of Bt's require careful safety testing, particularly with new strains (Glare and O'Callaghan 2000).

2.4.5 Useful Information on Fungi

There are many papers on the safety of fungi which can be used in statements. Excellent reviews on safety of *Beauveria bassiana. B. brongniartii* and *Metarhizum anisopliae* to vertebrates and to the environment have recently been published by Zimmermann (2007a, b). These publications could be a useful tool in the generation of a dossier on strains of these fungi. In fact, they could be the basis of a generic consensus dossier on these fungi. I will come back to this idea in my conclusions.

The main concerns with fungi relate to toxicology: pathogenicity and the production of metabolites, to allergenicity, and to environmental effects (host range). Many authors have addressed these issues with entomopathogenic fungi in relation to safety, particularly on non-target organisms. In general, there are no unacceptable effects on human and animal health and on non-target organisms (Strasser et al. 2000; Lynch and Thomas 2000; Goettel and Hajek 2001a; Goettel et al. 2001b; Vestergaard et al. 2003; Jaronski et al. 2003; Zimmermann 2007a, b). The impact on earthworms and Collembola was reviewed by Brownbridge and Glare (2007) and they concluded that applications of fungal products do not pose a significant risk to those non-target organisms. The effects of the application of *M. anisopliae* on soil microbiota (Kirchmair et al. 2008) and on the microbial community populations in growing media (Shah et al. 2009) was also minor and transient. In a review on persistence data of applied entomopathogenic fungi Scheepmaker and Butt (2010) reported that levels of these released fungi decline over time to natural background levels, and thus do not pose a major risk to soil biota.

2.4.6 Useful Information on Baculoviruses

The safety of baculoviruses to human and animal health and to the environment has been thoroughly studied and baculoviruses are generally recognized as safe (Burges et al. 1980a; Groner 1986, 1990; Cory 2003; Lacey 2008). The OECD published a consensus document on information used in the environmental risk assessment of baculoviruses (OECD 2002). The *raison d'être* for consensus documents was to deliver information that was mutually acceptable among OECD member countries.

The document contains information on the group of baculoviruses such as organism characteristics, behaviour in the environment, and safety considerations as well as many references. This document has formed the basis of most baculovirus dossiers.

Baculoviruses do not produce any metabolites or toxins, and cannot reproduce outside the host. This is a major difference when compared to bacteria and fungi. The generation of an active substance dossier for any baculovirus species is therefore relatively easy, and to obtain approval of a baculovirus as an active substance should not be difficult. Specific information is still requested for the formulated end-product. The concerns of the regulators are predominantly related to (1) the safety of the formulation additives, (2) microbial contamination, (3) the hairs of caterpillars as potential allergens, (4) stability of the product, and (5) efficacy. These issues should be properly addressed in the product dossier.

2.4.7 Additional Sources of Useful Information

With respect to applications of new active substances in the EU, the European Food Safety Agency (EFSA) coordinates the peer review of the Draft Assessment Report (DAR) with all MS and the European Commission. EFSA publishes a public version of the DAR during the evaluation process in order to give the public the possibility to submit comments. Furthermore, the conclusion of the peer review of the risk assessment by all MS is made available on the internet including the conclusions, the summary, several background documents with comments on the DAR, comments from expert meetings, and evaluation tables (http://dar.efsa.europa.eu). A study of these documents in detail provides a better understanding of the critical aspects of a risk assessment for a certain organism or group of organisms.

Some national authorities publish a report on the decision of a PPP on the internet. This can range from short notices on the decision and the approved uses and label to full evaluation reports describing all relevant risk assessments and conclusions thereof. Such a report is published by the Ctgb, the competent authority in the Netherlands (www.ctb-agro.nl), and the BVL (Bundesamt für Verbraucherschutz und Lebensmittelsicherheit), Germany (www.bvl.bund.de). Other authorities may provide them upon request (J. Meeussen, Ctgb, personal communication).

The US EPA also publishes evaluation reports and conclusion details on PPPs, called Biopesticide Active Ingredient Fact Sheets and a Biopesticide Registration Action Document (BRAD) (www.epa.gov). In Canada, the Pest Management Regulatory Agency (PMRA) publishes A Proposed Regulatory Decision Document and a Evaluation Report which contains information on the submitted dossier and the evaluation (www.pmra-arla.gc.ca).

2.5 Data Requirement Issues

The generation of a dossier is a very elaborate and complex task. Clear questions, appropriate guidance documents, and experience are of paramount importance in order to set up a dossier that fulfils all the requirements. Some of these prerequisites are lacking for biopesticides. Applicants have complained about the lack of appropriate data requirements while regulators criticize the poor quality of submitted dossiers. Both points are true to some extent. There are numerous items that could be improved in the data requirements and in the guidance documents. I will mention a few from my personal experience that seriously hamper the generation of a dossier. Improvements will help both industry, and regulators in their task to perform a proper risk assessment. This refers mainly to the procedure in the EU.

2.5.1 Taxonomy

In the EU, microorganisms are included on Annex I at the strain level (SANCO/10754/2005 rev. 5). The same is true for the USA and Canada. Since the taxonomic classification of microorganisms is under continuous debate and new technologies are still being developed, it is difficult to present the proper data for identification, and authorities require different methods. Currently, applicants may be submitting different methods for identification of a strain. This renders reference to related taxa unreliable and this situation is problematic for a risk assessment. A guidance document on the use of taxonomy in risk assessment has been published by the OECD (2003b) with regard to bacteria. The OECD BPSG (OECD 2008) is developing further guidance to standardize methodology and it recommends the identification of the strain at the highest possible level of detail.

For baculoviruses it has been concluded that Annex I Inclusion will be on species level based on the view that "In other cases in which the species is known to be *relatively homogeneous* and well studied it may be decided by experts if certain questions may be handled on a species/subspecies rather than on a strain level." New isolates have to be applied for at Member State level, and for the application of a new isolate a Guidance Document (SANCO/0253/2008 rev.2) has been developed. Following Annex I Inclusion of the new isolate, mutual recognition is possible. It is probably best to submit the new isolate dossier to the RMS that wrote the DAR on the original isolate, as they know all the details and should be capable of handling the procedure quickly and with low costs.

2.5.2 Metabolites

In the registration process, data are required for (relevant) metabolites. The definition of a relevant metabolite has been refined for chemicals (Sanco/221/2000 –rev.10). In contrast, for microbials a metabolite is not clearly defined. For a chemical substance, metabolites are defined as "all (biotic or abiotic) reaction or breakdown products of an active substance of a plant protection product, formed in the environment". Under several data requirement sections, more precise definitions are given. For example, in a guidance document on groundwater, definitions are given for "relevant metabolites", for "metabolites of no concern" and for "non-relevant metabolites". A stepwise scheme is presented to determine whether a metabolite is relevant. For drinking water, acceptable concentration levels are provided. Regarding ecotoxicology, definitions are given in specific guidance documents for major and minor metabolites.

For a microbial agent, metabolites are defined in 2001/36/EC as "products resulting from degradative and biosynthetic reactions taking place within the microorganism". Relevant metabolites are defined as "metabolites that are of concern for human or animal health and/or the environment". In the section on residues there is mention of substances of concern, as well as of impurities and relevant impurities. There are, however, no precise descriptions available (no guidance documents) for the determination of whether a metabolite, substance of concern, or impurity is relevant, and if so, what levels are acceptable. What is precisely understood as "are of concern for human health"? The distinction between metabolites and impurities needs clarification. The definitions for metabolites and consequently for the data requirements for those substances are ambiguous and confusing in the case of microbial agents, in contrast to chemicals.

The production of secondary metabolites is often part of the pathogenesis in entomopathogenic bacteria and fungi. These compounds may pose a risk to humans and animals when they are exposed to them. Therefore, it is pivotal to know whether a bacterium or a fungus, and particularly the strain that is being developed as a bioinsecticide, produces any metabolites that may pose a problem. I will focus on fungi since it is more complex than Bt, but much of the comments are true for bacterial compounds as well. For many fungal species it is well known whether and what kind of metabolites they produce (Strasser et al. 2000; Vey et al. 2001) and this is essential information for a risk assessor. For each strain subject to registration, however, the following strain-specific information is required: "the nature and structure of this substance, its presence inside or outside the cell and its stability, its mode of action (including external and internal factors of the microorganism necessary to action) as well as its effect on humans, animals or other non-target species" (Section 2.8 of 2001/36/EC). Investigating the exact metabolite profile of a certain strain is an elaborate and costly research project. Developing validation methods for analytical use of each metabolite (a necessary tool for residue and fate questions of a metabolite) is a project in and of itself that could cost up to €1.2 million per compound (Seger et al. 2005a, b; Strasser et al. 2004). It is obvious that these data requirements could be a major barrier for further development of a product (Strasser and Kirchmair 2006). Therefore, the EU RAFBCA project (Risk Assessment of Fungal BioControl Agents) investigated whether simpler methods could be developed to assess the potential risks of metabolites (for details see www.rafbca.com). Effects of crude extracts were compared with effects of purified metabolites. Crude extracts amplify the signal for the main metabolites, and may also take into account interactions between compounds. The findings of RAFBCA suggest that crude extracts represent the "worst case scenario" (Skrobek and Butt 2005). One of the recommendations from RAFBCA was that metabolites should be investigated in a tiered approach (Strasser et al. 2008). Crude extracts of the fungus should first be tested for harmful effects. If no harmful effects appear, further information and testing is not required. If negative effects do appear, further elucidation of metabolites could be warranted. Risk assessment of crude extracts requires less time and fewer resources and consequently is more cost-effective for companies wishing to register fungal BCAs.

In RAFBCA, the fate and environmental effects of metabolites were studied as well as whether these metabolites enter the food chain. With the fungi *M. anisopliae* (Längle et al. 2004), *B. brongniartii* (Pernfuss et al. 2007), and *L. muscarium* (A. Skrobek, W.J. Ravensberg, N. Ben El Hadj, A. Vey and T. M. Butt, unpublished data) it has been demonstrated that these metabolites do not pose a risk to the environment and that they do not enter the food chain.

Metabolites also pose a problem with regard to genotoxicity. The guideline states: "if the microorganisms produce exotoxins according to Section 2.8 of 2001/36/EC (e.g. "If other strains belonging to the same microbial species as the strain subject to the application are known to produce metabolites (especially toxins) with unacceptable human health and/or environment during or after its application...") then these toxins and any other relevant metabolites in the culture medium must also be tested for genotoxicity. Such tests on toxins and metabolites should be performed using the purified chemical if possible". Often, genotoxicity studies are required even if the organism does not relate at all to this point in Section 2.8 in the sense that no unacceptable effects are known from related strains. Three studies are required which altogether cost a considerable amount of money. In order to conduct the studies with the purified chemicals, considerable quantities need to be available. Theses are hard to produce and thus very expensive. I propose to test on genotoxicity only when the organisms produce metabolites according to the questions as in Section 2.8, i.e. there are known unacceptable effects from related strains, and with a tiered approach.

Because of the lack of a proper definition of a relevant metabolite, regulators tend to over-ask for data on metabolites. Only when "other strains belonging to the same microbial species as the strain subject to the application are known to produce metabolites (especially toxins) with unacceptable human health and/or environment during or after its application..." data on metabolites shall be provided (Section 2.8 of 2001/36/EC). The biocontrol industry does not work with these kinds of organisms, but only with Group 1 organisms: "unlikely to cause human disease" (EC/2000/54 (EC 2000a)). Still, most regulators require detailed information on metabolites. To improve this situation the OECD BPSG (OECD 2008) has initiated the development of a guidance document with regard to metabolites.

2.5.3 Various Other Data Requirement Issues

A number of other data requirements call for improvements. I will not go into detail but will briefly refer to them. These aspects are:

- storage stability: there are no guidelines specific to microbials. The test methods for chemicals must be used. Acceptable decline levels after a certain storage period have not been established. Currently, 10% loss of potency is considered unacceptable while bio-assay variability alone is often as much as 20%;

- microbial contamination: acceptable maximum limits of contaminants are lacking, and especially limits for each species or group of species of contaminant;
- genotoxicity: three tests are required and there is no tiered approach. One of the test methods, the Ames test which is based on mutant bacteria lacking the capability of the synthesis of histidine, is not appropriate for microbial pesticides because the microorganism may overgrow the mutagenicity-indicator bacteria on the agar plate. When a homogenate of the microorganism is used presence of histidine may interfere with the indicator bacteria which makes the test unreliable;
- sensitization, and classification and labelling: MPCPs are classified by default as potential sensitizers since there are no appropriate testing methods available. As a consequence they are labelled Xn R42/43 with a danger symbol with the wording "harmful" and risk phrases R42/43: "may cause sensitization by inhalation" and "may cause sensitization by skin contact". However, in the peer review report on *L. muscarium*, EFSA concluded in June 2009 in the PRAPeR expert meeting M03 that "Classification is not relevant for microorganisms according to classification rules for chemicals." The following labelling phrase was agreed: "microorganisms may have the potential to provoke sensitizing reactions" (http://www.efsa.europa.eu/en/scdocs/scdoc/1446.htm). This conclusion still needs to be officially included in the data requirements for microorganisms;
- ecotoxicology: data on earthworm toxicity are required although there are no pathogens known to earthworms. For this reason, it is highly unlikely that well-known microorganisms will affect earthworms, so testing is superfluous. Data on effects on the microbial community are required while applied microorganisms fall back to background levels and effects are minor and transient, in general;
- efficacy: levels of efficacy that are equivalent to efficacy of chemicals are often required. Biopesticides act differently and result assessments should reflect this. Nevertheless, test methods are designed for chemicals and chemical reference products must be used.

Many of these topics have been discussed in the REBECA Action and recommendations for improvements have been drawn up. The OECD BPSG is planning to translate these recommendations into guidance documents (see below) which are expected to be accepted for registration procedures.

2.6 Regulatory Authorities and Procedures in the EU

2.6.1 The Rapporteur Member State

For the registration of a new active substance, an applicant can choose in which EU country he wants to submit the dossier. The Rapporteur Member State (RMS) acts as the evaluation competent authority and as the contact point for the company in the EU during the evaluation process. The RMS performs the completeness check and writes the Draft Assessment Report (DAR), which is the basis for the peer review in the EU. Not every country, however, is willing to accept such an assignment, although legally they are obliged to accept it. For some applicants it has been difficult to find a country to act as a RMS for a new MPCA. Some countries state that they are too busy, or claim that they do not have enough experience. This demonstrates that an applicant is only to a certain extent free to choose a RMS. In appointing a country as RMS, the first consideration is the amount of the administrative fees. The fee between countries differs enormously, and these costs can be considerable. Another pivotal factor is the ability to have direct and efficient communication with the RMS, in which each others' language plays a role. Further, the willingness to have a pre-submission meeting and possibly follow up-meetings is determinative too. The experience of a RMS with a certain microorganism is a relevant factor to consider. Market aspects also play a role since the first provisional authorization for the product it is likely to be obtained in the RMS country. The choice of the RMS should take these aspects into account, although it is hard to know all of this in advance. Often, this experience is only gained during the process.

Only a number of MSs have gained experience with biopesticides. In the fourth stage of the review programme the following seven countries have acted as a RMS for a microorganism: Denmark, Estonia, France, Germany, Italy, Netherlands and Sweden. Most of the afore-mentioned also have evaluated "new" active substances. Austria, Belgium, Finland and the UK have also experience as RMS for "new" microbials. About ten EU countries now have experience with one up to three microorganisms as an active substance. Participating members of the OECD BPSG are Austria, Denmark, Germany, Netherlands, Sweden, Switzerland and the UK. I recommend selecting a RMS for their experience with a certain MPCA and for the ability of direct and open communication, and not only based on fees. The government-backed initiative to facilitate registration of microbials should also be weighed in the decision, see the paragraph on "national initiatives" below. This may save costs and time in the long run. Krause et al. (2006) came to similar conclusions for disease-suppressive organisms.

In the USA, the EPA, the regulatory authority, has set up a special division called the Biopesticides and Pollution Prevention Division (BPPD) that takes care of the registration of biopesticides. The BPPD is active since 1995 and has gained a lot of experience in evaluating the risks of biopesticides. In Canada, PMRA has a separate staff that reviews microbial and biochemical pesticides. The experience level is not as high as in the EPA due to much less registered microbials.

2.6.2 The Road to Annex I Inclusion

The procedure for a MPCA from application to Annex I inclusion is complex. The applicant submits a dossier with one or more representative uses to a chosen RMS. The first step is a completeness check by the RMS. After the evaluation of the dossier by the RMS and the writing of the DAR, a peer review by the European Food and Safety Authority (EFSA) follows. All MS are asked to comment on the DAR and they have the opportunity to raise questions. EFSA reviews these issues and consults experts in various meetings. Hereafter EFSA drafts a conclusion for a vote

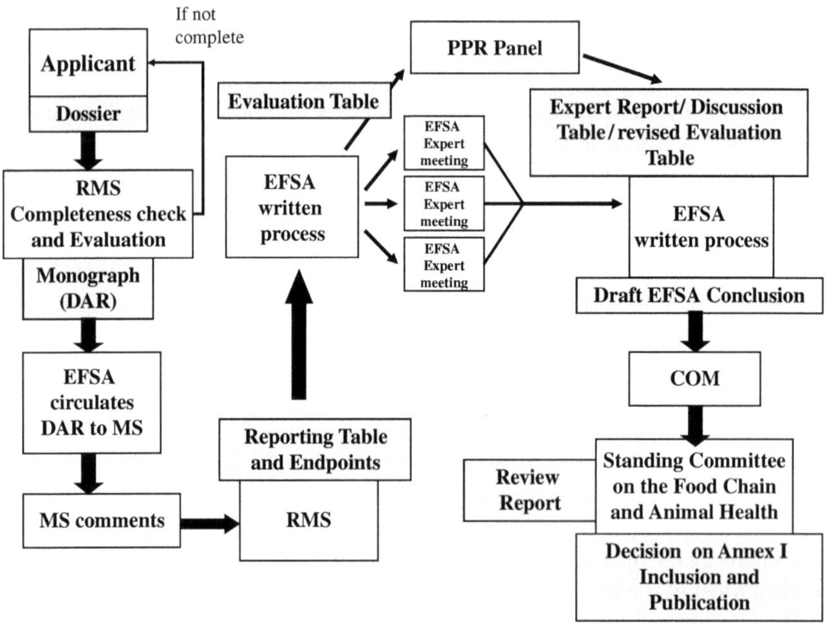

Fig. 5.1 The EU registration procedure for a new microorganism

by all MS that decide by qualified majority. Then the European Commission (COM) decides on Annex I inclusion and publication hereof in the Official Journal makes it official. The process (Fig. 5.1) should take 3 years (Table 5.1), but with MPCAs it has taken more then twice as long in most cases. The new Regulation (EC) No. 1107/2009 provides much clearer time lines for the authorities for the various steps in the procedure, and this should improve the registration process considerably.

2.6.3 National Product Authorizations

In some countries, registration is handled by an authority that is specifically established to evaluate and process registrations for plant protection products

Table 5.1 Official time lines of EU registration procedure for a new microorganism

Steps in registration process	Time period (91/414/EC)	Time period ((EC) 1107/2009)[a]
Completeness check by RMS	6 months	4.5 months
Preparation DAR by RMS	12 months	12 months
Peer review by EFSA	12 months	6 months
Decision Annex I listing	6 months	6 months
Total	36 months	28.5 months

[a]Additional time periods and clock stop periods are provided for submission of additional information and assessment thereof.

(and sometimes biocides). Examples are the Ctgb in the Netherlands, the CRD (formerly the PSD) in the UK, KEMI in Sweden, AFSSA in France. In other countries, registrations are handled by departments of Ministries, either of Agriculture, Environment or Health; sometimes even four to five Ministries are involved. In the case of a special authority for registration experience is built up over the years within the organization. It is evident that such an organization is geared up to process registrations in an appropriate way. Nevertheless, from personal experience I can say that even in such organizations the procedures are not flawless and delays are frequent.

Where a Ministry coordinates the evaluation process of a dossier, external experts are hired in to evaluate sections of the dossiers. Examples are experts for human toxicology, for environmental effects, for efficacy, etc. This is, for instance, the case in Belgium, Italy, Poland and Spain. These experts could be academia or experts from research institutes. These scientists are often overloaded with work, including some registration dossiers waiting on their desk for a risk assessment before a certain deadline. Meetings are only periodically organized, but when the work has not been carried out, the applicant can only wait for the next meeting. Decisions have been on hold for many months because external experts had left the assignments and Ministries were unable to replace them by new experts. The applicant's only resource is to lobby for solutions, but in fact is utterly powerless. The system with external experts is obviously an inefficient system with many problems. Each country has a different organization for registration, and for biopesticide companies that want to register their products in many different EU countries, these processes are a challenge to understand and to undergo. This is one of the most difficult aspects of registering a product in the EU. Once the active substance has been approved on a European level, it still takes many years and great effort to register a product in ten MSs. As an example, it took over 4 years to get a product authorization for Mycotal in Spain. This shows that this system is just not functioning. It is completely unpredictable when national registrations can be expected and what the costs are. Communication is the key word and obviously the language can be a barrier. Therefore, direct contact with the help of an experienced consultant is imperative for a timely registration.

2.6.4 National Initiatives to Improve Product Authorizations

Most of the discussion in the international scientific community has been focused on the data requirements for microbial agents. National authorization of products, however, has been hardly considered in international workshops and in the REBECA Action. Procedures and fees form a serious barrier to products coming on the market in many EU countries. In the UK, there was concern about the lack of microbial products on the market and a pilot project was launched in 2003 to encourage and facilitate registration of alternative control measures such as pheromones, plant extracts and biological organisms (Chandler et al. 2008). Reduced fees were an essential part of the project and the fee for national registration of a microbial organism was reduced from £60,000 to £22,500. Initial success led to the creation of a permanent Biopesticide Scheme in 2006. The key elements were the appointment of a "biopesticide champion" for all initial contacts and support, and specific guidance (pre-submission meetings). Further, applicants were and are encouraged to contact the Pesticide Safety Directorate (PSD; called Chemicals Regulation Directorate (CRD) since April 2009) early in the product development. The reduced fee structure is kept in place too. The fee for a new microorganism application in the EU is as low as £7,500. The Biopesticide Scheme has been a successful project since its foundation and has led to the registration of a moderate number of new biopesticides.

In Belgium, a special procedure was launched in 2007 in the framework of the "programme for reduction of pesticides and biocides" to improve the availability of biopesticides on the market (see www.fytoweb.fgov.be). Biopesticides are defined as plant extracts, microorganisms, pheromones and others. The project aims to give special consultancy to the applicant, a separate fast-track procedure for biopesticides, lower fees, and improved communication. One contact person has been assigned for the communication with applicants. Fees for a new active have been reduced from €100,000 for a traditional pesticide to €10,000 for a biopesticide and €3,000 for a national product authorization.

In the Netherlands, a similar situation as in the UK occurred. Decades of discussions and workshops, etc., had not led to tangible improvements and few new products were entering the market. The project GENOEG was set up in 2001 by the Ministry of Agriculture and the Ctgb, and was lead by the Centre for Agriculture and Environment (CLM) (see www.genoeg.net; and Chandler et al. 2008). The goal was to register more sustainable and low-risk products with a small dossier which would be quicker, and with lower costs. This should lead to the availability of sufficient non-chemical crop protection agents for the farmer, a policy goal of the Dutch government. The project aim was to offer consultancy to applicants in building a dossier and to provide financial support for the evaluation costs and additional required studies. There was no reduced fee for biopesticides established in the project as in the UK. The project ended in 2008 and achieved registration of ten new products. There has, however, not been a structural change in the procedure and no special contact point has been established for inquiries on biopesticides. The current fee structure of the Ctgb is complicated and based on actual expenses. For an application of a new substance for Annex I inclusion there is no differentiation between microorganisms and chemical substances, and the maximum amount is €242,720 (for 2009). When the evaluation appears to be less extensive, the lower fee of €164,682 will be charged. The application fee, however, is lower for a microorganism: €5,712 versus €11,424 for a chemical substance. For a national registration of a product, the fee is based on the actual studies that need to be evaluated and since the data requirements differ between a chemical and a microbial product, this results in fees of approximately €200,000 respectively €40,000. An applicant will receive a quotation after submission based on the size of the dossier. This makes an estimation of the costs beforehand almost impossible. A help-desk has been set up for general purposes, but answers are very formal and legal and do not really help potential biopesticide applicants in most cases. A more structural solution as set up in Belgium and in the UK is highly desirable.

These initiatives have lead to real progress in the procedures by the national authority, and this has led to the launch of new products on the market. Pre-submission meetings, reduced fees, and efficient and detailed communication have been the cornerstones of this success. It is obvious that there is a need for such systems in most EU countries. Governments that take the development of alternatives for synthetic chemical pesticides seriously have to implement such a system to improve registration of biopesticides.

In the USA, the IR-4 Project has been founded with the objective to provide safe and effective pest management solutions for specialty crop growers. A special programme for biopesticides has been established, and the primary objective of the IR-4 Biopesticide Research Program is "to further the development and registration of biopesticides for use in pest management systems for specialty crops or for minor uses on major crops" (Hartman and Markle 1999; www.ir4.ruthers.edu). Funds are available for efficacy research. The project provides regulatory assistance to applicants, can submit applications itself, on behalf of biocontrol companies, and is in close contact with the regulators of the EPA.

In Canada, PMRA has also introduced a program that facilitates registration of reduced risk pesticides, including biopesticides. As part of this initiative, the Agency offers a reduced fee structure, and has shortened review timelines for these products. Furthermore, the Pest Management Centre of the Ministry of Agriculture and Agri-Food Canada provides regulatory support to companies for biological pest control products that address priority pest management issues identified in consultations with growers (www4.agr.gc.ca). Budgets are available for research projects on efficacy and safety subjects.

2.7 The EU Policy Support Action REBECA

Registration of MPCAs and their products has been difficult and slow in the EU ever since the development of the first products. Despite more appropriate data requirements as established in 2001/36/EC, few biopesticides have reached the market in the EU. Registration of active substances was and is a lengthy process. The whole procedure from submission of the dossier to Annex I Inclusion took on average more than 75 months for "new" microbial agents (Hokkanen 2007). This period includes waiting for additional information from the applicant though. The details per microorganism are provided in Table 5.2. By comparison, approval in the USA only took about 28 months. As a consequence, companies hesitate to initiate product development. Research results remain on the shelf of the scientists and considerable research resources have been wasted in this way. Despite many EU-funded research projects on the development of MPCAs, the situation remained the same. In contrast, many MPCAs have been registered in the USA: 78 in the USA (Schneider

Table 5.2 Registration period for microorganism as new active substance from submission of the dossier until publication of the Annex I inclusion (Directive 91/414/EC)

Microorganism and strain ("...")	Applicant and product name (in capitals)	RMS	Submission date to RMS	Annex I inclusion: publ. date	Period (months)[a]
Paecilomyces fumosoroseus "PFR 97"	Certis/Biobest PREFERAL	BE	18-5-1994	25-6-2001	85
Pseudomonas chlororaphis "MA 342"	BioAgri CEDOMON	SE	15-12-1994	29-4-2004	113
Ampelomyces quisqualis "AQ10"	Ecogen-Intrachem AQ10	FR	12-4-1996	22-1-2005	105
Spodoptera exigua nucleopolyhedrovirus	Biosys-Certis SPOD-X	NL	12-7-1996	3-08-2007	133
Coniothyrium minitans "CON/M/91-08"	Prophyta CONTANS	DE	10-9-1997	13-8-2003	71
Gliocladium catenulatum "J1446"	Kemira-Verdera PRESTOP	FI	19-5-1998	22-1-2005	80
Bacillus subtilis "QST 713"	Agraquest SERENADE	DE	19-4-2000	14-2-2007	82
Pseudomyza flocculosa "PF-A22 UL"	Plant Products SPORODEX	NL	6-3-2001	Pending	> 108
Paecilomyces lilacinus "251"	Prophyta BIOACT	BE	15-9-2002	5-4-2008	67
Adoxophyes orana granulovirus "BV001"	Andermatt CAPEX	DE	29-11-2004	Pending	> 64
Paecilomyces fumosoroseus "Fe 9901"	FuturEco NOFLY	BE	4-2-2005	Pending	> 62
Candida oleophila "O"	BioNext NEXY	UK	12-7-2006	pending	> 45
Helicoverpa armigera nucleopolyhedrovirus	Andermatt HELICOVEX	EE	9-08-2006	Pending	> 44
Spodoptera littoralis nucleopolyhedrovirus	Andermatt LITTOVIR	EE	2-01-2007	Pending	>39
Pseudomonas sp. "DSMZ 13134"	Sourcon_Padena PRORADIX	NL	28-07-2007	Pending	> 32
Trichoderma atroviride "I-1237"	Agrauxine ESQUIVE	FR	28-8-2007	Pending	> 31
Aureobasidium pullulans	Bio-ferm BLOSS.PROTECT	AT	17-04-2008	Pending	> 23
Trichoderma asperellum "T 34"	Biocontrol Technologies; T34	UK	22-04-2009	Pending	> 11

[a]Updated until 31 March 2010; see European Pesticides Database: http://ec.europa.eu/sanco_pesticides/public/index.cfm?event=activesubstance.selection&a=1

2006) versus 21 in the EU (Kiewnick 2007). And, in 2006 the number of MPCPs available on the market was 225 in the USA versus 57 in the EU (Fjelsted 2006). Scientists and the IBMA have consistently attempted to improve this situation by means of lobbying activities directed at regulatory authorities, policy makers and to politicians (see the "Position Papers" on www.ibma.ch).

Eventually, the EU realized that there was a large problem concerning registration of microbial pesticides and other alternative pesticides such as pheromones and plant extracts. In 2004 the EU DG RESEARCH opened a Call for Proposals regarding improvements in the regulation of biological control agents (BCAs) in the Sixth Framework Programme. The REBECA (Registration of Biological Control Agents) proposal was written by Ehlers and others, and granted as a Specific Support to Policy Action. The objective was to: "review possible risks of biocontrol agents, compare regulation in the EU and the USA and to propose alternative, less bureaucratic and more efficient regulation procedures maintaining the same level of safety for human health and the environment but accelerating market access and lowering registration costs" (www.rebeca-net.de). Within the project, which ran in 2006 and 2007, many workshops were organized where research scientists, regulators and the biocontrol industry met and discussed critical issues related to data requirements, risk assessments, and procedures. This was the very first time in the EU that such a dialogue took place on such a scale. It did lead to a better understanding of the positions and responsibilities of all the parties and the difficulties each faced. There was a positive atmosphere where solutions were sought in order to resolve the stagnant position of biopesticide registration. Informal contacts and building relationships were valuable aspects for all. Unfortunately, there were few regulators present from southern European countries.

The project dealt with invertebrate BCAs, microbial BCAs, botanical extracts and semio-chemicals. Current regulations were reviewed, and proposals were made to improve procedures, harmonize risk assessments, reduce costs and accelerate procedures while maintaining safety to workers and consumers. A wealth of information was presented and reviewed. Intense discussions between stakeholders then took place in an attempt to design appropriate data requirements and risk assessment methods. It is beyond this book to present an elaborate review of the topics and the outcome of the project. The interested reader can find detailed reporting of the results on the project's website. I will briefly report the main recommendations for entomopathogens (cf. REBECA Deliverable 10 and 22) that are important for product developers and applicants and that may lead to a more balanced and quicker registration of BCAs.

1. Recommendations for improvement of the registration process:
   - improve communication between applicants and regulators, e.g. by a pre-submission meeting; a draft application form and an information package have been designed to facilitate and harmonize such meetings; a formal guidance document should be made. Applicants should also be invited for expert meetings at EU level;

- improve communication among regulators to harmonize evaluation approaches and to speed up the process. An expert group for microbials should be established that would meet regularly. A forum for communication could also be the OECD BPSG;
- use the lessons learned in the fourth stage of the review programme with "existing" microbial agents to facilitate new applications;
- lower fees substantially. Lobbying by all stakeholders at member state level is needed;
- develop and accept a generic approach for a certain group of active substances;
- keep to short and strict timelines for EU and national procedures which will improve predictability of the length of the process;
- develop guidance on efficacy evaluation by an expert group in order to harmonize decisions among member states. Assessments should become flexible and lower efficacy accepted.

2. Improvement on data requirements:

   - baculoviruses are considered safe and should be registered at species level. This has been implemented in the meantime (see above);
   - develop threshold levels for microbial contaminants in baculovirus products and others. The OECD BPSG has taken this up in their work programme;
   - develop more specific data requirements and guidance documents with respect to genetic stability, human infectivity studies, exposure data, sensitization, metabolites, persistence in the environment, and ecotoxicity and effects on non-target organisms.

Most of these items have been taken up by the OECD BPSG for further development.

An important document is the "Strategy for implementation of results" (Deliverable 30) that recommends how, by whom and when to obtain implementation of the project's proposals and recommendations. The EU microbial expert group, the OECD BPSG, EFSA and national regulators must continue to work on these issues. Industry should be involved through the IBMA, and scientists through IOBC too. A crucial topic has been whether or not to remove BCA regulation from 91/414/EEC and to design specific legislation for BCAs. There was no consensus reached on this point. The opinion of DG SANCO (The Health and Consumer Protection Directorate General of the European Commission) is that BCAs can be handled within the current legislation with adaptations in data requirements and the provision of appropriate guidance documents. Many regulators shared this opinion. The biocontrol industry desires new legislation and saw opportunities for this since 91/414/EEC in undergoing revision. The IBMA activities with regard to this aspect are mentioned below.

2.8 Harmonization Efforts by the OECD BioPesticides Steering Group

The OECD's Pesticide Programme strives to facilitate and improve registration procedures of crop protection products in member countries by harmonization and work sharing (Sigman 2005). The OECD BioPesticides Steering Group was established by the OECD Working Group on Pesticides (WGP) in 1999. The aim was to help member countries harmonize the assessment of biopesticides. The definition of biopesticides includes microbials, invertebrate biocontrol agents and semiochemicals. The first task was to review data requirements and to develop guidance documents, and a format for dossiers to be used in all member states (Richards and Kearns 1997). This was achieved in 2004, and guidance documents have been developed for the industry and for regulators (OECD 2003a, 2004a, b, c).

Currently, the BPSG is working on several issues related to data requirements for microbials (Meeussen 2007). The goal is to develop a document that will help evaluators in the assessment of microbial pesticides. The first set of issues under discussion concerned: taxonomic identification, genetic toxicity assessment, human exposure and risk assessment, metabolite residues in treated food crops, and efficacy evaluation. A number of meetings have been held with risk assessors from EU countries and from Canada and the USA. The OECD also invited representatives of the biocontrol industry (IBMA and the Biopesticide Industry Alliance (USA)) to participate in a workshop on the regulation on biopesticides in April 2008 in Arlington, USA, and in July 2009 in Paris. The industry appreciated this inclusive approach of stakeholders in this debate over difficult and sometimes controversial aspects of microbials. Ideas and problems from both groups were discussed in an open and constructive manner (OECD 2009). A "Working Document on the evaluation of microbials for pest control" (OECD 2008) has been published where these issues are treated elaborately, and guidance is given on how to fulfil the data requirements, both for industry and regulators.

The BPSG has established a working programme for 2009–2012 where remaining problems related to the registration of biopesticides will be resolved. These issues are data waivers, microbial contaminant limits, storage stability, environmental risk assessment, fungal metabolites, classification and labelling, and a checklist for pre-submission meetings. The intention is to produce guidance papers for each subject. Recommendations from REBECA will be taken forward and implemented where appropriate in these guidance documents. The same will be carried out with "lessons learned" from the evaluation of the review programme of the EU of "old" microbial agents, and with experiences from other countries, like the USA. The IBMA will be consulted for the exchange of information. The working programme of the OECD BPSG should lead to facilitation of the registration of biopesticides, and the shortening of the procedures through work sharing and harmonization between member countries. The outcome will benefit governmental evaluators as well as the industry if the EU and national authorities accept those guidance documents. Ultimately, one DAR should suffice for the whole world.

2.9 Registered Microbial Pest Control Agents in the EU

MPCAs that were on the market in Europe before 26 July 1993, the enforcement date of 91/414/EEC, had to go through a review programme for re-registration of the active substances. This was called the fourth stage of the review (Reg. 2229/2004) in which, among others, so-called "existing" microorganisms were re-evaluated. Annex II dossiers had to be submitted before 30 November 2005 according to the latest data requirements (2001/36/EC). Twenty-two species of microorganisms had been on the market in various MSs and sixteen of them have been defended by one or more notifiers, e.g. companies that submitted dossiers. The dossiers were processed by the EU-appointed Rapporteur Member States (RMS) and DAR's were written and evaluated. Based on the conclusion of the RMS that for microorganisms for which: "there are clear indications that it may be expected that they do not have any harmful effects on human or animal health or on groundwater or any unacceptable influence on the environment to be included in Annex I to Directive 91/414/EEC without detailed scientific advice from the European Food Safety Authority (EFSA) having been sought.", microorganisms were placed on Annex I by 1 May, 2009 (2008/113/EC). Peer reviews by EFSA were delayed and need to be finalized by December 31, 2012. Additional information could be required by EFSA later to keep the active substance on Annex I. The "existing" entomopathogens placed on Annex I are four subspecies of Bt: *aizawai, kurstaki, israelensis* and *tenebrionis*, and *B. bassiana, Cydia pomonella* GV, *Lecanicillium muscarium* (formerly *V. lecanii*) and *M. anisopliae* var. *anisopliae*. Further, nine disease control agents have been included on Annex I. One or more strains have been included per organism. In the re-registration procedure, ten MPCAs disappeared from the market for various reasons. Entomopathogens that were not further supported by the industry were the bacterium *B. sphaericus*, the fungi *Aschersonia aleyrodis, B. brongniartii* and *V. lecanii* strain Ve2 against aphids, and the baculoviruses *Agrotis segetum* GV, *Mamestra brassica* NPV and *Neodiprion sertifer* NPV. Three disease control agents were not supported.

Since 91/414/EEC has been implemented, only two "new" entomopathogenic strains have been included in Annex I: *Isaria fumosorosea* (*P. fumosoroseus*) strain "Apopka 97" and *Spodoptera exigua* NPV. Six other microorganisms have been included. Entomopathogenic microorganisms in the process of evaluation at the time of writing (March 2010) are the fungus *I. fumosorosea* (*P. fumosoroseus*) strain Fe 9901 and three species of baculoviruses (*Adoxophyes orana* GV, *Helicoverpa armigera* NPV and *Spodoptera littoralis* NPV). Microbial agents for control of fungi and bacteria under evaluation are one bacterium (*Pseudomonas* sp. "Proradix"), two fungi (*Trichoderma atroviride, T. asperellum*) and three yeasts (*Pseudomyza flocculosa, Aureobasidium pullulans, Candida oleophila*). There may be more applications filed, and due to confidentiality I cannot be sure whether the above list is complete. The entomopathogens registered in the EU in March 2010 (16 on strain level) and the ones in the process of registration are given in Table 5.3 (SANCO 2008). The number of entomopathogens (on strain level) registered in the USA is 37 in 2007 (EPA 2007b). Most of them (22) are Bt strains. Canada has 14 registered

Table 5.3 Entomopathogenic species and strains registered or under evaluatio in the EU in March 2010

Bacteria	Fungi	Baculoviruses
Bt aizawai (2)	*Beauveria bassiana (2)*	*Spodoptera exigua NPV (1; 2 p)*
Bt kurstaki (5)	*Lecanicillium muscarium (1)*	*Cydia pomonella GV (1)*
Ffd Bt israelensis (1)	*Metarhizium anisopliae (1)*	*Adoxophyes orana GV (1 p)*
Bt tenebrionis (1)	*Paecilomyces fumosoroseus (1; 1 p)*	*Helicoverpa armigera NPV (1 p)*
		Spodoptera littoralis NPV (1 p)

(. . .) – Number of registered strains.
(p) – Registration pending.
P. fumosoroseus is now called *Isaria fumosorosea.*

entomopathogenic strains (PMRA 2008). With regard to microorganisms which can be used in organic agriculture in EU Member States, Council Regulation 2092/91 (EC 1991b) defines that all microorganisms are allowed provided that they are not genetically modified. Of course PPPs based on microorganisms need a regular national approval too before they can be used.

2.10 Registration Costs

Costs for registration are considerable. It is, however, not possible to give a fixed cost figure for a microbial agent and/or product. Costs depend on the organism and its field of use, and on the country of submission. It is also difficult to distinguish precisely developmental costs from registration costs. These overlap to a certain extent. Research conducted in the process of product development provides valuable infor mation for the dossier. In the literature, many authors mention global figures on the costs of registration, but few are concrete. Gelernter (2005) estimated the costs for safety tests and registration of a biocontrol product to be US $1–2 million. Marrone (2007) stated that the registration of *Bacillus subtilis* in the EU exceeded US $2 million. Hökeberg (2006) mentioned that regulatory expenses for a biofungicide amounted to €1,6 million; Andermatt (2006) estimated the registration costs of a baculovirus to range between €1–€2 million. Costs in the USA are estimated to range between $500,000–1,000,000 (Evans 2004). Clearly, costs are much higher in the EU compared to the USA. Many authors have stated this fact, but few exact figures are available.

A survey among biocontrol companies revealed that registration costs for a microbial pesticide in the EU are about €1,890,000 (Hokkanen 2007). More than half of these costs are external costs (experimental studies, consultant costs, administrative fees) and the breakdown of the total costs showed that toxicological tests accounted for the largest share (43%). These figures refer to all types of microorganisms, and the authors do not specify the field of use nor the number of countries in which approval was sought. The greater the number of label uses requested, the greater the number of efficacy trials that will have to be conducted.

The broader the label, the more complicated the risk assessment becomes, and all these factors contribute to the total costs.

I will try to provide a more concrete figure for registration of a pathogen as an active substance in the EU and an authorization in one country for a limited field of use. It is customary to separate the costs for the active substance dossier (Annex II dossier) from the product dossier (Annex III dossier). The latter contains data on the product (chemical-physical properties, product toxicology data and efficacy data). Fees differ enormously between countries for acting as a RMS as well as for national approvals. Fees for a RMS for registration of an active substance range from €5,000 (Spain) and €240,000 (NL; this is the maximum fee, actual fee is based on cost recovery base) in 2009. Fees per product authorization depend on the requested uses (crops and targets) and vary between €5,000 and €50,000 and possibly higher, depending on the country. I presume that these costs will be on average €100,000–150,000 for obtaining an active substance approved in the EU and a product authorization with one target crop and pest in one EU member state. Below I will give an estimate of registration costs for a new active substance, in other words for the generation of an Annex II dossier, without administrative fees.

- Bt's are the only entomopathogenic bacteria from which experience exists. Cost estimates are not available in the literature. An estimate is about €300,000–€500,000 per new Bt strain. This amount depends very much on whether a company already registered a Bt strain before and has access to ample information. If a company has to register its first Bt strain, costs may be more than double. For a new entomopathogenic bacterium that is not a Bt, costs could be much higher, particularly when toxins are involved in the mode of action. Costs may accumulate to around 2 million Euro;
- Entomopathogenic fungi vary considerably in their biology and this leads to differences in dossier costs. At the present time, strains of four species have been registered in the EU: *B. bassiana, M. anisopliae, L. muscarium* and *Isaria fumosorosea* (*P. fumosoroseus*). I estimate that minimal costs for a fungal insecticide are around €500,000. This becomes much higher when unknown metabolites need to be analysed and tested. Expenses may extend 2 million Euro in that case;
- Baculoviruses are considered safe to human and animal health and to the environment in general. Moreover, they do not produce metabolites and cannot reproduce outside the host. Extrapolation from information of the baculovirus family, from which information is widely publicly available, can therefore be used to build up a species dossier, the taxonomic level that is accepted for baculovirus registration. Costs of a registration dossier are therefore considerably less than for fungi and bacteria. I expect that for less than half a million Euros a full dossier can be made. Consequently, registration fees can be lower since the dossier requires less time to evaluate;
- EPNs need to be registered is some countries and a limited dossier needs to be presented (see below). When efficacy trials are required, this adds substantially to the costs. The estimate of €20,000 per country (see below) seems to be the

highest reported thus far for an approval at species level. If approval is sought for more targets and/or crops, this may increase the costs for efficacy trials. I presume this will not be much higher in the future. Data may be used in several countries which makes the average costs for approval per country in the EU acceptable.

Registration costs are often reported as the main hurdle to the development of a MPCP. Therefore, it is essential to have a reliable estimate of the costs beforehand. This is a very difficult task for a product developer and each case is different. Experience and consultation with experts will help, as will contact with contract laboratories who conduct toxicological tests. A pre-submission meeting with the authorities may clarify which data and studies are required and this may improve the estimate. In many cases data requirements can be answered by statements. Costs may be much lower if an applicant can refer to data of a closely-related approved strain by means of a letter of access. For a less well-studied organism, costs may be much higher than indicated above.

During the generation of the dossier new and unexpected issues may appear which could significantly raise the expenses. For example, toxicological studies may reveal negative effects, where after higher tier studies are necessary. Expenses will depend on whether the microorganism is well-known in the literature or whether it is a newly-discovered organism with limited or no available data. The latter will make the building of a registration dossier much more expensive and time-consuming. The generation of an Annex II and an Annex III dossier will cost a certain amount. In contrast, the applicant's choice of the RMS determines the total administrative fees.

The costs for a product dossier for a typical greenhouse use, for one vegetable crop and one target pest, amount to about €50,000–100,000, without the administrative fees. The fee tariffs are sometimes made up by the fees for each (risk) assessment and it is complicated to ascertain what the total amount of fees will be for a product authorization. Each country has its own fee structure and finding out what the costs will be is quite difficult.

2.11 Data Protection

Reports of experimental studies as well as statements are considered data. Data protection is regulated in Article 13 of Directive 91/414/EEC. Annex II data are protected for 10 years following Inclusion in Annex I. When it concerns an "existing" microbial agent, data protection lasts 10 years from the date of the first registration in each member state; this was not harmonized, and was regulated per country. For new data on an "existing" active ingredient under the review programme, data protection lasts 5 years from the decision on Annex I Inclusion. For Annex III data, protection lasts 10 years from the first registration for new products (harmonized), and for old products, 10 years from the date of the first registration in each country (not harmonized). In the new Regulation 1107/2009 data protection is regulated in article 59. It will be 10 years, but for plant protection products based on low risk substances the period is extended to 13 years, thus also for most microorganisms. Once data protection has lapsed, national authorities should, upon request, inform applicants which data are unprotected. Reference to unprotected studies accelerates the building of dossiers for new products and reduces costs considerably. In the case of microbial agents and products this has hardly been done, and generic products (products based on the same strain) are not currently found on the market, in contrast to chemicals where this is commonplace. Moreover, national enforcement of data protection and track records of protected and unprotected data are often not in place. In addition, the provisions on general data sharing, the data sharing of vertebrate studies due to test animal protection, and data compensation are not clear. The use of unprotected data for generic products may upset relationships in the market which could lead to fierce competition and other undesired effects. In the field of biopesticides this is yet not the case.

Registration and the concomitant data package are considered protection from competitors in the field of MPCPs where patents are not always possible or not applied for (the possibility of a patent for a MPCA will be discussed below). Therefore, it is useful to know the data protection rules and consider them in the development of a MPCP. When the development of a biopesticide is conducted in collaboration with a research institute, data protection should be considered from the early phases of the research.

3 Entomopathogenic Nematodes and Regulations

3.1 Entomopathogenic Nematodes and Safety

The safety of nematodes to human health and to the environment has been the subject of many studies and publications. Overviews of this subject have been given by Bathon (1996), Ehlers and Hokkanen (1996), Akhurst and Smith (2002) and Ehlers (2003, 2005, 2006). Much information can also be found in presentations and papers given within the COST Action 850 "Biocontrol Symbiosis", a European network that organized the scientific cooperation in the field of nematode use for biological plant protection (www.cost850.ch). The safety of the symbiotic bacteria of the genera *Photorhabdus* and *Xenorhabdus* was reviewed by Boemare et al. (1996a, b) and Boemare (2004). Various animal toxicity studies have been conducted with the bacteria, independently from the nematodes, and the bacteria did not show any pathological effects on mammalian vertebrates. An overview of these tests is given by Akhurst and Smith (2002). There is one reported case of allergenicity to *Xenorhabdus*, but the risk of allergenicity is considered very low (Akhurst and Smith 2002). There is concern, however, about *P. asymbiotica* which is sometimes found in *Heterorhabditis indica*. This bacterium is an opportunistic human pathogen and can cause severe infections (Akhurst and Smith 2002; Gerrard et al. 2004). Therefore, this bacterial species must be avoided in nematode production and an accurate identification at the start of research on *H. indica* and its symbiotic bacterium is indispensible. In general, the nematode-bacterium complex is considered to be safe to human health and to the environment which allows production and use as a biocontrol agent. Nematodes have a long history of safe use and there are no reports of lasting harmful effects.

3.2 Entomopathogenic Nematodes and the EU Directive 91/414/EEC

Nematodes are subject to various regulations which are quite different from each other in European countries (Ehlers 1996, 2003, 2005; Richardson 1996; Loomans 2007). They are not covered by the European legislation on plant protection products (Directive 91/414/EEC). However, when the new contours of the revision of this Directive became clearer in 2005, it turned out that there was an attempt made by DG SANCO to include nematodes in the new EU Regulation (1107/2009) as active substances of plant protection products. In a draft amendment (SANCO/10159/2005, dated 6th April 2005) a proposal was made stating that: "For the purposes of this regulation, nematodes used in or as plant protection products will be assimilated to microorganisms". Individual companies that produce entomopathogenic nematodes, the IBMA, and scientists working within COST Action 850 objected to this intention. The arguments used were the following:

- EPNs are broadly regarded and accepted as macro-organisms and as such are also regulated in various countries in the EU and in other countries;
- EPNs are classified by the EPPO as macro-organisms in their EPPO Standards for "Safe use of biological products" PM 6/1-2. EPPO also included EPNs on the "List of Biological Control Agents" that have been widely used and are considered to be safe (PM 6/3(3)) (EPPO 2008). The EPPO list contains six species of entomopathogenic nematodes;
- The OECD considers EPNs invertebrate biocontrol agents and not microorganisms (OECD 2004d);
- The FAO also categorizes EPNs in their guidance documents (ISPM No. 3) under invertebrate BCAs and not as microorganisms (biopesticides) (FAO 2005);
- Many countries have adopted these guidelines and EPNs are often regulated according to these guidelines;
- In the USA, the EPA has exempted nematodes from any kind of registration.

From the above it is clear that EPNs are regarded as macroorganisms by many different authorities and that many of these authorities consider EPNs safe. If the EU regarded EPNs as microorganisms within the renewed Regulation on plant protection products, these BCAs would fall under both European legislation and various national regulations and would become over-regulated. All these regulations would have made it extremely confusing and very difficult to obtain approval for the use of EPNs. Later that year, in 2005, the EC agreed to withdraw the proposal to include nematodes within the scope of the new regulation (Agrow 2005).

3.3 Registration and Data Requirements in Various Countries

Several European countries have a regulation in place which requires a registration procedure for nematodes. Some, however, do not have any regulations for invertebrate biocontrol agents (IBCAs). A list of countries with requirements can be found on the website of the COST 850 Action under the subject "Legal & safety" (www.cost850.ch). A brief overview of data requirements for approval of the release of nematodes is given by Ehlers (2005) for European and various other countries. In general, a dossier has to be presented on safety to human health and to the environment. Critical aspects are non-target effects, dispersal, and persistence. Some countries, such as Austria and Switzerland, also require efficacy data. Requirements for countries outside the EU are reviewed in detail by Bedding et al. (1996) for Australia and New Zealand, and by Rizvi et al. (1996) and Akhurst and Smith (2002) for the USA. Details on British legislation regulating non-indigenous nematodes were described by Richardson (1996). The context of the Dutch legislation for IBCAs, including nematodes, is explained by Loomans and Sütterlin (2005). The authors compare and contrast the Dutch Flora and Fauna Act that regulates releases of native and non-native animals with international legal frameworks. This perspective illustrates the complexity and confusion of the legislation around the use of IBCAs. Recommendations for regulations of nematodes and harmonization were already made in 1996 by researchers and the biocontrol industry (Ehlers and Hokkanen 1996). It was concluded that there should be no regulation for native species. On the other hand, tailored regulation was recommended for non-native species. A set of data requirements was proposed and a harmonized approach was recommended. However, nematode regulation across Europe is still a country by country approach and varies considerably.

At the present time, Austria has the most elaborate registration requirements for nematodes in Europe. They entail information on identification, mode of action, dispersal and establishment, application, efficacy and side-effects. Details on production, formulation, contaminants, toxins, human pathogenicity, storage, and waste disposal are also required as well as a MSDS (Strauch 2004). Registration of an EPN species took more than 2 years and costs were over €20,000 (R.-U. Ehlers, personal communication). This is a considerable cost factor in the marketing of a biocontrol agent for a small market. Although this particular approval was for broad use, registration costs would be prohibitive if all countries required similar approval procedures. Belgium, Finland, France, Greece, Italy, and Portugal currently do not mandate registration of nematodes. Denmark, Germany, Hungary, Ireland, Spain and the UK do require a permit for non-native species. Usually, a dossier can be generated based on public information. Spain has implemented a new regulation for exotic IBCAs in 2007, but data requirements and criteria are unclear.

Registration of nematodes is often complicated by the fact that their release must comply with several different regulations, particularly for exotic species. Examples are wildlife acts, conservation acts, biodiversity acts, and quarantine acts, in addition to the regulations on release. The authorizing agency usually must ensure that their regulation complies with all these others, but sometimes the applicant must deal with this. This renders the design of procedures and decisions complicated and confusing.

3.4 The Struggle with Regulations for Import, Export, and Use of Macrobial Biocontrol Agents

I will present the Dutch situation at the end of 2009 as an example to illustrate the current struggle with regulations for IBCAs. Improvements and adaptations were still being carried out at the time of writing. The release of animals has been regulated by the Flora and Fauna Act since 2002. Without approval, animals cannot be released into the environment. However, at the time of implementation of the act, no measures were taken with regard to biocontrol and so from one day to the next, biocontrol was illegal, and growers were breaking the law. A procedure was then developed in 2005 to allow the continuation of the release of beneficial organisms for biocontrol. In order to release a new beneficial invertebrate organism, one needs to apply for derogation for release through submission of a dossier. Importation, on the other hand, is not regulated in this Flora and Fauna Act, but one cannot release animals without the approval. The dossier requires information on the identification, origin of the organism, and the purpose of the release. The second section asks information on the biology and ecology, which are the basis of a risk assessment in section three. The procedure follows the environmental risk assessment (ERA) developed for IBCAs (Van Lenteren et al. 2003, 2006). For native organisms, the effect of a mass release on flora and fauna near the point of release is evaluated. Generally, this is a transient and limited effect and not a concern. For non-natives, the focus is on potential establishment and effects on non-targets and a risk assessment must be presented. Costs of the procedure are €100, and the decision should made within 8 weeks. In 2008, however, decisions were about 2 years overdue. The "Application Form" (REBECA 2007a) and the "Guidance Document" (REBECA 2007b) developed within the REBECA project are used from 2009 onwards. The procedure is still in development and is not yet officially enforced by law. In 2005, six species of entomopathogenic nematodes obtained derogation for release based on a safe history of use (cf. EPPO criteria) and they were placed on a generic list.

In addition to the approval for release, the Netherlands have implemented the International Standards for Phytosanitary Measures (ISPM) No. 3 of the FAO (the first revision of the *Code of conduct for the import and release of exotic biological control agents* and thus regulates import and export (FAO 2005)). For importation into the European Community, veterinary certificates are needed according to the EU Act on veterinary checks on animals entering the Community (91/496/EEC (EC

1991c) and 97/78/EC (EC 1998a)). This act regulates border inspections of animals, regarding diseases and well-being, by a veterinarian. IBCAs fall under the broad category "other living animals". In the Netherlands this responsibility is delegated to the Food and Consumer Product Safety Authority ("Voedsel en Waren Autoriteit" (VWA) in Dutch). This organization has no expertise at all in the area of IBCAs. Criteria for inspection are unclear for beneficial arthropods, except for bees (92/65/EC (EC 1992)). A veterinary certificate from the exporting country is necessary as well. Some countries only provide phytosanitary documents while others furnish veterinary documents only. Phytosanitary certificates are accepted too and as long as the arthropods are not quarantine pests (according to 2000/29/EC (EC 2000b)), shipments will be released. An importation certificate should be released within 48 h. Once these documents have been made, customs handles of import taxes and inspection of safety of goods, after which shipments are handed over to the importer. For importation from EU countries, no border inspections are required.

For exportation within the EU, export certificates are sometimes required. For exportation outside the EU, phytosanitary and/or veterinary certificates may be required depending on the importing country. Veterinary certificates are provided by the VWA and must confirm that no animal diseases occur in the environment around the production facility and that the purpose of the goods is "biological control of plant pests". Phytosanitary certificates must state that the shipment contains no "harmful organisms". These certificates are provided by the Plant Protection Service ("Plantenziektenkundige Dienst" (PD) in Dutch). The main requirements are identity and purity. These need to be ascertained by production protocols, records of identification by experts, and QC data of the production process and the product. Twice a year, the production facility and the records are audited by the PD. From 2009 onwards a production facility can be certified based on their own in-house quality control system. Upon approval of the production facility, certificates are given for every shipment. The chance is great that all these import and export procedures will become more and more complicated.

Furthermore, with regard to air freight of goods the regulations on safety of civilian air freight require that only authorized personnel be allowed to work in the shipping department of the BCA producer. For export to the USA even stricter rules apply.

3.5 The Confusing Burden of Procedures

For many countries, it seems more important that procedures be followed and the "paperwork" be in order than that criteria be appropriate and proper inspection have taken place. There are few countries that really understand the matter in detail and have the expertise to evaluate the certificates properly. Proper inspections on the critical issues are hardly conducted due to the lack of expertise. As a general rule certificates are issued as long as procedures are set up and followed. With regard to all the required documents, shipping within the EU is the least complex. Exportation to many African and South American countries and Japan and South-Korea is the most difficult. Usually, customs are also involved which often means further delays. The above picture illustrates how complex and bureaucratic import and export of IBCAs has become. Paperwork takes time and increases costs, and causes delays in transports and releases of shipments. Certificates are issued within two to three work days. This last aspect is critical with products with a shelf-life of only a few days. These procedures also impact EPNs, although in general there is less concern with transport of EPNs than of arthropods.

The IBCA manufacturers and distributors face a number of problems that stem from these regulations. The administrative burden is increasing, and authorities often have no real expertise on this field, thus delays of shipments are common. This may jeopardize the quality of the products and timely releases at the users' sites. The design of the procedures and their adaptation to the biocontrol industry diverts many efforts from many stakeholders, and often the results are far from clear. Uncertainties and ambiguities are omnipresent and regulations keep changing. This has become a real burden and is difficult to manage within a small industry. Koppert, for example, exports EPNs to more than thirty countries, each with their own regulations. It is clear that there is a need for the harmonization of legislation and for the development of appropriate procedures and protocols, and procedures should be handled by one competent authority per country only. In the Netherlands the authorities responsible for veterinary checks and for phytosanitary checks announced in December 2008 that they will merge into one authority. Combining the veterinary and phytosanitary check into one certificate is also anticipated. This may facilitate procedures in future.

3.6 The EU Policy Support Action REBECA and Nematodes

Within REBECA nematodes were considered IBCAs. A separate working group of nematode experts discussed whether the ERA developed for insects and mites (REBECA 2007c) could be used for nematodes. The working group concluded that the same ERA could be used for nematodes, but data on host range, establishment and dispersal of EPNs "would not normally be required". The following recommendations on data requirements for nematodes were drawn up (REBECA 2007d):

- accurate identification;
- for *H. indica,* the absence of the bacterium *P. asymbiotica* should be ascertained;
- no further data are needed for indigenous species;
- non-indigenous species to be deposited in a recognized collection;
- data on origin, distribution and host range should be provided for exotic species.

An application form and a guidance document were developed in order to apply for approval for the import, shipment, rearing and release of IBCAs in European countries (REBECA Deliverable 22). The form refers to native and non-native organisms, and the information required can easily be supplied for nematodes when literature and data waivers are accepted. The information is required at species level, but information on a lower taxonomic level may be required where appropriate. Product information is requested, but the approval is given for the organism, not the product. So, many product formulations and product sizes may be based on this approval. The procedure is intended to be used in any EU country. In this way REBECA recommends a harmonized approach to the approval of the use of a new macrobial organism for biocontrol.

Furthermore, REBECA does not recommend developing a new and central legislation within the EU (REBECA 2007e: deliverable 22). DG SANCO concurs. But in this way, national authorities will continue to regulate IBCAs. In order to harmonize regulation, REBECA recommended that an Expert Group be established that will give advice on the release of new species. Adoption of the "standard procedure" as developed within REBECA is a highly desirable instrument to facilitate standardization. A joint EPPO-IOBC Panel has been created in March 2008 which adapted and modified the REBECA proposal and updated the procedure and criteria for updating the "positive list" (see IOBC Commission on "Harmonized regulation of biological control agents (CHIBCA)" at www.iobc-wprs.org). The adoption of the "Application Form" developed within REBECA as a standard is under discussion within EPPO. A species can be added to the "positive list" of EPPO when it is either indigenous and widespread in the EPPO region, or established and widespread in the EPPO region, or has been used for 5 years of use in at least five EPPO countries. The agents are listed on the basis of an expert judgement and when there is "sufficient knowledge to indicate the absence of significant risks, or the availability of reliable risk management measures" (EPPO 2008). Inclusion of new species should be evaluated by the Panel on a yearly basis. This regulatory process is still under development. Adoption of such a regulatory system by individual countries would be on a voluntary basis. Most biocontrol companies favour this option, rather than a central EU regulation.

3.7 Intellectual Property Rights and Data Protection Related to Nematode Registration

Protection of data is a critical issue in registration. When dossiers have to be submitted for evaluation by the authorities, companies want to see an independent approval system per applicant where R & D and registration efforts are protected. A generic approval of an organism after this has been cleared following the dossier evaluation of the first applicant leads to less R & D, unfair competition, and a loss in competitive advantage for the first applicant. It is often said that the registration requirement may create more difficulties for small companies than for larger companies, but this should be looked at on a product-specific basis rather than on a company-specific basis. A smaller company specialized in a certain BCA may have a greater market share with that particular beneficial than a larger company in which the product is minor. Although company strategies may vary, a product is generally valued on its own merits and in the end needs to be profitable. As a solution to duplication of efforts and evaluations, data could be shared between companies and submitted to authorities with a letter of access and an arrangement for data compensation. Furthermore, data protection legislation needs to be developed and implemented. This approach is also recommended by REBECA for IBCAs (REBECA 2007c). Approval could be made general, for instance, by the placing the species on the "positive list" of EPPO as mentioned above. A period of 5 years should be long enough for a company to recover the initial research and registration costs.

4 Exotic Organisms and Microbial Pesticides

4.1 Biodiversity Regulations

The collection, export and import, and the use of microorganisms (species or strains) may be subject to biodiversity acts. The Convention on Biological Diversity (CBD) (Rio de Janeiro 1992) has now been implemented by 190 countries (Bigler 2008). Biological control agents are more and more an issue, also within the context of the Access and Benefit Sharing (ABS) guideline that has been developed in 2000 (Biber-Klemm and Martinez 2006). In general, the CBD and ABS promote conservation of biodiversity and sustainable use of its components. At the same time, they regulate ownership and the sharing of economic benefits from genetic resources in a bilateral way. These regulations require researchers and companies to seek approval for the collection of genetic resources from the competent authorities, and to arrange benefit sharing from the potential use of the material. The focus is mainly on pharmaceutical products. The ABS, however, includes microorganisms and nematodes. Implications for biocontrol are not yet clear and discussions are ongoing to avoid inappropriate regulations from hampering or impeding biocontrol (Cock et al. 2009). Up to today research and the development of biopesticides has not been hindered by the CBD as far as I know. Within the Lubilosa project, however, a trust fund has been set up within the context of the CBD and royalties are shared between African countries (Lomer et al. 2001; EC 2002).

The ABS procedure will become implemented by countries and may make exploration and exploitation of organisms complex and liable to negotiations on "fair and equitable benefit sharing". The "IOBC Global Commission on Biological Control and Access and Benefit Sharing" has been established that will try to clarify concepts, terms, working definitions and sectoral approaches with regard to scientific research and utilization (IOBC 2008; Cock et al. 2009). Recommendations are provided to governments and to the CBD in 2010 for implementation of an ABS regime that ensures arrangements acceptable to all parties. For the moment it is difficult to predict the impact on biocontrol, but at some point in time the biocontrol industry will need to be involved in order to avoid an unworkable situation.

4.2 Registration and Exotic Organisms

The use of exotic species or strains of microbial entomopathogens is not specifically regulated within registration laws. There are no specific data requirements given in the guidelines. Regulators will generally require sufficient information on the host range, effects on non-target organisms, and on fate and behaviour in the environment for a non-native organism in order to be able to make a reliable risk assessment. If the outcome complies with the Uniform Principles an approval will be given.

With regard to nematodes, specific approval procedures need to be followed for non-indigenous species in some countries. This is discussed above. There is, however, no European-wide regulation on this aspect. Responsibilities laid down in ISPM3 are also applicable as explained above. EPPO has published to guidelines on safe use of exotic BCAs. The term "exotic" excludes organisms which are indigenous to the country. The first guideline (EPPO 1999) describes the required information with regard to the first import for research under contained conditions. The second describes the procedure for import and release of exotic BCAs (EPPO 2001). It provides guidance for the preparation of a dossier which will be evaluated by the competent authority. These procedures may be used on a voluntarily basis per country.

5 Safe Handling of Microorganisms

It is vital to know in detail the microorganism under study, from early discovery up to commercial use, since microorganisms may present challenges to health and safety of the people handling it. First of all, proper identification is needed, followed by an initial risk assessment based on the taxonomy of the organism and existing knowledge. Few studies have been published on exposure to MPCAs by workers and applicators. Strasser and Kirchmair (2006) reviewed the literature on this subject and concluded that unacceptable effects on human health have not been documented. Working with microorganisms is regulated by several legislations. National legislation for protecting workers should be followed as well as international legislation. Furthermore, Directive 2000/54/EC (EC 2000a) concerning the protection of workers from risks related to exposure to biological agents at work, is applicable. Classification on the basis of hazard is the first step, and only Group 1 species ("unlikely to cause human disease") should be accepted as biocontrol agents. An example where work with a microorganism was abandoned due to this supposition was a project where *Serratia marcescens* strains (Group 2 organism: "may cause human disease") had demonstrated potential insecticidal properties (De Kogel and Ravensberg 2006). Deliverable 12 of the REBECA project identified a Positive list of "low risk" MPCA candidates.

Several regulations apply to the handling of microorganisms such as collecting, packaging, shipping, and storage. These could be, for instance, postal and transport (by road and by air) regulations. An overview of these regulations and when they are applicable is given by Smith (2000). It is worthwhile to take notice of these regulations. Generally, Group 1 organisms are not subject to many of these regulations. Details on laws, regulation, and restrictions on transport of biological material are described by Weihs and Rohde (2005b). This mainly refers to transport in and between the USA and Europe by postal service and by air, including safe packaging instructions. With regard to transport by road, microbial biopesticides are generally exempt of ADR rules (The European Agreement concerning the International Carriage of Dangerous Goods by Road). Only biological products with microorganisms that can cause human or animal diseases are classified as infectious substances and need to comply with ADR protocols (www.unece.org/trans/danger/publi/adr.html). Regulations only tend to become stricter for transport and import as a result of the threat of bioterrorism (Foster 2007). ISPM3 also describes responsibilities of the exporter and importer for proper packaging, labelling and storage during shipment, although details are not given on requirements.

A material safety data sheet (MSDS) needs to be established as soon as the initial risk assessment has been conducted, and this should be updated once more data become available during the generation of the registration dossier. The MSDS is a mandatory requirement; it offers a synoptic overview of the most important health and safety aspects of the organism/product for professional users, including what to do in case of an emergency (2001/58/EC (EC 2001b)). The developer of a MPCP is liable for any harmful effects resulting from exposure to the product and he has a duty to inform others of the potential risks of the product when using and handling of the product is done according to conventional practices.

6 Intellectual Property Rights

Intellectual property rights can be divided in two categories of legal property rights: copyrights and industrial property rights. The latter comprises patents, plant variety rights, industrial design rights, trade secrets and trademarks. For a biopesticide manufacturer patents could be interesting. Many aspects, however, need to be considered in case patent application is pursued. Trademarks are easily required for a product name; the product name is registered and this offers protection. The other intellectual property rights (IPR) are not very relevant for a biopesticide manufacturer and are not further discussed here. Whether biological material is eligible for patent protection has been the subject of many debates. The patenting of naturally occurring organisms as such is not possible, but the patent law distinguishes between discovery and invention. Inventions can be patented, whereas discoveries cannot; the legal definition is still under debate and evolving alongside new technologies. For more details, I refer to Westerlund (2002). The patent law in Europe is laid down in the Convention on the Grant of European Patents of 5 October 1973, commonly known as the European Patent Convention (EPC). This is a multilateral treaty that laid the groundwork for the European Patent Organisation and provides an autonomous legal system according to which European patents are granted. Later, the Directive 98/44/EC (EC 1998b) of the European Parliament and of the Council of 6 July 1998 on the legal protection of biotechnological inventions was established. This Directive was meant to harmonize the laws of Member States regarding the patentability of biotechnological inventions. A broader overview of laws and regulations for the protection of biological material and its history is provided by Fitzner (2005).

6.1 Patentability of an Entomopathogenic Microorganism

Novelty, inventive step and industrial applicability are the criteria for patentability. These three criteria are briefly discussed in relation to entomopathogens:

1. *Novelty.* The mere discovery of an organism is not patentable. If the microorganism (species or strain) has been isolated and selected by a specific method for a particular purpose, however, and it is properly identified and characterized, and its specific features for the intended use are newly described, then it may be patentable. A sample of the organism must then be deposited in a recognized culture collection (37 authorities). These are listed in the 1977 Budapest Treaty on the International Recognition of the Deposit of Microorganisms for the Purpose of Patent Procedure (Budapest Treaty 1977). This treaty stipulates that only one deposit is necessary to cover protection in the 72 contracting countries. Details on the procedure of the deposit of biological material are described by Weihs (2005a);
2. *Inventive step.* The invention must be new and non-obvious to the skilled person. Not only should it be new, it should have an element of innovation, a novel use. This is often a difficult requirement in patent evaluation. If anything of the invention has been disclosed, it is considered to be "state of the art" and not new anymore, and therefore not patentable. If it is a new use with something that already exists, this may also be patentable. A new use of a known microorganism could thus be patentable;
3. *Industrial applicability.* This criterion is usually not so hard to satisfy. It must be "technically applicable" and be made for or used in any kind of industry, including agriculture.

6.2 Patent Application

Inventors have three options for a patent application. This can be submitted nationally, to a national patent office; to the European Patent Office (EPO), which comprises 34 contracting states, for a European patent; and as a PCT (Patent

Cooperation Treaty) application which covers 94 countries. If the EPO grants a patent, it still must be registered in each designated country. A PCT application covers all participating countries, but eventually one has to choose the countries, and patents are applied for per country and granted per country. Patents are nationally granted and are essentially independent nationally-enforceable and nationally-revocable patents.

The date of application to the patent authority, often called the filing date, is generally the date for determining whether the invention is new. If a patent application is a first filing, the filing date is generally the same as its priority date. Until publication of the application, generally after 18 months, the patent is confidential. An agreement on confidentiality must always be signed when material and information is shared with and disclosed to other parties before the publication of the patent application.

6.3 Patent Protection Period

Patents are granted for a period of 20 years, from the priority date on. This period can only be extended for a maximum of 5 years (Regulation (EC) No 1610/96 (EC 1996)) in the case of medicinal drugs and plant protection products. The patent term extension has been established because product development and registration procedures consume a large proportion of the patent period, and the protection period in the market is largely shortened by then. A supplementary protection certificate (SPC) can be requested on a national basis only, and has to be applied for within half a year following official authorization to release the product on the market. Market exclusivity, however, cannot exceed 15 years. A SPC comes into force only after the corresponding patent expires. It is essentially an extension of the lifetime of a patent, giving a company a better chance on return of investment.

6.4 Patents on Microbial Pest Control Agents

A useful publication on the patenting of microbial insecticides has been produced by Cresswell (1997) that treats the principles of the patenting of microorganisms. He also mentioned the differences between applying for a patent in Europe and in the USA. Many patents have been granted concerning biopesticides. The subject of the patents generally is the method of production, formulation, or use of the product, or combinations thereof. An overview of patents up to 2000 in this area has been presented by Montesinos (2003). He studied 215 patents on strains of microorganisms with potential as a biopesticide or plant growth promoter. Most patents were deposited in the USA and owned by research institutes. The majority concerned bacteria and fungi, but few patents were granted for nematodes and viruses. The number of patents does not, however, reflect the number of registered products. I have not executed a detailed search on patents since Montesinos' search, but the situation has not really changed since 2000. Companies or research institutes expect to receive benefits from patents, such as a competitive edge, or income from patent licenses. In few cases are royalties paid for isolates, however (Cherry and Gwynn 2007). And almost none of the microbial control products currently on the market are protected by patents (Gelernter 2007). A small number of biocontrol manufacturers do have patents on their production and/or formulation technology, mostly concerning fungi and nematodes.

6.5 Considerations Regarding the Patenting of Microbial Pesticides

The advantages and disadvantages of applying for a patent need careful consideration (Table 5.4). Several aspects could be advantageous to a company. Patents are property and contain a certain financial value in the market. They may give a company a unique market position. Other parties are prohibited from using the protected technology or product for a long period. Licensing the technology to others may generate extra income. They may also strengthen a company's negotiation position in several situations.

There are also a number of disadvantages to patenting. All details of the invention must be fully disclosed. Competitors are able to track the inventions that are being made and what kind of products may be marketed in the near future. The costs of filing a patent and of maintaining it for 20 years can be considerable. In the first years there are high application costs. After the patent has been granted, yearly increasing taxes have to be paid during the full patent term. For one EU country the total costs amount to about €20,000; for a European patent this is approximately €50,000; it can accrue to over €300,000 if a world wide patent is pursued. Countries have their own policies in granting patents and some countries do not allow patents on crop protection as is the case in Poland. This makes the whole process complex. Infringement of the patent needs to be monitored by the patent-holder himself and he is responsible for bringing the violator to court. Disputes that need to be solved in

Table 5.4 Considerations on when to or not to apply for a patent

Apply for a patent when:	Do not apply for a patent when:
• Competitors can be prevented from using this technology or product	• Cost are higher than benefits
• A large market exist	• A short product life cycle is foreseen
• A long product life cycle is expected	• The invention is easy to circumvent
• It defends R & D advances	• Violation is difficult to prove
• The invention cannot be kept secret	• The invention can be kept secret
• It is key technology of the company	• The country patent system is weak
• It delivers market position advantages	• Infringement procedures are too expensive
• It influences market price negotiations	• A small market and margins exists
• It offers the potential to sell knowledge	• A head start in time is sufficient
• It strengthens the financial position	• Registration offers enough protection

court can be long-lasting and expensive. There are also other disadvantages that are beyond the scope of a single manufacturer. Concern about the freedom of research in relation to entomopathogens is discussed by Gelernter (2007). A related example is the debate over the use of neem products for crop protection (Kocken and van Roozendaal 1997).

A patent on a microorganism or a product formulation is enforceable, but a patent on a certain production technology is difficult to evaluate and check. Sometimes it is better to keep production methods secret, and apparently most manufacturers follow this approach. Small changes in processes may negate the violation of a patent, whereas full disclosure of the invention may give the other party valuable ideas. These are critical aspects to consider before a patent is applied for. Moreover, registration offers considerable protection as does a registered brand name. Applying for a patent needs careful consideration before any time and money is spent. Lisansky (CPL 2006a) considered IPR of minor importance to the success of a biopesticide company and stated that "Companies should have patents on relatively few key aspects of their technology and should focus instead on having know-how". Further, he warned that "It too, feeds expectations by giving the illusion of value". When patents are held by a research institute, a license can be obtained by a biopesticide producer in exchange for a royalty payment arrangement. This can be an exclusive or non-exclusive license, and in this case careful considerations need to be made to decide whether all the arrangements and costs are worthwhile.

In one case, the advantage of a patent on the nematode *Phasmarhabditis hermaphrodita* held by Becker Underwood (WO 93/00816) has clearly led to a monopoly position in the market. The discovery of the nematode in combination with its symbiotic bacterium and the novel use against slugs appeared to be patentable. For this reason other companies have refrained from producing and selling this nematode. To my knowledge, this is the only example where a patent has kept others from becoming active in a certain field, where alternatives have not been easy to develop. In a similar case, the University of Florida holds a patent on an isolate of *S. scapterisci*, and an exclusive license for production and marketing has been given to Becker Underwood.

7 Conclusions and Recommendations

7.1 Obstacles and Improvements in the Registration of Microbial Pesticides

The implementation of 91/414/EEC was supposed to bring about a harmonized system for the registration of plant protection products within the EU. In reality it has not improved registration, at least not for MPCAs. An average time period of more than 7.5 years for the Inclusion on Annex I of a microbial active substance is not acceptable for any stakeholder. The wait for return on investment is in many cases too great a challenge for a small biocontrol company; they either go bankrupt or refrain from starting a registration procedure. Research results are not developed into products, or products are sold illegally. The procedures for national authorizations are also complicated, long and costly. It may take more than 10 years from the submission of the application for approval of an active substance at EU level to the launch of the product on the market in a number of countries. From the outset of the development of the product until actual sales, the situation in the market is likely to have changed considerably. In the meantime the pest problem may have changed and/or other control products may have solved it by then. The example of Preferal for control of whitefly where registration took more than 10 years illustrates this problem (Ravensberg and Sterk 2004). What is left of the market for which the product was initially developed? This is a great risk that is worsened by the lengthy registration period.

My perspective on the six greatest obstacles in the registration process and suggestions for improvement are summarized below.

7.1.1 Many Data Requirements Are Not Appropriate for Microbials

The data requirements for metabolites form the largest hurdle. The lack of a proper definition of a relevant metabolite and the lack of expertise in this area leads to unnecessary concerns, and regulators tend to over-ask data on metabolites. According to the guideline, data on metabolites need only to be provided when unacceptable human health and/or environmental effects are likely to occur (Section 2.8 of 2001/36/EC). However, the biocontrol industry does not work with these kinds of organisms, but only with Group 1 organisms ("unlikely to cause human disease") (2000/54/EC). Still, most regulators require detailed information on metabolites. With the current data requirements it is hard to fulfil the analysis of metabolites. For a new MPCA where information on metabolites may be null, costs for a full metabolite analysis will be a serious limit to further development. Two EU funded research projects BIPESCO (ww.bipesco.com) and RAFBCA (www.rafbca.com) investigated the effects of metabolites of fungal biocontrol agents on residue and environmental effects. Results indicated that metabolites pose no risks and that the risk assessments could be simplified (Strasser and Pernfuss 2005). These project results were not yet considered by DG SANCO or EFSA. For the assessment of metabolites, the REBECA project recommends studying crude extracts for the toxicological and genotoxicological assessment (Strasser et al. 2008). Clear definitions and standardized test protocols coupled with a guidance document need to be developed.

The production of generic documents for the main groups or species of entomopathogens could facilitate the writing of registration dossiers and their evaluation. An efficient example is the OECD consensus document on the safety of baculoviruses. It should be accepted that core data will be provided at species level. Obviously, strain-specific information will still be required. At the same time, it would be helpful to all stakeholders if accepted waivers became public to prevent duplication of work and this would help harmonization of decisions by regulators. Furthermore, a tiered system needs to be established.

The REBECA project reviewed most deficiencies and ambiguities in the data requirements and the OECD BPSG has taken many of these items up into their work package for the next 4 years (OECD 2008). This suggests a revision of 2001/36/EC (EC 2001a) is necessary.

7.1.2 Guidance Documents for Applicants as Well as Regulators

Applicants have indicated that more guidance documents are needed to address the specific features of microbials. Also, risk assessors have indicated that they need more tools for a proper risk evaluation. Mensink and Scheepmaker (2007) designed risk decision trees and summary tables as tools to distinguish acceptable from unacceptable risks. These tools should facilitate and harmonize environmental safety assessments for regulatory purposes. In the Dutch Policy Note on Sustainable Crop Protection (LNV 2004) it is stated that the government is committed to improving the European harmonization on crop protection products and to developing guidance documents with regard to the evaluation of those products, with special reference to metabolites and plant strengtheners. This illustrates that governments recognize the lack of proper guidance. The OECD BPSG planned to develop new guidance documents and to consider including recommendation from REBECA (OECD 2008). Perhaps it would be better if EFSA as the evaluating authority would develop such documents and that these would be enforced as legal guidelines for the EU rather than OECD guidance documents.

7.1.3 Testing Methods Adapted to Microbials

The existing test protocols (OECD, EPPO and others) are developed for chemical substances. As an example, testing shelf-life according to the protocol for accelerated storage stability at 54°C makes no sense for a microbial. New or properly adapted methods have to be developed for microbial products. This process needs to be taken up by the OECD in consultation with IBMA members.

7.1.4 Regulators Lack Experience with Microbials

Regulators have only dealt with a limited number of microbial dossiers and have not been able to build experience. Many are not educated or trained in this area. Often, new and young employees fill this position; sometimes external experts are hired for short periods. This does not allow the regulators to focus on microbials and to get acquainted with the specific field, and learning is slow in this way. In contrast, the BBPD of the EPA has regulators that only deal with microbials. A general European agency for evaluation of microbials could be a solution, within EFSA for instance. Another idea is the appointment of a few specialized MSs that will carry out this task for the EU, just like the appointment of two lead Rapporteurs in the fourth list re-registration programme for microorganism. There needs to be enough confidence among regulators to be able to make this work.

7.1.5 Procedures Are the Most Serious Obstacle

The procedure for Inclusion on Annex I is complex and is too long. First, the RMS performs the completeness check (up to 6 months). Then, after the RMS has produced a DAR (1 year), the other twenty-six MS, the EC and EFSA make comments and several experts' review rounds take place (2–3 years). The applicant has the opportunity to reply to comments twice. Finally, EFSA writes a conclusion and the EC makes an official decision which it publishes. This can take many years, deadlines are not always met, and the various steps are not very transparent to an applicant.

Provisional national authorization may be given after completeness check, but many MS do not accept this and wait for the DAR or even until Annex I Inclusion. Many countries have their own administrative requirements in their language and procedures are not very transparent. Evaluation can take years and more data and explanations can be requested. All regulators face a log-jam of dossiers and biopesticides are not treated in any kind of fast-track system. Fees are high in most cases and differ considerably per country. Often it is hard to find out the amount of the fees. Fees structures should be harmonized; a reduction in fees for biopesticides will promote more registrations. The initiative of Belgium and the UK should be followed widely in the EU. An option may be to replace fees with a tax on sales.

In my experience, the biggest hurdle to the commercialization of biopesticides is the process of national registrations. It takes great effort, some years, and a considerable amount of money to obtain approval in ten countries, for instance. I estimate that the total expenses of Annex I Inclusion and approval in ten countries could reach approximately €2 million. One starts with the registration procedure, but the timing and amount of expenses of approvals is unknown. This "black box" system must be greatly improved. This could be done by a wide implementation of mutual recognition, by the establishment of zonal registrations, and by work-sharing between countries. In my opinion there is only one way out of this complicated process of national registrations and that is to create a centralized European authority that will evaluate registrations in an efficient way. EFSA could execute this role, for instance. National registrations should accordingly be a short procedure. The EPA in the Unites States is a model where federal approval is first given, and then state approvals are easily obtained. The EPA also has a special branch (the BPPD) that assesses biochemical products, including microorganisms. It is obvious that such a set-up allows for an accumulation of experience with biopesticides. As a result, risk assessors understand this field very well. From my personal experience working with both the EU and the USA systems, it is clear to me that experience with biopesticides is a key aspect of proper and timely registration. Unfortunately, the European system does not allow for this since the few applications over the last 30 years were scattered among all the countries.

It is a political decision to change the current system and design a new system. This is a complicated matter and why it has failed until now has been elaborately described by Chandler et al. (2008). Apparently, the young biocontrol industry lacks a strong supporting policy network, and therefore it has not been able to sufficiently influence decision-makers. Unfortunately, there does not seem to be much support for a central European agency at the present time. Even REBECA did not recommend a central authority given its poor prospects. I consider this a missed opportunity since the REBECA project was supposed to provide "scientific support to policies". The project clearly concluded that the current legislation is severely limiting registration of new products and that data requirements need to be adjusted. This does, however, not improve the difficult process of national product authorizations and I believe that only a new legal framework can improve this.

If the current system remains in place, the hope of the biocontrol industry is vested in the revision of the EU law as laid down in the new Regulation (EC) No. 1107/2009 which foresees improvements such as zonal registrations and mutual recognition. This must become a well-functioning system in order to improve the situation. Within countries, special divisions within the regulating authorities should be set up to assess the biopesticides in order to build experience and confidence with these products. Deadlines must be met and costs reduced. If the system were more transparent and open to communication, this could lead to predictable timeframes which are essential to companies with regard to return on investment. With improved transparency at every step of the evaluation process, companies are more likely to understand the system and to accept delays and costs.

The EU registration process took more than 5 years for chemicals as well. EFSA and the EC have recognized the need for improvement and in January 2009 a new body was established, the Pesticide Steering Committee, to manage the overall process. Improvements should be made in efficiency, work-sharing, communication, meeting deadlines, procedures and guidance documents. Unfortunately, there is no specific mention of biopesticides (EFSA 2008).

7.1.6 The Biocontrol Companies Are Small and Inexperienced

Companies should also become more self-critical and improve their ways of working as well as the quality of their dossiers. Employ people with experience in registration of biopesticides; when needed, hire a consultant who knows the procedures and the details and who has easy access to regulators. The extra expenses will pay for themselves via a quicker registration. Ensure that the dossier is complete, and use communication with the authorities as much as needed. It is unwise to start discussions on the better safety profile of biopesticides in comparison to chemicals; this has little impact on regulators (CPL 2006b). On the other hand, regulatory authorities should reach out to the applicant and offer assistance. The appointment of a contact person as the CRD (PSD) has done with the "biopesticide champion" deserves imitation by all countries.

The REBECA project also identified the main obstacles in the registration process and these are similar to the ones I described above (REBECA 2007f). Within the EU, the new registration process in accordance with 91/414/EEC and the scientific peer review of the evaluations prepared by MSs is considered as follows: "overall the programme has been extremely successful" (Fay et al. 2007). No

Table 5.5 Recommendations for improvements in the registration of microbial pesticides

Data requirements	Regulatory aspects
• Develop appropriate data requirements: revision of 2001/36/EC • Develop appropriate guidance documents • Design a tiered approach • Establish category of low risk substances accompanied by relevant data requirements • Generate guidance documents on accepted data waivers and statements • Generate generic dossiers of the main species or groups of species • Develop specific testing methods adapted to microorganisms and microbial products • Allow for tests conducted by the producer, non-GLP • Adapt efficacy criteria and allow for a broader extrapolation across crops • Establish clear defined end-points	• Establish a special European group of regulators with experience with microorganisms • Organize training workshops for regulators when new developments take place • Develop harmonized procedures • Develop a harmonized fee system • Reduce fees • Establish and keep to known and short decision periods • Improve transparency of evaluation process, decision periods and costs • Provide guidance and support system to applicants ("biopesticide champion") • Install pre- and post-submission meetings with applicant • Optimize mutual recognition system • Develop harmonization worldwide via OECD

doubt, major improvements have been made in terms of joint reviews, new guidance documents, harmonization and administrative support to the programme by the ECCO-Team. This may be true for chemicals. Only a fraction of the more than thousand active substances were microorganisms and the process of evaluation of microorganisms is still problematic, complex, and lengthy. The above illustrates that there is much work to be done before this programme can be called successful for microorganism. Full commitment from regulators, scientists and the industry is required to improve the registration of microbials. A new and separate registration process and more appropriate data requirements must be developed: innovative products require innovative regulations. Suggestions for improvements are listed in Table 5.5.

7.2 *Recommendations for a New Registration Procedure for Nematodes*

Currently, there is no uniform regulation of nematodes across EU member states; the use of EPNs is regulated at the level of individual member states. The variation in existing regulations and procedures is unworkable for the biocontrol industry. It also creates confusion and lack of credibility for regulation when neighbouring countries have different regulations in force, or even no regulations. There is a need for a harmonized registration of entomopathogenic and molluscidal nematodes. Proposed steps in the set-up of a new regulation are given in Table 5.6 (see

Table 5.6 Proposed steps for a simplified and harmonized regulation for entomopathogenic nematodes

Data requirements	Regulatory aspects
• Establish limited set of data requirements • Develop simple test methods • Demand data requirement at species level only, not at strain level • No requirements at product level • Accept data waivers and statements • Exclude presence of P. asymbiotica • Exclude efficacy data requirements • Only notification for native species • Environmental risk assessment exclusively for non-native species • Tests conducted by producer, non-GLP • Establish end-points • Pre-submission meeting to establish required data package	• Develop harmonized procedures • Develop guidance documents • Include all stakeholders • Low procedural costs • Known and short decision periods • Each applicant must supply its own dossier or have a letter of access referring to another dossier • Develop data protection legislation • Establish zonal system and mutual recognition system • Update the EPPO "positive list" every 5 years • Harmonization worldwide via OECD • Import/export inspections: minimal for in vitro-produced nematodes

also Ravensberg 2004). Similar conclusions were drawn up by scientists and the industry during a round table discussion within the IOBC WG "Insect pathogens and entomopathogenic nematodes" in 2003 (Blum et al. 2003). The procedure should be the same for every applicant and data protection legislation needs to be in place. The procedure should not be expensive and decisions should be made within short time periods. Nor should it become a barrier for research, commercialization, and release of new species.

The recommendations from the REBECA project are a result of discussions between regulators, scientists, and industry. This does, however, not mean that they are based on full consensus on all requirements. Whether or not these recommendations will be implemented on a national level or a European level, and to what extent, is uncertain at the present time. Were that so, it would be a voluntary system. There are currently no concrete plans for this. If the recommendations are accepted, it would mean that some of the requirements would have to be withdrawn by a number of countries such as efficacy data. It remains to be seen whether countries are willing to do that.

The IBMA and scientists, and organizations like EPPO, IOBC, OECD and the EU must all work together to harmonize legislation and make registration simpler and workable, within Europe and in other countries. Test protocols and other guidance documents need to be developed for data generation, for instance, by the OECD, the IOBC and the IBMA. The EPPO should update its "positive list" and continue to evaluate new species for placement on the list as recommended in deliverable 21 of REBECA. In 2008, following the REBECA recommendation, the Joint EPPO/IOBC Panel on Biological Control Agents has been re-activated and the "Positive List" has been revised (EPPO 2008). The Panel intends to regularly update the lists. If regulation of nematodes remains a national process, ideally some umbrella organization will take the lead in the procedure in order to receive commitment from countries that are not familiar with IBCA regulations at the present. The building of a network and of confidence between regulators will be helpful to the broad acceptance of this new approach. The approval process should be similar to the one envisaged for other IBCAs.

Since nematodes are different from the beneficial insects and mites, nematode experts should be asked to join the by REBECA proposed EPPO/IOBC Expert Group in case new nematodes are applied for. So far, only one nematode specialist participates in the EPPO/IOBC Panel. It will be beneficial to the industry if nematodes are regulated all over Europe within a harmonized system of application and evaluation. Dossier development will become easier if one standard dossier can be used, and costs will be acceptable. Fees should be kept low and decision-making within short time frames. Officially approved use of nematodes will favour the image of biocontrol.

Another unsolved problem is the import and export regulations and the required certificates for export outside the EU or Europe, and for import into the EU. When the REBECA approach is accepted by a broader range of countries, this will most likely facilitate these procedures. This requires further action from world-wide operating organizations such as OECD, FAO, IOBC and IBMA. All stakeholders should be included and participate actively. The IOBC Commission on "Harmonized regulation of biological control agents (CHIBCA)" has adopted as their task the periodic updating and harmonization of regulation beyond the two years of the REBECA project and in the whole IOBC WPRS region, including countries that did not participate in the EU project REBECA (Bigler 2007). I recommend that this commission take the lead here, together with the IBMA.

7.3 New Regulations Which May Affect the Registration and Use of Microbial Pesticides

The EU has revised 91/414/EEC, and replaced it with a Regulation that will regulate the placing of PPP on the market: Regulation (EC) No 1107/2009. The new legislation must be implemented by the Member States by 14 June 2011. Principal objectives of the new law are to maintain a high level of protection of humans, animals and the environment, to reduce administrative burden linked to inclusion and authorization, and to increase harmonization between MS. The Regulation also promotes the use of IPM and non-chemical methods in recital 35: "In order to ensure a high level of protection of human health and the environment, plant protection products should be used properly, in accordance with their authorization, *having regard to the principles of integrated pest management and giving priority to non-chemical and natural alternatives wherever possible*". These changes may influence the registration and use of MPCAs/MPCPs positively. Other aspects that may facilitate the process for microbials are a positive list of safeners and synergists, a negative list for co-formulants, and simplified procedures for basic substances ("foodstuff") and a category of low risk substances and products. Further, obligatory mutual recognition of PPP authorizations granted by MSs in the same zone will be implemented. There will be three ecological zones and a single zone of contained environments: for greenhouses and for post-harvest storage facilities. This will significantly reduce procedural burdens and accelerate product releases into a larger market. Decisions would be taken within 90 days.

Low risk substances and products will be defined, and their registration will be for a longer period (15 years instead of 10 years standard). A fast track authorization and a reduced dossier is proposed. Initially biopesticides would not qualify for this category due to the fact that they are considered sensitizing and persistent. In a later version of the proposal, however, the exclusion criterion referred to "sensitizing chemicals" only. It seems that the lobbying activities of the IBMA have paid off in this case. A provision of comparative assessment and substitution of products may be installed with a view to encouraging the substitution of dangerous substances by safer alternatives, which could stimulate biopesticides.

Another new EC legislation is Directive 2009/128/EC (EC 2009b). The aim of this Directive is "to achieve a sustainable use of pesticides" and it covers, among others, the development of national action plans to reduce risks of and dependence on pesticides, and to promote IPM. Furthermore, biological control measures will be included in training programmes. This Directive is likely to promote the use of microbials and other non-chemical pest control products. Member States shall bring into force the laws, regulations and administrative provisions necessary to comply with this Directive by 14 December 2011.

A new Regulation 1185/2009/EC (EC 2009c) has been adopted on collection on statistics of sales and use of pesticides for assessing policies and risks related to pesticides. Currently, there are no official figures on the scope of use of biological plant protection products and this keeps politicians from making decisions in this area. The biopesticide industry has been reluctant to release any data because of competition. Policymakers need to have insight into the relevance of biocontrol in order to be able to justify support to the biocontrol sector. It will become obligatory to provide data to the EC on the use of biopesticides. This will ultimately benefit the biocontrol industry when official data demonstrate the need to facilitate registration and promotion of IPM.

7.4 Snake Oils and Illegally Sold Microbial Pesticides

Some companies refrain from seeking official approval thanks to the difficult and lengthy registration procedures and start selling products without registration. In almost any country in Europe dozens of unregistered biopesticides are sold by a conservative estimate. Often these products are of low quality and efficacy may be minimal. Safety is also a concern when these products are used without proper advice and knowledge. As a consequence, growers may develop a negative attitude towards biopesticides in general, which serves no one's interest. An easier registration system may encourage more companies to register their products and to advertise them openly. The current registration system demands an enormous amount of resources, while very little attention is devoted to enforcement of regulations. More efforts are needed in the monitoring and control of PPPs in general, but certainly of illegally used biopesticides also. There is a vital role for national enforcement agencies; however, all stakeholders should participate in this effort and fight this problem.

7.5 Position and Lobbying Actions of IBMA

IBMA has been very active with regard to registration issues since its foundation in 1995. In fact, one of the common goals of the member companies was to improve and facilitate registration. In the first congress of the organization in 2003 "Bringing science to practice" in Beziers, France, legislation was one of the main topics. The EC was represented, and it was recognized that registration was a major hurdle for commercialization and that methods and guidance needed to be developed for companies as well as regulators. However, progress in that direction has been very slow, hence the situation presented above. Lately, the IBMA has undertaken lobbying activities with regard to the proposal for the new PPP regulation. The main objective was to keep the option for a separate and specific regulation for BCAs open by changing the current text of article 2.2: "This Regulation shall apply to substances, including microorganisms and viruses, having general or specific action against harmful organisms or on plants, parts of plants or plant products, … ." with the addition: "*It shall, however, cease to apply to microorganisms, viruses, pheromones and biological products once a specific regulation on biological control products has been adopted.*" Unfortunately, this has been unsuccessful, the new Regulation does not mention anything on a specific regulation for biologically-based products.

The IBMA also asked that MPCAs be exempted from inappropriate exclusion criteria for low risk substances (art. 22). Two criteria, sensitizing and persistence (half life in soil > 60 days), exclude them. MPCAs should benefit from the favourable status of low risk substances. The criterion "sensitizing" has been changed and is only referring to "sensitizing chemicals", but the persistence criterion remained and this may exclude some microbial control agents for this low risk category.

The IBMA is also actively contacting members of Parliament, officials in DG SANCO, DG ENVIRONMENT and others. Contacts at member state level are approached to try to increase support for the position and wishes of the biocontrol industry, and to favour the position of BCAs in general. For instance, the IBMA UK has developed a good working relationship with the CRD (PSD) that helps applicants to prepare submissions (Whittaker 2007). Through the foundation of IBMA, a representative body of the biocontrol industry has been established that has become a contact for interactions and negotiations with official organizations, and a spokesman for the companies. This has been very helpful and IBMA must continue to play an active role in improving and promoting the position of biological control. Members should be willing to commit resources to this purpose that will benefit all stakeholders.

7.6 Future Outlook

After more than five decades of slow progress in the adaptation of registration for microbial pesticides, the situation is gradually improving. The EU REBECA Action has brought together regulators with scientists and the biocontrol industry representatives, and a new dialogue has been established. All seem committed to improving the situation and ready to work actively on solutions. The work programme of the OECD BPSG is very promising in this respect (OECD 2008). The IBMA is also actively expanding its influence on politicians and decision-makers. The role of farmers is still too limited and it is up to the biocontrol industry to engage them more and to encourage them to exert political pressure. The same is true for the supermarkets. With GlobalGAP, a private sector body of supermarkets that sets voluntary standards for the certification of agricultural products, IPM is becoming more and more standard procedure (www.globalgap.org). This will ultimately offer more opportunities for biopesticides. New regulations in the EU will put chemical pesticides in a more difficult position. Ideally, biopesticides will be promoted more and registration eased. All stakeholders must remain active in this field and continue to assign resources to this task. Changes in regulatory processes tend to be slow, but I believe that the future for biopesticides in terms of registration is promising.

References

Agrow, 2005. EU revises registration proposals on nematodes. *Agrow* 476, July 22, 2005: 9.

Akhurst, R. and K. Smith, 2002. Regulation and safety. In: R. Gaugler (ed), *Entomopathogenic nematology*. CAB International, Wallingford. pp. 311–332.

Andermatt, M., 2006. Virus. *Paper presented at the REBECA workshop on risk assessment of microbials*, April 12–13, 2006. Innsbruck, Austria. www.rebecea-net.de. Cited November 20, 2008.

Bale, J.S., J.C. van Lenteren, and F. Bigler, 2008. Biological control and sustainable food production. *Phil. Trans. R. Soc. B* 363: 761–776.

Bathon, H., 1996. Impact of entomopathogenic nematodes on non-target hosts. *Biocontrol Sci. Technol.* 6: 421–434.

Bedding, R.A., S. Tyler, and N. Rochester, 1996. Legislation on the introduction of exotic entomopathogenic nematodes into Australia and New Zealand. *Biocontrol Sci Technol.* 6: 465–475.

Betz, F.S., 1986. Registration of baculoviruses as pesticides. In: R.R. Granados, B.A. Federici (eds), *The biology of baculoviruses. Vol. II. Practical application for insect control*. CRC Press, Boca Raton. pp. 203–222.

Biber-Klemm, S. and S. Martinez, 2006. Access and benefit sharing. *Good practice for academic research on genetic resources*. Swiss Academy Science, Bern.

Bigler, F., 2007. Commission on "Harmonized regulation of invertebrate biological control agents (CHIBCA)" activity report 2003–2005. *IOBC/WPRS Bull.* 30(3): 63–65.

Bigler, F., 2008. The Presidents page. *IOBC/WPRS Profile* 45: 2.

Bigler, F., J.S. Bale, M.J.W. Cock, H. Dreyer, R. Greatrex, U. Kuhlmann, A.J.M. Loomans, and J.C. Van Lenteren, 2005. Guidelines on information requirements for import and release of invertebrate biological control agents in European countries. *Biocont. News Inform.* 26(4): 115 N–123 N.

Blum, B., R.-U. Ehlers, S. Haukeland-Salinas, H. Hokkanen, K. Jung, U. Kuhlmann, I. Menzler-Hokkanen, W.J. Ravensberg, H. Strasser, P. Warrior, and M. Wilson, 2003. Letter to the editors – biological control agents: safety and regulatory policy. *BioControl* 48: 477–484.

Boemare, N., 2004. Safety of nematode associated bacteria. *Paper presented at the COST Action 850, WG 3 and 5: Evaluation of risks related with the use of EPN and discussion on requirements for registration of invertebrate biocontrol agents,* 5–7, November 2004. Kiel, Germany, www.cost850.ch. Cited September 20, 2008.

Boemare, N., C. Laumond, and H. Mauleon, 1996a. The entomopathogenic nematode-bacterium complex: biology, life cycle and vertebrate safety. *Biocontrol Sci. Technol.* 6: 333–345.

Boemare, N., R.-U. Ehler, A. Fodor, and A. Szentirmai, 1996b. COST 819 Entomopathogenic nematode – symbiosis and pathogenicity of the nematode-bacterium complex. Office for official publications of the EC, Luxembourg, EUR 16727 EN.

Brownbridge, M. and T. Glare, 2007. Impact of entomopathogenic fungi on soil-dwelling invertebrates. In: S. Ekesi, N.K. Maniania (eds), *Use of entomopathogenic fungi in biological pest management.* Research Signpost, Kerala, India. pp. 295–312.

Budapest Treaty, 1977. Budapest Treaty on the International recognition of the deposit of microorganisms for the purposes of patent procedure Budapest Treaty (1977) (Budapest Union). www.wipo.int/treaties/en/registration/budapest/index.html. Cited November 12, 2008.

Burges, H.D., 1981. Safety, safety testing and quality control of microbial pesticides. In: H.D. Burges (ed), *Microbial control of pests and plant diseases 1970–1980.* Academic Press, London. pp. 737–767.

Burges, H.D., G. Croizier, and J. Huber, 1980a. A review of safety tests on baculoviruses. *BioControl* 25(4): 329–339.

Burges, H.D., J. Huber, and G. Croizier, 1980b. Guidelines for safety tests on insect viruses. *Entomophaga* 25(4): 341–348.

Burges, H.D., A. Krieg, P. Lüthy, and H. de Barjac, 1982. Guidelines for safety tests and registration of bacterial pesticides. *BioControl* 27(3): 225–235.

Chandler, D., G. Davidson, W.P. Grant, J. Greaves, and G.M. Tatchell, 2008. Microbial biopesticides for integrated crop management: an assessment of environmental and regulatory sustainability. *Trends Food Sci. Technol.* 19: 275–283.

Cherry, A.J. and R.L. Gwynn, 2007. Perspectives on the development of biological control agents in Africa. *Biocontrol Sci. Technol.* 17(7): 665–676.

Cock, M.J.W., J.C. van Lenteren, J. Brodeur, B.I.P. Barratt, F. Bigler, K. Bolckmans, F.L. Cônsoli, F. Haas, P.G. Mason, J. Roberto, and P. Parra, 2009. Do new access and benefit sharing procedures under the convention on biological diversity threaten the future of biological control? *BioControl* 55(2): 199–218.

Cook, R.J., W.L. Bruckart, J.R. Coulson, M.S. Goettel, R.A. Humber, R.D. Lumsden, J.V. Maddox, M.L. McManus, L. Moore, S.F. Meyer, P.C. Quimby, J.P. Stack, and J.L. Vaughn, 1996. Safety of microorganisms intended for pest and plant disease control: a framework for scientific evaluation. *Biol. Control* 7: 333–351.

Cory, J.S., 2003. Ecological impacts of virus insecticides: host range and non-target organisms. In: H.M.T. Hokkanen, A.E. Hajek (eds), *Environmental impacts of microbial insecticides*. Kluwer, Dordrecht. pp. 73–91.

CPL, 2006a. *Biopesticides 2007.* CPL Business Consultants, Wallingford.

CPL, 2006b. *Biopesticides 2007, Vol. 5. A how to do it guide to biopesticides.* CPL Business Consultants, Wallingford.

Cresswell, R.C., 1997. Microbial insecticides and patent law. In: H.F. Evans (ed), *Microbial insecticides: novelty or necessity? BCPC Symposium Proceedings* No. 68, Coventry, April 16–18, 1997. BCPC, Farnham. pp. 235–242.

Cross, J.V. and D.R. Polonenko, 1996. An industry perspective on registration and commercialization of biocontrol agents in Canada. *Can. J. Plant Pathol.* 18: 446–454.

De Kogel, W.J. and W.J. Ravensberg, 2006. Bacterie losgelaten op plagen. *Groeten en Fruit* 35: 13.

EC, 1991a. Council Directive 91/414/EEC of 15 July 1991 concerning the placing of plant protection products on the market. *Official Journal of the European Union L* 230, 19.08.1991: 1–32.

EC, 1991b. Council Regulation (EEC) No 2092/91 of 24 June 1991 on organic production of agricultural products and indications referring thereto on agricultural products and foodstuffs. *Official Journal of the European Union* L 198, 22.07.1991: 1–15.

EC, 1991c. Council Directive 91/496/EEC of 15 July 1991 laying down the principles governing the organization of veterinary checks on animals entering the community from third countries and amending directives 89/662/EEC, 90/425/EEC and 90/675/EEC. *Official Journal of the European Union* L 268, 24.9.1991: 56–68.

EC, 1992. Council Directive 92/65/EEC of 13 July 1992 laying down animal health requirements governing trade in and imports into the community of animals, semen, ova and embryos not subject to animal health requirements laid down in specific community rules referred to in Annex A (I) to Directive 90/425/EEC. *Official Journal of the European Union* L 268, 14.9.1992: 54–72.

EC, 1996. Regulation (EC) No 1610/96 of the European Parliament and of the council of 23 July 1996 concerning the creation of a supplementary protection certificate for plant protection products. *Official Journal of the European Union* L 198, 08.08.1996: 30–35.

EC, 1998a. Council Directive 97/78/EC of 18 December 1997 laying down the principles governing the organisation of veterinary checks on products entering the community from third countries. *Official Journal of the European Union* L 24, 30.1.1998: 9–30.

EC, 1998b. Directive 98/44/EC of the European Parliament and of the council of 6 July 1998 on the legal protection of biotechnological inventions. *Official Journal of the European Union* L 213, 30.7.1998: 13–21.

EC, 2000a. Directive 2000/54/EC of the European Parliament and of the council of 18 September 2000 on the protection of workers from risks related to exposure to biological agents at work. *Official Journal of the European Union* L 262, 17.10.2000: 21–45.

EC, 2000b. Council Directive 2000/29/EC on protective measures against the introduction into the Community of organisms harmful to plants or plant products and against their spread within the Community. *Official Journal of the European Union* L 169, 10.7.2000: 1–112.

EC, 2001a. Commission Directive 2001/36/EC of 16 May 2001 amending council Directive 91/414/EEC concerning the placing of plant protection products on the market (Data requirements micro-organisms). *Official Journal of the European Union* L 164, 20.6.2001: 1–38.

EC, 2001b. Commission Directive 2001/58/EC of 27 July 2001 amending for the second time Directive 91/155/EEC defining and laying down the detailed arrangements for the system of specific information relating to dangerous preparations in implementation of Article 14 of European Parliament and council Directive 1999/45/EC and relating to dangerous substances in implementation of Article 27 of council Directive 67/548/EEC (safety data sheets). *Official Journal of the European Union* L 212, 7.8.2001: 24–33.

EC, 2002. Second report of the European community to the convention on biological diversity. Thematic report on access and benefit sharing. Box 10. Benefit sharing by the International Locust Control Programme (LUBILOSA). http://biodiversity-chm.eea.europa.eu/convention/cbd_ec/. Cited September 5, 2008.

EC, 2003. Commission Directive 2003/82/EC of 11 September 2003 amending Council Directive 91/414/EEC as regards standard phrases for special risks and safety precautions for plant-protection products. *Official Journal of the European Union* L 228, 12.9.2003: 11–28.

EC, 2005. Council Directive 2005/25/EC of 14 March 2005 amending Annex VI to Directive 91/414/EEC as regards plant protection products containing micro-organisms. *Official Journal of the European Union* L 90, 8.4.2005: 1–34.

EC, 2009a. Regulation (EC) No 1107/2009 of the European Parliament and of the council of 21 October 2009 concerning the placing of plant protection products on the market and repealing council Directives 79/117/EEC and 91/414/EEC. *Official Journal of the European Union* L 309(52): 1–50.

EC, 2009b. Directive 2009/128/EC of the European Parliament and of the Council of 21 October 2009 establishing a framework for community action to achieve the sustainable use of pesticides. *Official Journal of the European Union* L 309(52): 71–86.

EC, 2009c. Regulation (EC) No 1185/2009 of the European Parliament and of the council of 25 November 2009 concerning statistics on pesticides. *Official Journal of the European Union* L 324, 10.12.2009: 1–22.

EFSA, 2008. Evaluation of the European Union pesticide safety review process. *The EFSA J.* 157: 1–110.

Ehlers, R.-U., 1996. Current and future use of nematodes in biocontrol: practice and commercial aspects in regard to regulatory policies. *Biocontrol Sci. Technol.* 6: 303–316.

Ehlers, R.-U., 2003. Biocontrol nematodes. In: H.M.T. Hokkanen, A.E. Hajek (eds), *Environmental impacts of microbial insecticides*. Kluwer, Dordrecht. pp. 177–220.

Ehlers, R.-U., 2005. Forum on safety and regulation. In: P.S. Grewal, R.-U. Ehlers, D.I. Shapiro-Ilan (eds), *Nematodes as biocontrol agents*. CAB International, Wallingford. pp. 107–114.

Ehlers, R.-U., 2006. Safety and regulation of entomopathogenic nematodes. http://www.cost850.ch/legal_safety. Cited November 8, 2008.

Ehlers, R.-U. and H.M.T. Hokkanen, 1996. Insect biocontrol with non-endemic entomopathogenic nematodes (*Steinernema* and *Heterorhabditis* spp.): conclusions and recommendations of a combined OECD and COST workshop on scientific and regulatory policy issues. *Biocontrol Sci. Technol.* 6: 295–302.

EPA, 2007a. Pesticides; data requirements for biochemical and microbial pesticides. Federal register: 40 CFR Part 158, October 26, 2007.Vol. 72, Nr. 207: Section 158.2100.

EPA, 2007b. Biopesticide active ingredients and products containing them. March 2007. http://www.epa.gov/pesticides/biopesticides/. Cited October 24, 2009.

EPPO, 1999. EPPO Standards – safe use of biological control. PM 6/1(1): first import of exotic biological control agents for research under contained conditions. *EPPO Bull.* 29: 271–272.

EPPO, 2001. EPPO Standards – safe use of biological control. PM 6/2(1): import and release of exotic biological control agents. *EPPO Bull.* 31: 33–35.

EPPO, 2008. List of biological control agents widely used in the EPPO region. PM 6/3(3) – version 2008. http://archives.eppo.org/EPPOStandards/biocontrol_web/bio_list.htm. Cited February 6, 2009.

EPPO/CABI, 1997. EPPO/CABI Workshop on safety and efficacy of biological control in Europe. Proceedings of workshop held in Streatley-on-Thames, 1996-03-26/28. *EPPO Bull.* 27: 1.

Evans, J., 2004. Shifting perceptions on biopesticides. *Agrow* No. 455, September 3, 2004: 18–21.

FAO, 2005. Guidelines for the export, shipment, import and release of biological control agents and other beneficial organisms. ISPM No. 3, FAO, Rome (available at www.ippc.int).

Fay, E., D.J. Flynn, J.R. Lundehn, P.J. Chapman, and R.D. Mason, 2007. The joint evaluation procedure for active substances contained in plant protection products within the European community – 10 years of the ECCO-Project. *J. Verbr. Lebensm.* 2: 61–77.

Federici, B.A. and J.P. Siegel, 2007. Safety assessment of Bacillus thuringiensis and Bt crops used in insect control. In: B.G. Hammond (ed), *Food safety of proteins in agricultural biotechnology*. CRC Press, Boca Raton. pp. 45–102.

Fitzner, U., 2005. Laws and regulations for the protection of biotechnological inventions. In: J.L. Barredo (ed), *Microbial processes and products*. Humana Press, Totowa. pp. 465–493.

Fjelsted, A., 2006. Current regulation system and data requirements within the EU. *Paper presented at the REBECA workshop on "Potential risks of microbials"*. Innsbruck, Austria, 12–13 April 2006. www.rebeca-net.de. Cited November 8, 2008.

Foster, J.E., 2007. Bioterrorism: past, present and preparedness in plant protection. *Proceedings of the XVIth BCPC Intern. Congress – Plant protection 2007*, October 15–18, 2007, Glasgow, Scotland. pp. 426–427.

Franceschini, S. and F. Jondini, 2004. EU registration problems and possible solution. *IOBC/WPRS Bull.* 27(8): 423–426.

A. Garibaldi (ed), 1995. Microbial control agents in sustainable agriculture: field experience, industrial production and registration. *Proceedings of the congress,* October 18–19, 1995 Saint Vincient (Aosta) Italy.

Gelernter, W.D., 2005. Biological control products in a changing landscape. *Proceedings of the XIVth BCPC Internnational Congress – Crops science & technology 2005*, October 31–November 2, 2005, Glasgow, Scotland. pp. 293–300.

Gelernter, W.D., 2007. Microbial control in Asia: a bellwether for the future? *J. Invertebr. Pathol.* 95: 161–167.

Gerrard, J., N. Waterfield, R. Vohra, and R. ffrench-Constant 2004. Human infection with *Photorhabdus asymbiotica*: an emerging bacterial pathogen. *Microbes Infect.* 6(2): 229–237.

Glare, T.R., and M. O'Callaghan, 2000. *Bacillus thuringiensis: biology, ecology and safety.* Wiley & Sons, Chichester.

Glare, T.R. and M. O'Callaghan, 2003. Environmental impacts of bacterial biopesticides. In: H.M.T. Hokkanen, A.E. Hajek (eds), *Environmental impacts of microbial insecticides.* Kluwer, Dordrecht. pp. 119–149.

Goettel, M.S. and A.E. Hajek, 2001a. Evaluation of non-target effects of pathogens used for management of arthropods. In: E. Wajnberg, J.K. Scott, P.C. Quimby (eds), *Evaluating indirect ecological effects of biological control.* CAB International, Wallingford. pp. 81–97.

Goettel, M.S., A.E. Hajek, J.P. Siegel, and H.C. Evans, 2001b. Safety of fungal biocontrol agents. In: T.M. Butt, C. Jackson, N. Magan (eds), *Fungi as biocontrol agents: progress, problems and potential.* CAB International, Wallingford. pp. 347–375.

Groner, A., 1986. Specificity and safety of baculoviruses. In: R.R. Granados, B.A. Federici (eds), *The biology of baculoviruses.* CRC Press, Boca Raton. pp. 178–202.

Groner, A., 1990. Safety to nontarget invertebrates of baculoviruses. In: M. Laird, L.A. Lacey, E.W. Davidson (eds), *Safety of microbial insecticides.* CRC Press, Boca Raton. pp. 135–147.

Hajek, A.E., M.L. McManus, and I.D. Junior, 2007. A review of introductions of pathogens and nematodes for classical biological control of insects and mites. *Biol. Control* 41: 1–13.

Hall, R.A., G. Zimmermann, and A. Vey, 1982. Guidelines for the registration of entomophagous fungi as insecticides. *Entomophaga* 27(2): 121–127.

Hartman, C.L. and G.M. Markle, 1999. IR-4 Biopesticide program for minor crops. In: F.R. Hall, J.J. Menn (eds), *Biopesticides: use and delivery.* Humana Press, Totowa. pp. 443–452.

Hauschild, R. and B. Speiser, 2007. A comparison of legislation and current practice regarding evaluation and registration of plant protection products based on micro-organisms, plant extracts and pheromones in the EU, USA, Canada and Australia. *Report prepared in the course of WP 2 of the REBECA project, December 2007.* www.rebeca-net.de. Cited September 13, 2008.

Hokkanen, H., 2007. Survey among industry concerning biological plant protection products. *Paper presented at the REBECA workshop on "balancing the benefits and costs of regulating biological plant protection products"*, May 6–8, 2007 Porvoo, Finland. www.rebeca-net.de. Cited November 20, 2008.

Hunt, E.J., U. Kuhlmann, A. Sheppard, T.-K. Qin, B.I.P. Barratt, L. Harrison, P.G. Mason, D. Parker, R.V. Flanders, and J. Goolsby, 2008. Review of invertebrate biological control agent regulation in Australia, New Zealand, Canada and the USA: recommendations for a harmonized European system. *J. Appl. Entomol.* 132(2): 89–123.

Hökeberg, M., 2006. Bacteria against plant pathogens. *Paper presented at the REBECA workshop on "potential risk of microbials"*, Innsbruck, Austria, April 12–13, 2006. www.rebeca-net.de. Cited November 20, 2008.

Ignoffo, C.M. and T.L. Couch, 1981. The nucleopolyhedrosis virus of Heliothis species as a microbial insecticide. In: H.D. Burges (ed), *Microbial control of pests and plant diseases 1970–1980*. Academic Press, London. pp. 329–362.

IOBC, 2008. IOBC Global commission on biological control and access and benefit sharing. *IOBC Global Newslett.* 84, October 2008: 5–7. www.iobc-global.org. Cited March 29, 2009.

Jackson, T.A., 2003. Environmental safety of inundative application of naturally occurring biocontrol agent, Serratia entomophila. In: H.M.T. Hokkanen, A.E. Hajek (eds), *Environmental impacts of microbial insecticides*. Kluwer, Dordrecht. pp. 169–176.

Jaronski, S.T., M.S. Goettel, and C.J. Lomer, 2003. Regulatory requirements for ecotoxicological assessments of microbial insecticides – how relevant are they? In: H.M.T. Hokkanen, A.E. Hajek (eds) *Environmental impacts of microbial insecticides*. Kluwer, Dordrecht. pp. 237–260.

Kiewnick, S., 2007. Practicalities of developing and registering microbial biological control agents. *CAB Rev.: Perspectives Agric. Vet. Sci. Nutr. Nat. Resour.* 2(No. 013): 1–11.

Kirchmair, M., S. Neuhauser, L. Huber, and H. Strasser, 2008. The impact of soil treatment on soil mycobiota. *IOBC/WPRS Bull.* 31: 239–244.

Klier, A., 2000. Bacillus thuringiensis: risk assessment. In: J.-F. Charles, A. Delécluse, C. Nielsen-LeRoux (eds), *Entomopathogenic bacteria: from laboratory to field application*. Kluwer, Dordrecht. pp. 485–503.

Kocken, J. and G. van Roozendaal, 1997. The neem tree debate. *Biotechn. Develop. Monitor* 30: 811.

Krause, M.S., A.C.R.C. Vanachter, and T.J.J. de Ceuster, 2006. Commercial research and development of disease-suppressive microorganisms. *IOBC/WPRS Bull.* 29(2): 63–66.

Lacey, L.A., 2008. Codling moth granulovirus: a comprehensive review. *Biocontrol Sci. Technol.* 18(7/8): 639–663.

Lacey, L.A. and R.W. Merritt, 2003. The safety of bacterial microbial agents used for black fly and mosquito control in aquatic environments. In: H.M.T. Hokkanen, A.E. Hajek (eds), *Environmental impacts of microbial insecticides*. Kluwer, Dordrecht. pp. 151–168.

Lacey, L.A. and J.P. Siegel, 2000. Safety and ecotoxicology of entomopathogenic bacteria. In: J.-F. Charles, A. Delécluse, C. Nielsen-LeRoux (eds), *Entomopathogenic bacteria: from laboratory to field application*. Kluwer, Dordrecht. pp. 253–273.

Laird, M., 1981. Repository for data on the safety of insect pathogens. In: H.D. Burges (ed), *Microbial control of pests and plant diseases 1970–1980*. Academic Press, London. pp. 913–914.

Laird, M., L.A. Lacey, E.W. Davidson (eds), 1990. *Safety of microbial insecticides*. CRC Press, Boca Raton.

Längle, T., M. Kirchmaier,, T. Bauer, J. Raffalt, C. Seger, B. Pernfuss, and H. Strasser, 2004. Environmental risk assessment of soil-applied fungal biological control agents with respect to European registration. *IOBC/WPRS Bull.* 27(8): 197–200.

Libman, G.N. and S.C. MacIntosh, 2000. Registration of biopesticides. In: J.-F. Charles, A. Delécluse, C. Nielsen-LeRoux (eds), *Entomopathogenic bacteria: from laboratory to field application*. Kluwer, Dordrecht. pp. 333–353.

Lisansky, S.G., 1986. The registration of microbial products: industry's viewpoint. In: R.A. Samson, J.M. Vlak, D. Peters (eds), *Fundamental and applied aspects of invertebrate pathology. Proceedings of the fourth international colloquium fundamental and applied aspects of invertebrate pathology,* 18–22 August 1986, Veldhoven, The Netherlands. pp. 637–640.

Lisansky, S., 1994. International harmonization in biopesticide registration and legislation. *Proceedings of the BCPC International Congress – Pests and diseases,* Brighton 1994. pp. 1397–1402.

LNV, 2004. Nota: Duurzame gewasbescherming – Beleid voor gewasbescherming tot 2010. *Rapport van Ministerie van Landbouw, Natuur en Visserij*, 25.5.2004.

Lomer, C.J., R.P. Bateman, D.L. Johnson, J. Langewald, and M. Thomas, 2001. Biological control of locusts and grasshoppers. *Ann. Rev. Entomol.* 46: 667–702.

Loomans, A., 2007. Regulation of invertebrate biological control agents in Europe: review and recommendations in its pursuit of a harmonised regulatory system. Report EU project REBECA, February, 2007. www.rebeca-net.de. Cited October 20, 2008.

Loomans, A. and S. Sütterlin, 2005. Regulation of invertebrate biological control agents: international context and situation in The Netherlands. *IOBC/WPRS Bull.* 28(1): 179–182.

Lynch, L.D. and M.B. Thomas, 2000. Nontarget effects in the biocontrol of insects with insects, nematodes and microbial agents: the evidence. *Biocont. News Inform.* 21(4): 117 N–130 N.

Marrone, P.G., 2007. Barriers to adoption of biological control agents and biological pesticides. *CAB Rev.: Perspectives Agric. Vet. Sci. Nutr. Nat. Resour.* 2(No. 051): 1–12.

McClintock, J.T., 1999. The federal registration process and requirements for the United States. In: F.R. Hall, J.J. Menn (eds), *Biopesticides: use and delivery*. Humana Press, Totowa. pp. 415–441.

McCoy, C.W., 1996. Pathogens of eriophyoid mites. In: E.E. Lindquist, M.W. Sabelis, J. Bruin (eds), *World crop pests. Eriophyoid mites. Their biology, natural enemies and control*, Vol. 6. Elsevier, Amsterdam. pp. 481–490.

Meadows, M.P., 1993. Bacillus thuringiensis in the environment: ecology and risk assessment. In: P.F. Entwistle, J.S. Cory, M.J. Bailey, S. Higgs (eds), *Bacillus thuringiensis, an environmental biopesticide: theory and practice*. Wiley & Sons, New York, NY. pp. 193–220.

Meeussen, J.J., 2007. OECD biopesticide steering group. Past, presence and future. *Paper presented at the annual biocontrol industry meeting*, October 22–23, 2007, Lucerne. www.abim-lucerne.ch. Cited November 20, 2008.

Mensink, B.J.W.G. and J.W.A. Scheepmaker, 2007. How to evaluate the environmental safety of microbial plant protection products: a proposal. *Biocontrol Sci. Technol.* 17(1): 3–20.

Miltenburger, H.G. (ed), 1980a. *Safety of baculoviruses as biological insecticides. Proceedings of the Symposium* 13–15 November 1978. Federal Ministry for Research and Technology, Bonn.

Miltenburger, H.G., 1980b. Viral pesticides: hazard evaluation for non-target organisms and safety testing. In: B. Lundohlm, M. Stackerud (eds), *Environmental protection and biological forms of control of pest organisms. Ecol. Bull.* 31: 57–74.

Montesinos, E., 2003. Development, registration and commercialization of microbial pesticides for plant protection. *Int. Microbiol.* 6: 245–252.

Neale, M.C., 1997. Biopesticides-harmonization of registration requirements within the EU Directive 91/414 – an industry view. *EPPO Bull.* 27(1): 89–94.

Neale, M., 2000. The regulation of natural products as crop-protection agents. *Pest Manage. Sci.* 56: 677–680.

Neale, M. and P. Newton, 1999. Registration/regulatory requirements in Europe. In: F.R. Hall, J.J. Menn (eds), *Biopesticides: use and delivery*. Humana Press, Totowa. pp. 453–471.

OECD, 1997. Consensus document on information used in the assessment of environmental applications involving *Pseudomonas*. *Series on harmonization of regulatory oversight in biotechnology*, No. 6. OCDE/GD(97)22. OECD, Paris (available at www.oecd.org).

OECD, 1998. OECD Workshop. Sustainable pest management: safe utilization of new organisms in biological control. *Phytoprotection* 79, Supplement. OECD, Paris (available at www.oecd.org).

OECD, 2002. Consensus document on information used in the assessment of environmental applications involving baculoviruses. *Series on harmonization of regulatory oversight in biotechnology*, No. 20. ENV/JM/MONO(2002)1. OECD, Paris (available at www.oecd.org).

OECD, 2003a. Guidance for registration requirements for microbial pesticides. *Series on pesticides* No. 18. ENV/JM/MONO(2003)5. OECD, Paris (available at www.oecd.org).

OECD, 2003b. Guidance document on the use of taxonomy in risk assessment of micro-organisms: bacteria. *Series on harmonization of regulatory oversight in biotechnology*, No. 29. ENV/JM/MONO(2003)13. OECD, Paris (available at www.oecd.org).

OECD, 2004a. Guidance for information requirements for regulation of invertebrates as biological control agents (IBCAs). *Series on pesticides* No. 21. ENV/JM/MONO(2004)1. OECD, Paris (available at www.oecd.org).

OECD, 2004b. OECD guidance for country data review reports on microbial pest control products and their microbial pest control agents (Monograph Guidance for microbials), February 2004. *Series on pesticides* No. 22. OECD, Paris (available at www.oecd.org).
OECD, 2004c. OECD guidance for industry data submissions for microbial pest control products and their microbial pest control agents (Dossier guidance for microbials), February 2004. *Series on pesticides* No. 23. OECD, Paris (available at www.oecd.org).
OECD, 2004d. Guidance document on methods for detection of micro-organisms introduced into the environment: bacteria. *Series on harmonization of regulatory oversight in biotechnology*, No. 30. ENV/JM/MONO(2004)7. OECD, Paris (available at www.oecd.org).
OECD, 2008. OECD working document on the evaluation of microbials for pest control. *Series on pesticides* No. 43. ENV/JM/MONO(2008)36. OECD, Paris (available at www.oecd.org).
OECD, 2009. Report of workshop on the regulation of biopesticides: registration and communication issues, 15–17 April 2008, EPA, Arlington, USA. *Series on pesticides* No. 44. ENV/JM/MONO(2009)19. OECD, Paris (available at www.oecd.org).
Otvos, I.S., H. Armstrong, and N. Conder , 2005. Safety of *Bacillus thuringiensis* var. *kurstaki* applications for insect control to humans and large mammals. *Proceedings of the 6th Pacific rim conference on the biotechnology of Bacillus thuringiensis and its environmental impact,* Victoria BC, 2005. pp. 45–60.
Panetta, J.D., 1999. Environmental and regulatory aspects: industry view and approach. In: F.R. Hall, J.J. Menn (eds), *Biopesticides: use and delivery*. Humana Press, Totowa. pp. 473–484.
Pernfuss, B., T. Längle, R. Kron Morelli, and H. Strasser, 2007. Fungal BCAs in the European Union: *Beauveria brongniartii* (Sacc.) Petch as the model organism to address key questions. *IOBC/WPRS Bull.* 30(1): 111–116.
PMRA, 2008. Product information. http://pr-rp.pmra-arla.gc.ca. Cited November 21, 2008.
Quinlan, R.J., 1990. Registration requirements and safety considerations for microbial pest control agents in the European economic community. In: M. Laird, L.A. Lacey, E.W. Davidson (eds), *Safety of microbial insecticides*. CRC Press, Boca Raton. pp. 11–18.
REBECA, 2007a. Guidance document, methods for environmental risk assessment for macrobial (Invertebrate) biocontrol agents. *Regulation of biological control agents specific support action,* EU Project no. SSPE – CT – 2005-022709. www.rebeca-net.de. Cited November 21, 2008.
REBECA, 2007b. Proposal for regulatory system and testing guidelines for macrobial BCAs based on retrospective case studies. REBECA, Deliverable 23, Appendix A: recommendations for regulation requirements for entomopathogenic nematodes (EPNs). *Regulation of biological control agents specific support action,* EU Project no. SSPE – CT – 2005-022709. www.rebeca-net.de. Cited November 21, 2008.
REBECA, 2007c. Proposals on how to accelerate regulation and reduction of fees. REBECA deliverable 27. *Regulation of biological control agents specific support action,* EU Project no. SSPE-CT-2005-022709. www.rebeca-net.de. Cited November 28, 2008.
REBECA, 2007d. Application form for the import, shipment, rearing and release of invertebrate biological control agents in European countries. REBECA deliverable 23. *Regulation of biological control agents specific support action*, EU Project no. SSPE-CT-2005-022709. www.rebeca-net.de. Cited November 28, 2008.
REBECA, 2007e. Guidance document: methods for environmental risk assessment for macrobial (Invertebrate) biocontrol agents. REBECA deliverable 23. *Regulation of biological control agents specific support action*, EU Project no. SSPE-CT-2005-022709. www.rebeca-net.de. Cited November 28, 2008.
REBECA, 2007 f. Proposals on how to accelerate regulation and reduction of fees. REBECA deliverable 27. *Regulation of biological control agents specific support action,* EU Project no. SSPE-CT-2005-022709. www.rebeca-net.de. Cited November 28, 2008.
Ravensberg, W.J., 2004. Industry view on regulation of IBCAs. *Paper presented at the COST Action 850, WG 3 and 5: "Evaluation of risks related with the use of EPN and discussion on requirements for registration of invertebrate biocontrol agents"*, Kiel, Germany, 5–7 November 2004. www.cost850.ch. Cited December 15, 2008.

Ravensberg, W.J. and Y. Elad, 2002. The reality – a commercial perspective to plant disease biocontrol. *IOBC/WPRS Bull.* 25(10): 125–130.

Ravensberg, W.J. and G. Sterk, 2004. Experiences in registration of microbial BCAs in Europe: a biocontrol industry perspective. *Paper presented at the EU-RAFBCA-IBMA-IOBC workshop 30th September, 2004, centre Albert Borschette, Brussels, "New Insights into risk assessment and registration of fungal biocontrol agents in Europe"*. www.rafbca.com. Cited December 15, 2008.

Richards, J. and P. Kearns, 1997. Work on micro-organisms within the OECD's environmental health and safety programme. In: H.F. Evans (ed), *Microbial insecticides: novelty or necessity? BCPC Symposium Proceedings* No. 68, Coventry, April 16–18, 1997. BCPC, Farnham. pp. 223–226.

Richardson, P.N., 1996. British and European legislation regulating Rhabditid nematodes. *Biocontrol Sci. Technol.* 6: 449–463.

Rizvi, S.A., R. Hennessey, and D. Knott, 1996. Legislation on the introduction of exotic nematodes in the US. *Biocontrol Sci. Technol.* 6(3): 477–480.

SANCO, 2008. Summary table of the current status of active substances. http://ec.europa.eu/food/plant/protection/evaluation/3010_rev_040808_2008.xls. Cited January 20, 2009.

Scheepmaker, J.W.A. and T.M. Butt, 2010. Natural and released inoculum levels of entomopathogenic fungal biocontrol agents in soil in relation to risk assessment and in accordance with EU regulations. *Biocontrol Sci. Technol.* 20(3): 503–552.

Schneider, W., 2006. US EPA Regulation of biopesticides. Microbial and biochemical pesticide regulation. *Paper presented at the REBECA workshop on current risk assessment and regulation practice,* Salzau, Germany, September 18–22, 2006. www.rebeca-net.de. Cited November 28, 2008.

Seger, C., S. Sturm, T. Längle, W. Wimmer, H. Stuppner, and H. Strasser, 2005a. Development of a sensitive high-performance liquid chromatography-diode array detection assay for the detection and quantification of the *Beauveria* metabolite oosporein from submerged culture broth and bio-control formulations. *J. Agric. Food Chem.* 53: 1364–1369.

Seger, C., T. Längle, B. Pernfuss, H. Stuppner, and H. Strasser, 2005b. High-performance liquid chromatography-diode array detection assay for the detection and quantification of the *Beauveria* metabolite oosporein from potato tubers. *J. Chromatogr. A* 1092: 254–257.

Shah, F.A., C. Greig, S. Hutwimmer, H. Strasser, P. Dyson, B. Carlile, and T.M. Butt, 2009. Evaluation of the effects of the insect pathogenic fungus *Metarhizium anisopliae* on microbial populations of disparate plant growing media. *Fung. Ecol.* 3(3): 185–194.

Siegel, J.P., 2001. The mammalian safety of *Bacillus thuringiensis*-based insecticides. *J. Invertebr. Pathol.* 77(1): 13–21.

Sigman, R., 2005. OECD's vision for the regulation of crop protection products. *Proceedings of the BCPC international congress – crop science & technology 2005,* 31 October–2 November 2005, Glasgow, Scotland. pp. 329–336.

Skrobek, A. and T.M. Butt, 2005. Toxicity testing of destruxins and crude extracts from the insect-pathogenic fungus *Metarhizium anisopliae*. *FEMS Microbiol. Lett.* 251: 23–28.

Smith, D., 2000. Legislation affecting the collection, use and safe handling of entomopathogenic microbes and nematodes. In: A. Navon, K.R.S. Ascher (eds), *Bio-assays of entomopathogenic microbes and nematodes*. CAB International, Wallingford. pp. 295–313.

Smits, P.H. and J.M. Vlak, 1994. Registration of the first viral insecticide in the Netherlands: the development of Spod-X, based on *Spodoptera exigua* nuclear polyhedrosis virus. *Med. Fac. Landbouw. Rijksuniv. Gent.* 59(2a): 385–392.

Strasser, H., 2000. A researcher's obstacle race: report on the registration of a new bioinsecticide. *IOBC/WPRS Bull.* 23(2): 9–13.

Strasser, H. and M. Kirchmair, 2006. Potential health problems due to exposure in handling and using biological control agents. In: J. Eilenberg, H.M.T. Hokkanen (eds), *An ecological and societal approach to biological control*. Springer, Dordrecht. pp. 275–293.

Strasser, H. and B. Pernfuss, 2005. What have BIPESCO and RAFBCA achieved that could help with risk assessment and registration? *IOBC/WPRS Bull.* 28(2): 189–192.

Strasser, H., B. Pernfuss, C. Seger, and T. Längle, 2004. Case study 1. *Beauveria brongniartii* for pest control in potatoes. *Paper presented at the EU-RAFBCA-IBMA-IOBC workshop 30th September, 2004, Centre Albert Borschette, Brussels, "New insights into risk assessment and registration of fungal biocontrol agents in Europe"*. www.rafbca.com. Cited September 20, 2008.

Strasser, H., M. Typas, C. Altomare, and T.M. Butt, 2008. REBECA proposal on the assessment of microbial metabolites. *IOBC/WPRS Bull.* 31: 21–26.

Strasser, H., A. Vey, and T.M. Butt, 2000. Are there any risks in using entomopathogenic fungi for pest control, with particular reference to the bioactive metabolites of *Metarhizium, Tolypocladium* and *Beauveria* species? *Biocontrol Sci. Technol.* 10: 717–735.

Strauch, O., 2004. EPN registration. *Paper presented at the COST Action 850, WG 3 and 5: "Evaluation of risks related with the use of EPN and discussion on requirements for registration of invertebrate biocontrol agents"*, 5–7 November 2004, Kiel, Germany. www.cost850.ch. Cited September 20, 2008.

Summers, M., R. Engler, L.A. Falcon, P.V. Vail (eds), 1975. Baculoviruses for insect pest control: safety considerations. *Selected papers from EPA-USDA Working Symposium,* Bethesda, Maryland. Am. Soc. Microbiol., Washington, DC.

Tooby, T.E., 1997. A regulatory perspective: is there a framework for encouraging use of biopesticides? In: H.F. Evans (ed), *Microbial insecticides: novelty or necessity? BCPC Symposium Proceedings* No. 68, Coventry, April 16–18, 1997. BCPC, Farnham. pp. 215–222.

Van Lenteren, J.C., 1996. Regulatory issues related to biological control in Europe. *IOBC/WPRS Bull.* 19(1): 79–82.

Van Lenteren, J.C., D. Babendreier, F. Bigler, G. Burgio, H.M.T. Hokkanen, S. Kuske, A.J.M. Loomans, I. Menzler-Hokkanen, P.C.J. van Rijn, M.B. Thomas, M.G. Tommasini, and Q.-Q. Zeng, 2003. Environmental risk assessment of exotic natural enemies used in inundative biological control. *BioControl* 48: 3–38.

Van Lenteren, J.C., J. Bale, F. Bigler, H.M.T. Hokkanen, and A.J.M. Loomans, 2006. Assessing risks of releasing exotic biological control agents of arthropod pests. *Ann. Rev. Entomol.* 51: 609–634.

Vestergaard, S., A. Cherry, S. Keller, and M. Goettel, 2003. Safety of hyphomycete fungi as microbial control agents. In: H.M.T. Hokkanen, A.E. Hajek (eds), *Environmental impacts of microbial insecticides*. Kluwer, Dordrecht. pp. 35–62.

Vey, A., R.E. Hoagland, and T.M. Butt, 2001. Toxic metabolites of fungal biocontrol agents. In: T.M. Butt, C. Jackson, N. Magan (eds), *Fungi as biocontrol agents: progress, problems and potential*. CAB International, Wallingford. pp. 311–346.

Vilas-Bôas, G.T., A.P.S. Peruca, and O.M.N. Arantes, 2007. Biology and taxonomy of *Bacillus cereus, Bacillus anthracis*, and *Bacillus thuringiensis*. *J. Can. Microbiol.* 53: 673–687.

Waage, J. 1997. Global developments in biological control and the implications for Europe. *EPPO Bull.* 27(1): 5–13.

Waage, J., 2007. The sustainable management of biodiversity for biological control in food and agriculture: status and needs. *FAO Background Study Paper* No. 38. ftp://ftp.fao.org/ag/cgrfa/bsp/bsp38e.pdf. Cited November 21, 2008.

Weihs, V., 2005a. How to deposit biological material for patent purposes. In: J.L. Barredo (ed), *Microbial processes and products*. Humana Press, Totowa. pp. 451–464.

Weihs, V. and C. Rohde, 2005b. Safe dispatch and transport of biological material. In: J.L. Barredo (ed), *Microbial processes and products*. Humana Press, Totowa. pp. 437–449.

Westerlund, L., 2002. Invention or discovery. In: L. Westerlund (ed), *Biotech patents: equivalency and exclusions under European and U.S. patent law*. Kluwer, New York, NY. pp. 23–58.

Whittaker, M.S., 2007. Regulatory innovation and the biopesticide industry. *Proceedings of the XVIth International Plant Protection Congress 2007*, October 15–18, 2007, Glasgow, Scotland. pp. 600–601.

Zimmermann, G., 2007a. Review on safety of the entomopathogenic fungi *Beauveria bassiana* and *Beauveria brongniartii*. *Biocontrol Sci. Technol.* 17(5/6): 553–596.

Zimmermann, G., 2007b. Review on safety of the entomopathogenic fungus *Metarhizium anisopliae*. *Biocontrol Sci. Technol.* 17(9): 879–920.

Zimmermann, G., P. Lüth, E. Bode, M. Hommes, B. Selzam, S. Guske, and J. Huber, 2004. Report on the third meeting on biological plant protection "Commercialisation of microbial plant protection agents". *Nachtrichtenbl. Deut. Pflanzenschutzd.* 56(6): 131–136.

Chapter 6
Implementation of a Microbial Pest Control Product in an Integrated Pest Management Programme

Abstract Key factors to cost-effective implementation of a microbial pest control product are application strategy, compatibility, and knowledge transfer to the user. This requires the design of a comprehensive integrated pest management programme in which the microbial pest control product is to be incorporated. The first element is an optimal application strategy per targeted pest and cropping system. Determinative parameters such as dosage and spray volume are discussed as well as factors that influence these parameters such as the tritrophic plant-host-pathogen relationship. The second element is the incorporation of the microbial pest control product in an IPM system. Compatibility with chemical pesticides and with beneficial organisms must be established and critical aspects are mentioned for each type of pathogen. Microbial pest control products and natural enemies are often simultaneously used in IPM systems to enhance control. Synergistic, additive and negative interactions are reviewed. Combinations of microbial pest control products and combinations with other means are discussed. The risk of resistance will be briefly treated. The third element is the implementation of a microbial pest control product into practice. Adoption of new products needs knowledge transfer in the entire supply chain to the grower, and pivotal considerations including economic aspects will be reviewed. Requirements and recommendations will be provided on how to accomplish a successful implementation. Considerations will be presented with regard to the compatibility profile, the advantages and the disadvantages of biopesticides, the cost of the implementation process, and to the potential of combined use of biopesticides.

1 Introduction

In greenhouse crops, microbial pesticides are generally used as part of an integrated pest management programme (IPM). They are mainly used as an inundative release strategy from which direct effects are expected resulting in a considerable decrease of the pest population within a relatively short period. In greenhouse vegetable crops, the main biocontrol components of the IPM system are natural enemies. Microbial pest control products (MPCPs) are often used as selective and corrective measures, or as additional pest control tools. The same can be seen in ornamental

W.J. Ravensberg, *A Roadmap to the Successful Development and Commercialization of Microbial Pest Control Products for Control of Arthropods*, Progress in Biological Control 10,
DOI 10.1007/978-94-007-0437-4_6,

crops such as roses and chrysanthemum, but in other ornamental crops they are used in a more chemical oriented programme. In such a case, they are used as a stand-alone product against a particular pest, or as an additional control product. MPCPs can be used to control key pests, often in combination with other methods, or they can be used against secondary or minor pests. As justly stated by Hofstein and Fridlender (1994): "Design of a comprehensive disease control program is the main challenge of those who intend to introduce biological pesticides as an attractive alternative or addition to existing control programs". The same is of course true for a pest control programme. The concept of an IPM programme is to have an optimum exploitation of all the pest control options with a strong emphasis on preventative measures and biological control methods. This implies a good understanding of all components and their interactions, in any given cropping system.

The concept of integration of pesticides in IPM programmes in greenhouses has been reviewed by Blümel et al. (1999). They described the importance of selective pesticides, and sequential test methods on detecting side-effects on natural enemies. Furthermore, approaches were suggested on pesticide application methods and on how to minimize negative effects on beneficials. Little attention, however, was given to microbial pesticides. An introduction on compatibility of pesticides, including MPCPs, with biological control agents is given by Hassan and Van de Veire (2004). They briefly reviewed the use of pesticides, including microbials, in IPM systems in greenhouse crops.

Given the entomopathogenic character of MPCPs and their use together with beneficial arthropods, as well as the use with other pesticides, a "two-sided" compatibility profile of a MPCP with these other crop protection components must be established. This is indispensable for successful implementation and commercialization of a MPCP. In the literature compatibility of entomopathogens with other methods of pest control has been studied for many cases. This kind of research is usually aimed at investigating whether there are any adverse effects of chemical pesticides on MPCPs, or whether MPCPs have any adverse effects on natural enemies and pollinators. The latter is sometimes called biocompatibility. An early overview of research results of entomopathogenic bacteria, fungi, viruses, nematodes and protozoa and compatibility with chemicals and beneficial arthropods is given by Jacques and Morris (1981). They also reviewed the combined use of entomopathogens with pesticides. Examples were given of enhanced as well as of reduced effects. The option to combine a MPCP with other methods may give improved control through additive or synergistic effects. Compatibility studies and combination effects have been investigated by many scientists with all four types of pathogens treated in this book. It is not possible to review all these studies. I will highlight the most significant publications below with the focus on greenhouse crops.

The efficacy of entomopathogens depends on a number of factors such as environmental and agronomic factors. These relationships need to be studied. Some of the factors can be influenced so that entomopathogens can be used in a more effective way. Inglis et al. (2001) reviewed the importance of temperature, humidity, sunlight, soil type, rainfall, and interactions among these environmental factors, and their influence on the activity of fungal entomopathogens. Understanding these influences is vital for developing an effective IPM programme in various cropping systems. The authors illustrated this by examples from glasshouse crops (aphids) and field crops (Colorado potato beetle, European corn borer and whiteflies). Compatibility with chemicals as well as with beneficials is reviewed, and innovative strategies such as using semio-chemicals and trap crops, were discussed. These factors are also relevant for entomopathogenic bacteria, nematodes and baculoviruses, and vary per organism, target pest and cropping system.

In the development of a MPCP, following product development, the first step is to establish the optimal application strategy under commercial conditions. The product should be used in such a way that its pest control features are fully exploited within the given situation. The application strategy needs to take into account factors such as the delivery technology, the biology of the pest, the crop and how it is cultivated, and the environmental conditions. Each factor likely to influence the entomopathogens efficacy needs to be considered and studied, if deemed critical, in the application strategy. Secondly, the use of the MPCP must be implemented in the entire pest and disease control programme. An integrated programme needs to be established, with optimal effects on the pests and minimal harmful effects to all the biological pest control agents. These two aspects are the cornerstones of a successful implementation of a new MPCP in commercial horticultural crops. I will discuss how to consider these pivotal elements in the development and integration of a MPCP in an IPM system.

Furthermore, I will discuss other relevant aspects related to this topic such as the chance of resistance with pathogens, practical aspects of testing compatibility, the transfer of knowledge to the end-users, and the cost considerations concerning implementation of a MPCP within an IPM programme. Finally, I will provide recommendations and conclusions on the critical steps in the implementation of a MPCP in a multi-component pest and disease control system, and on the potential of combinations of control agents to enhance biological pest control.

2 Development of an Application Strategy

In order to achieve effective results with a MPCP against a certain pest, an optimal application strategy needs to be developed. For a "broad-spectrum" MPCP, this may differ per target species. First of all, the level of the pest determines when and how to intervene. Vital factors therein are the application method, the frequency of applications, intervals between applications, the optimal dosage, the spray volume used, and appropriate equipment. Furthermore, the strategy is strongly influenced by the crop, the cultivation methods, the type of greenhouse, the climate, etc. (Fig. 6.1). These parameters influence the target pest as well as the pathogen. The plant-host-pathogen is a tritrophic system that is influenced by a multitude of factors which results in complex study systems. To investigate all these relationships is impossible and it is obvious that the most influential factors should be determined and studied where they are expected to have a relevant effect on the performance of the

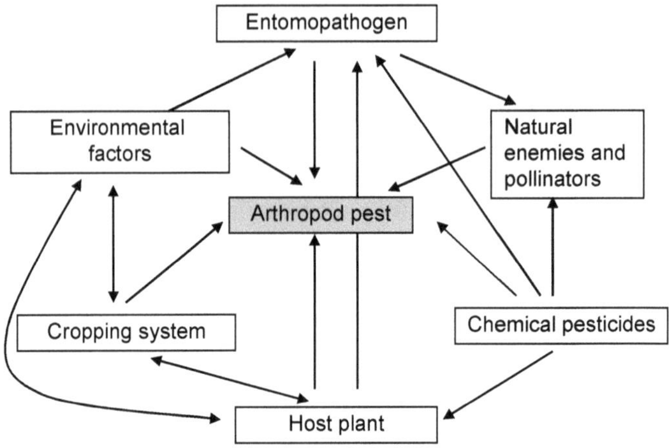

Fig. 6.1 Interactions that need to be considered in the development of an application strategy and in the implementation of a microbial insecticide in an integrated pest management system

biopesticide. In the literature, only few examples have been presented where the authors have investigated and established a complete and effective application strategy for a pathogen. Depending on the pathogen and its biology, particular aspects have to be considered, and I will discuss them below.

Factors influencing activity and persistence of Bt and its toxins are reviewed by Glare and O'Callaghan (2000). The fate of Bt is only known in general terms. Persistence can be years in soil, but only days in the phyllosphere. Little is known from greenhouse environments. Ingestion of Bt by the pest organism usually follows application quickly and therefore environmental factors influencing survival of Bt are less critical as with fungi. Generally, insects become less susceptible to Bt as larvae become older. The same is known for entomopathogenic fungi and baculoviruses. If possible, this should be taken into account with regard to timing of the treatments. Examples of greenhouse application strategies with Bt are hard to find in literature, most likely because Bt behaves almost like a chemical and is less influenced by crop and environmental parameters.

Entomopathogenic fungi may pose the greatest challenge of the pathogens treated in this book due to their mode of action. This is exemplified by a thorough study of application parameters on efficacy of *Beauveria bassiana* for control of *Frankliniella occidentalis* (Ugine et al. 2007a). The objective was to develop a spray application guideline for maximum efficacy of *B. bassiana* in a greenhouse ornamental crop. They investigated the application rate, the interval between applications, the water volume and the timing of the sprays during the full growing cycle of impatiens. Thrips population reduction was the highest when plants were sprayed multiple times with 5-day intervals with high rate and high volume applications. Early treatments in the plant cycle gave larger suppressive effects than later treatments. But Ugine et al. (2007a) also found great variation in the population reduction between spray programmes. This extensive study illustrates the complexity of establishing an effective spray programme and the variations that may occur. Another example is the use of the entomopathogenic fungus *M. anisopliae* for thrips control in plant-growing media in combination with chemicals (Ansari et al. 2007). The fungus proved to be a robust biocontrol agent giving good control independent of the growing media, the method of application (pre-mixing resp. drenching) and whether used alone or in combination with insecticides. Earlier work had shown similar results against the black vine weevil (Bruck and Donahue 2007; Shah et al. 2007a).

The use of *Spodoptera exigua* NPV in ornamental crops and tomato was investigated by Smits et al. (1987). Reduction of feeding damage was important in ornamental crops and virus application timing should be directed for young caterpillars as early as possible. Sprays should be directed to the underside of the leaves since early instars mainly feed there. This is difficult in dense crop canopies as with chrysanthemum and gerbera, but feasible in tomato. The authors also recommended using pheromones for monitoring the population and optimal timing of sprays.

Nematodes are very different from traditional pesticides as well as from microbial pesticides due to their size and active behaviour. It is essential to understand the attributes and the limitations of these biocontrol agents. An overview of the crucial application factors is given by Koppenhöfer (2007), Wright et al. (2005) and Shapiro-Ilan et al. (2006). An example of developing an effective application strategy is given by Ehlers (2003a) for control of black vine weevil in strawberry. Their study showed that application through the drip-irrigation system did not give an even distribution to all plants. Instead, a dipping method with young plants was developed and this proved successful for the control of black vine weevil. Other interesting examples of the development of an application strategy are the use of nematodes with insecticides to control leafminers in lettuce and Chinese cabbage (Head et al. 2003) and to control fungus gnats with nematodes in poinsettia and impatiens, grown in various potting media (Jagdale et al. 2004). Application strategies with nematodes for glasshouse pests are reviewed by Tomalak et al. (2005).

Above examples illustrate the complexity of an application strategy and which factors need to be taken into account when an optimal application strategy is being developed. I will discuss some of these parameters more in depth.

2.1 Delivery Methods for Microbial Pest Control Products

Efficient delivery of the infective propagules is crucial for a good efficacy. A number of authors have addressed the issue of optimal application of biopesticides (Evans 1999; Matthews 2000; Bateman et al. 2007; Chapple et al. 2007). Because of the particulate and living nature of biopesticides, they pose a very different challenge to efficient delivery through appropriate equipment than chemicals. The available delivery system, however, is usually the existing equipment developed for chemicals. MPCPs should be capable of application through standard available equipment with minimal special requirements. Growers will not readily change or buy equipment just to apply MPCPs. Nor will they easily accept a very different spray regime or more frequent applications than is normal practice. Only minor adjustments are possible and acceptable for growers. As an example, pre-treatment soaking of Mycotal already turned out to be hampering adoption of this product (see Chapter 3 under Section 5.2). Taking out sieves when nematodes are applied or changing nozzle types for products with small particles are about the maximum accepted modifications.

In protected crops, the most used type of technology for foliar treatment is the high pressure hydraulic nozzle system. This ranges from knapsack sprayers to handgun sprayers, up to automatic spraying boom systems. To some extent other equipment is used such as air-blast sprayers, hot and cold foggers, and low volume misting systems (LVM). Biopesticides are generally applied by means of the hydraulic nozzle system. With these systems good coverage can be obtained, and nozzle position can be adjusted to cover the upper side or underside of leaves or even both. Foggers and LVMs are sometimes used, but the deposit hardly reaches the underside of the leaves, and for many pests coverage is inappropriate then and good control is not achieved. Spinning discs are not used in protected crops since it is not possible to penetrate into the dense canopy with this technique. The methods used to deliver biopesticides have been reviewed by Gan-Mor and Matthews (2003), and they argued that often the method of application has not been adequately considered. They recommended more research in the area of formulation and engineering, but the challenge remains getting farmers to buy and use new technology. On the other hand, I have seen growers constructing their own spraying devices or dripping application systems when they are convinced that this will bring a solution to their problems. It depends on their motivation and the results they can achieve. In general, however, one cannot count on this from the outset of the development of a new product or a new programme.

The application of nematodes differs from the application of bacteria, fungi and viruses because they are much bigger. For soil applications, coarse droplets can be used and this has been working well. Critical aspects, also for the microorganisms, but even more for nematodes, are sedimentation in the tank, pump pressure and the size of sieves and nozzle openings. Many studies have highlighted the critical factors of the application technology used for nematodes (Nilsson and Gripwall 1999; Fife et al. 2003; Wright et al. 2005; Laczyński et al. 2006, 2007) and alternatives as spinning discs have been suggested for foliar treatments (Mason et al. 1999). Optimizing the parameters for application of EPNs is still a topic for research after 30 years of using nematodes for pest control (Brusselman et al. 2007).

2.2 Applications and Timing, Frequency, Intervals and Combinations

The timing of an application is essential in respect to controlling the pest and/or keeping the damage acceptable. The biology of the pest, the level of the pest population and the mode of action of the pathogen must be considered, and thus well known. Often, there is a difference in susceptibility between the life stages of a pest and the application should be targeted at the most susceptible life stage. This is possible when the insect population shows distinctive generations, but generally generations start to overlap later in the growing season and then this becomes increasingly difficult. In greenhouse crops with favourable conditions for a pest with a short life cycle, it takes repeated treatments before the pest population starts to decrease considerably. This has been demonstrated with whitefly and thrips (Ravensberg et al. 1990). Intervals between treatments and the frequency are determining factors then. Determination of the time of spraying is very important in MPCPs. The time of the application during the day can be determinative for the result of an application when environmental conditions influence the survival and activity of the pathogen. To avoid high levels of UV, high temperatures or low humidity, it is often advised to spray pathogens early in the day or late in the afternoon. It has even been suggested to spray shortly before shipping plants to customers in order to control insect pests (Osborne et al. 2008). Conditions during shipping allow the fungus *Isaria fumosorosea* (formerly *Paecilomyces fumosoroseus*) to infect and kill larval stages of *Bemisia tabaci* in poinsettia.

Combined treatments are often carried out by growers for reasons of convenience and cost savings. Most of the time, these combinations (tank-mixes) consist of two chemical pesticides and are targeted at different problems, like aphids and powdery mildew, or caterpillars and whitefly. Tank-mixes are not often carried out for the control of a single pest. Combined treatments including a MPCP, or of two MPCPs, has not been common practice, at least not in the Netherlands. Alternate use of products with different mode of actions to control a pest is more commonly applied; including MPCPs. Resistance management is one of the reasons for this. It can also be that older larval stages or adult stages of the pest require a different product. When a microbial pesticide is used in a tank-mix or in an alternating scheme, the compatibility must be known, either of direct effects or of residual effects. The number of combinations is endless and in order to achieve the optimal effect of an entomopathogenic product in an integrated programme, it requires a lot of knowledge on many interactions. This is discussed below. Further to using a biopesticide as a blanket treatment, biopesticides can also be used as a spot treatment, or as a band spraying (top or lower bands), creating separation in location and making it compatible in this way. For instance, a product may be sprayed on the top of the plants only, leaving natural enemies lower in the crop unharmed.

2.3 Environmental Effects

Abiotic factors are salient parameters that can influence the results of MPCPs. The most important one is temperature; others are relative humidity or availability of free water, solar radiation and wind. Soil or other growing medium conditions must be considered when pathogens are applied to the growing medium, as with nematodes and fungi. The environmental conditions can be controlled to some extent in greenhouse crops by computerized environmental control. In temperate regions growers use advanced techniques and modern greenhouse structures to control the environment as much as possible. Completely closed greenhouses are a new trend in which climate control is fully regulated. In Mediterranean greenhouses, climate control is much less advanced and temperature and humidity ranges are large. For an overview of greenhouse structure factors influencing biological control, I refer to Lindquist and Short (2004). Consequently, greenhouse technology in a broad sense has an impact on the activity of entomopathogens. Mediterranean greenhouses are also much more open for reasons of cooling, and wind can have a great effect on relative humidity, even at the microclimate at the leaf-boundary level which is critical for the efficacy of fungi (Fargues et al. 2003a). UV radiation is harmful to many pathogens, but under glass or plastic this effect is greatly reduced. Nevertheless, UV can still be harmful and various types of plastics can have different effects (Costa et al. 2001).

Butt (2002) and Jaronski (2010) reviewed the environmental factors that influence the success of fungi against insects. It has been often mentioned that epizootic effects of pathogens are very valuable and a marked difference from chemicals. I have witnessed wonderful epizootic effects in cucumber and chrysanthemum with *L. muscarium* on whitefly and with *L. longisporum* on aphids in the 1980s following treatments of the commercial products. This worked in low greenhouses with cucumbers where the RH was very high at certain periods of the year. A similar situation occurred under the plastic covers used for decreasing the day length period to induce flowering in chrysanthemum. But since these technologies have changed dramatically, such favourable conditions for fungi no longer occur. In using biopesticides, epizootics cannot be counted on, and moreover, they are unlikely to occur. I will come back to this later in this chapter.

2.4 Crop and Cultivation Effects

The cropping system, the crop structure and cultivation techniques also influence the efficacy of pathogens. The greenhouse climate is to some extent influenced by the spacing of the plants and by the density of the canopy. A grower will focus on production and decides on the basis of optimal yield for a certain plant density, and for growing techniques such as watering regime, fertilization, temperature settings, usage of screens for energy saving or for temperature control. For instance, modern Dutch greenhouses differ enormously from plastic covered crops in the Mediterranean basin in cultivation technology and climate. Still, surprisingly, Fargues et al. (2003b) found no difference in efficacy of *L. muscarium* (Mycotal) on the greenhouse whitefly between a sophisticated glasshouse and a polyethylene-covered greenhouse. Shipp et al. (2002) demonstrated that the canopy of the plant affects the infection levels of pest insects using *B. bassiana*. Whitefly and thrips infection levels were higher in the top of the plant than in the middle canopy, but this was less prominent for the cotton aphid. These differences were both found in greenhouses with a low and with a higher relative humidity.

For pathogens used against pests or life stages of a pest occurring in the growing medium, other aspects need to be checked such as the influence of the growing medium on the efficacy of the product. Shah et al. (2007b) demonstrated that both the application method and growing medium influenced conidial leaching when conidia of *M. anisopliae* were applied as a drench or premixed into the medium. Inoculum losses were greater following drench application than premixing irrespective of media type. Ansari et al. (2008a) demonstrated that mortality of soil-dwelling life stages of the western flower thrips with fungal pathogens was similar in different growing media such as peat, coir, bark and mixtures of these with green waste compost. Spores of the fungi *M. anisopliae, B. bassiana* and *I. fumosorosea* (*P. fumosoroseus*), either applied as a drench or incorporated as a mix, caused a high mortality irrespective of the growing medium and the method of application, indicating that these fungi can be used as effective biocontrol agents in a range of growing media. These examples show that factors that are expected to negatively affect efficacy are not always that important, illustrating that research is required to show what factors are critical.

2.5 Host Plant-Mediated Effects

Host plants may directly or indirectly influence survival and efficacy of pathogens. Plants produce a wide range of phytochemical substances which may directly influence the pathogen's survival on the leaf. Plants also may influence the pest's susceptibility to the pathogen. These tritrophic interactions between host plants, host insects and entomopathogenic bacteria, fungi and viruses are reviewed by Cory and Hoover (2006), and by Cory and Ericsson (2010) specifically for fungal entomopathogens. Very little is known about nematodes and plant-mediated effects. Plant structure, plant morphology and leaf characteristics (leaf size and shape, wax layer, density of hairs, number of stomata) will have an effect on the microclimate which will influence the conditions for survival and germination, and thus efficacy. Inbar and Gerling (2008) reviewed these host plant interactions between whiteflies and natural enemies, including entomopathogens. Most studies on plant-mediated effects are carried out with fungi, most likely due to their mode of action and their dependency on the environmental conditions.

It has been shown (reviewed by Navon 1993, and by Glare and O'Callaghan 2000) that host plants can strongly affect the insect's susceptibility to Bt. Even between host plant cultivars and within a season, susceptibility can vary. A brief overview of these host plant effects in entomopathogenic fungi is given by Butt (2002). An example to illustrate the extent of host plant factors is a study by Ugine et al. (2005) who showed that the susceptibility of western flower thrips reared and exposed on kidney beans was six times greater to *B. bassiana* than thrips reared and exposed on impatiens. The mechanism behind this effect was a difference in conidial acquisition between treated bean and impatiens leaves (Ugine et al. 2007b). The underlying mechanism of differential rates of conidia pick up has not been determined. Differences in susceptibility to fungal pathogens that were dependent on the host plant on which the insect pest was reared were also demonstrated for *B. tabaci* (Poprawski and Jones 2001) and for *Trialeurodes aleyrodis* (Poprawski et al. 2000) for the two fungi *B. bassiana* and *I. fumosorosea*, and for *B. tabaci* to *B. bassiana* (Olleka et al. 2009). Other studies revealed differences between host plants and persistence of spores as well as mortality by fungal insecticides. Survival of spores of *Aschersonia aleyrodis* was influenced by the host plant, as was mortality of whitefly larvae (Meekes et al. 2000). Yet, Vidal et al. (1998) did not find significant differences in mortality of *B. argentifolii* by *I. fumosorosea* between cucumber, tomato, cabbage, sweet pepper, nor between three tomato cultivars. Comparative trials with *S. feltiae* or *L. muscarium* against *Thrips palmi* revealed no differences in mortality with either of the biocontrol agents on chrysanthemum, tomato or sweet pepper leaves in laboratory trials (Cuthbertson et al. 2005a).

The activity of baculoviruses can also be influenced by the host plant. An elaborate overview of theses effects is given by Cory and Myers (2003). Host plant chemicals can negatively influence the efficacy of baculoviruses (Hoover et al. 1998) and this can differ largely between plant species and even between plant parts (Ali et al. 1998).

2.6 Relevant Considerations for the Development of an Application Strategy

Developing an application strategy requires a case by case approach. The biology of the pathogen and its susceptibility to biotic and abiotic factors may affect its efficacy to a large extent. The application strategy needs to take these factors into account as much as possible (Table 6.1).

Applying the product with appropriate equipment is clearly crucial. This may look like an open door, but too often this element is underestimated. The equipment must be well adjusted so that the propagules are not damaged by the application and that good coverage is achieved. Because of the particular nature of the products, they work by contact or digestion, and optimal coverage determines the results. This is more critical with microbials than with chemicals which often have a translaminar or systemic uptake in the plant.

It is of paramount importance to know the influence of environmental effects on the performance of the MPCP. Growers generally are reluctant to adapt their climate to optimize the activity of the products, so one need to find to best window to apply

Table 6.1 Essential factors that need to be considered in the development of an application strategy of a microbial insecticide

- Mode of action of the entomopathogen
- Application method and adjustment of equipment
- Optimal dosage and spray volume
- Timing of application, depending on pest level and presence of susceptible life stages of the pest
- Treatment frequency and intervals
- Host plant and cropping system
- Environmental conditions

the product. Conditions and time of application should be chosen in such a way that it will result in a good activity of the pathogen. Crop, cultivation and host plant effects cannot be avoided, so it is important to know that they may play a role and to find a solution in that particular case. In ornamentals, various plant species and cultivars may be grown concurrently, and it cannot be assumed that the MPCP will work equally effective on all host plants.

It is necessary to know to which factors a pathogen is particularly sensitive. In fungi, the most critical factor is germination and the dependency on the relative humidity or available free water. In bacteria like Bt and in baculoviruses this is much less critical due to their mode of action via *per oral* uptake. The use of nematodes requires other considerations. The most relevant factors should be identified in each case and studied in order to develop a successful application strategy.

3 Development of an Integrated Pest Management Strategy

The goal of an IPM programme is to bring together all possible control options in order to achieve an efficient, sustainable and cost-effective control of the pest, or as in most cases, a complex of pests and diseases. For a MPCP that will be introduced as a new element in an existing IPM programme, many variables and a number of multi-trophic interactions have to be considered and studied. It is impossible, however, to investigate all these interactions, and the critical topics need to be selected based on knowledge of the host and the pathogen, and within a certain cropping system. The relevant aspects of such a tritrophic system with regard to the application strategy of the product are discussed above, but in an IPM system many other interactions occur (Table 6.2). Means of control of the same pest are often integrated, as well as control means against other pests and diseases in the same crop. The MPCP could potentially interfere with these, or the other way around. Therefore, it is required to identify and study these potentially interfering interactions such as the effects of chemical crop protection products on the MPCP, and the effects of the MPCP on natural enemies and pollinators. A "two-sided" compatibility profile of the MPCP needs to be determined. This is

Table 6.2 Important elements that need to be considered for a successful use of microbial pest control products in an IPM strategy

• Compatibility of chemical pesticides with MPCPs
• Compatibility of MPCPs with natural enemies and pollinators
• Interactions between MPCPs and natural enemies
• Combinations of different types of MPCPs
• Combined use of MPCPs with other pesticides
• Combined use with other control techniques
• Knowledge transfer and training of technical consultants, extension workers and growers
• Inclusion of a MPCP in an existing IPM system
• Costs of a MPCP and its application

a large research programme. The most relevant interactions have to be identified and, initially, only they need to be studied. Once a MPCP is used on a larger scale, feedback from its use may reveal new topics of concern which then need to be studied in order to give proper advice to limit negative interactions. Ultimately, an IPM approach should be developed in which MPCPs can obtain maximum effects, without interfering with the effectiveness of other practices. This is an ideal scenario. In practice negative effects may occur. Intervention is possible, and, for instance, new releases of natural enemies can be made when necessary. Monitoring after applications is needed to check whether control of the target pests has been achieved and is maintained, and how the populations of the biocontrol agents develop.

To my knowledge, no studies have been published covering this topic in full for a specific MPCP in a greenhouse crop. Nevertheless, it is generally well understood that proper incorporation of a biopesticide into an IPM programme is essential. In the early days of fungal insecticides, when Vertalec was developed for control of aphids in chrysanthemum, successful integration with fungicides was the prime concern. Many practical aspects and the compatibility were studied. Recommendations for use were adapted from "lessons learned" from the laboratory and the field (Quinlan 1988). Another example is given by Lomer et al. (1999) who reviewed the development of a strategy of incorporation of *M. anisopliae* in the control programme of locusts. An example of integration of pest and disease control methods for greenhouse tomato is given by Bardin et al. (2004) who studied control of botrytis, powdery mildew and whitefly, using three biological control agents. A review of IPM development with microbial pesticides, including interactions, combinations and mixtures is presented by Dent (1997). He concluded that, up to the mid 1990s, little research had been performed on true integration of control measures and that farmers were left to sort it out themselves. Since then this has received more attention and some examples will be given.

Entomopathogenic fungi used for foliar pests appear to be the most difficult with regard to application and integrated pest management strategies. First, these organisms are, due to their mode of action, more dependent on environmental conditions and plant-related aspects than are bacteria and baculoviruses. Second, fungi potentially have a broader host range including non-target organisms, which requires a better understanding of the relevant interactions. Third, chemical fungicides are often used in these IPM systems and their effects on the entomopathogenic fungi can be deleterious and therefore must be investigated. Most examples on side-effects of pesticides on microbial agents in the literature are focussing on fungi. With Bt and viruses these aspects are less critical and therefore these pathogens are easier to integrate in a crop protection system. For nematodes the same is true.

Testing compatibility of entomopathogens with pesticides is required to see if adverse effects of any given combination are occurring, and if so, to what degree and for how long these persist. Knowing these adverse effects, often called side-effects, gives the opportunity to avoid them or to intervene and repair the situation. Many interactions occur through unintentional sequential or simultaneous use of various plant protection products. On the other hand, an obvious approach is to control a pest by combined use of two different control agents. This is aimed at enhancing control, at giving quicker control of the pest, or at reducing the damage more rapidly. Combined use requires knowledge of their interactions and to what extent they contribute to improve control.

The optimum IPM strategy using a MPCP should be studied in the developmental phase of a product. There are, however, numerous pest control programmes exploited which differ per crop, per season, per country and region, etc. These programmes continuously change due to new cultivation techniques, new cultivars, new pests and diseases, and due to new pesticides and disappearance of pesticides. Also socio-economic aspects play a role in a pest control programme and these also give rise to changes. This means that initially only basic interactions can be studied which will help in designing an optimal pest control programme. Such programmes, however, need to be evaluated regularly using field experience and feedback from users and advisers in order to adapt them to current and new developments in a broad sense.

Interactions between MPCPs and other control methods can be studied in the laboratory. This often reflects a direct exposure scenario as optimal contact is achieved in bio-assays or in small scale greenhouse trials. In case there are no adverse side-effects in laboratory tests, it can be assumed that this will be the same in the field. However, negative side-effects of a certain combination of agents may be overestimated in such testing scenarios. Practical aspects of testing compatibility and combined use will be discussed below.

3.1 Compatibility of Chemical Pesticides with Microbial Pest Control Products

In the registration process of plant protection products, manufacturers are obliged to test their product for effects on non-target organisms, including natural enemies and pollinators. Data on compatibility with other plant protection products (chemical as well as biological) is only required when the use of certain tank mixtures is to be authorized and recommended on the label. Generally, this is not the case and side-effects of chemicals on biopesticides are thus not demanded by regulatory authorities, and tests are rarely conducted by the agrochemical companies. The same holds for the registration of biopesticides: compatibility tests with chemicals are not required. This leaves the task to the biopesticide manufacturer to test susceptibility of the product to a range of relevant chemicals so that proper advice can be given for use in IPM systems. The effect of biopesticides on chemical pesticides is usually not tested, and I do not know of an example where a biopesticide negatively influences the activity of a chemical pesticide.

In IPM programmes in greenhouse crops, chemical pesticides are often part of the control measures. The use of fungicides is standard in most greenhouse crops. Due to their mode of action, insecticides are generally less harmful to entomopathogens than fungicides. Compatibility of chemical fungicides needs to be investigated, particularly on fungal MPCPs. EPNs, Bt's and baculoviruses are less susceptible to fungicides. Chemical bactericides are hardly available and are rarely used in greenhouses. Anti-viral products are not available.

Theoretically, MPCPs may also have negative side-effects on other MPCPs, but I do not know of any case where an entomopathogen directly affects the survival and activity of another; this is also rarely tested. But for all cases it is true that if a product shows a reduced activity, one must never assume that another pest control product cannot have an effect, always test it when there is any doubt.

3.1.1 Side-Effects of Chemical Pesticides on Entomopathogenic Fungi

Numerous studies have focussed on the compatibility of chemical pesticides with MPCPs. For entomopathogenic fungi, overviews of these effects are given by Inglis et al. (2001) and (Shah et al. 2009). Inhibitory effects are reported, as well as additive and synergistic effects. Recently, Cuthbertson et al. (2005b, 2008a) investigated the compatibility of some chemical and natural insecticides with *L. muscarium*. Exposure to the chemical insecticide solution at the recommended rates for 24 h decreased spore germination dramatically for teflubenzuron, imidacloprid and nicotine, and for some natural insecticides. Germination was only moderately reduced by spiromesifen and polysaccharide/alginate, and not affected by buprofezin. When the fungus was applied on 24-h dry residue of the chemical, there seemed to be no inhibition. Generally, fungicides have adverse effects on fungal pathogens. Direct as well as persistent effects may occur and could impede the activity of the mycoinsecticide. Often the effect is prominent on foliage. In soil and other growing media the effects may be less deleterious as demonstrated for *Metarhizium anisopliae* and a range of fungicides (Bruck 2009). But all fungicides need to be tested, and depending on the outcome, the use of the MPCP and the fungicide must be well separated in time or space.

3.1.2 Side-Effects of Chemical Pesticides on Entomopathogenic Bacteria and Baculoviruses

Side-effects of chemicals are rarely tested on bacterial and viral insecticides. This is due to the nature of these organisms and their propagules which are not susceptible to chemical pesticides. There is no report of any problem of integration with chemical pesticides, and bacterial and viral insecticides are often tank-mixed with chemicals. Baculoviruses, however, cannot be tank-mixed with copper containing products. Further, they are inactivated at low and high pH values of a tank-mix ($5 < pH > 8.50$), and by disinfectants such as hypochlorite.

3.1.3 Side-Effects of Chemical Pesticides on Entomopathogenic Nematodes

Dauer juveniles of EPNs are relatively tolerant to many substances, including chemical pesticides, but they may be very susceptible to nematicides. In the IOBC Working Group "Pesticides and Beneficial Organisms" nematodes have been incorporated in the testing programmes since the early 1990s and a laboratory method has been developed to test side-effects of chemicals on nematodes (Vainio 1992). Results can be found in subsequent IOBC bulletins of this Working Group. More results are in publications of Rovesti (1991), Vainio (1994), Peters and Poullot (2004), Gutiérrez et al. (2008) and in a review on compatibility of EPNs with chemicals by Koppenhöfer and Grewal (2005). The effects of adjuvants and surfactants have also been tested and they can be toxic to EPNs (Peters and Poullot 2004). Natural products like neem, neem oil, cynnamaldehyde and soaps need to be tested too and some have shown a considerable toxicity to the dauer juveniles of EPNs (Krishnayya and Grewal 2002).

3.2 Compatibility of Microbial Pest Control Products with Natural Enemies and Pollinators

The compatibility of microbial pest control products with natural enemies often is a subject in academic research as part of host range and non-target effects testing. For regulatory requirements this aspect has achieved relatively little attention. This is in contrast to chemical pesticides. The compatibility of chemical pesticides with natural enemies has been extensively tested for regulatory purposes and for integration into IPM programmes with natural enemies. Testing the effects of pesticides on non-target arthropods is required for registration purposes in Europe according to the Council Directive 91/414/EC. It is obligatory to test side-effects on two model arthropods, the natural enemies *Typhlodromus pyri* and *Aphelinus rhopalosiphi*, and on honeybees. These natural enemies are, however, not relevant for greenhouse IPM programmes. In the past, agrochemical companies only performed these obligatory tests, but since IPM has become standard in greenhouses, the testing of important greenhouse natural enemies is considered more and more (Schnorbach 2006). An example is the testing of side-effects of spinosad in order to develop an IPM system (Miles 2006). Testing methods and evaluation procedures have been developed for chemical pesticides and non-target arthropods within or in conjunction with experts of the IOBC Working Group "Pesticides and Beneficial Organisms" and in the ESCORT Working Group (Barrett et al. 1994; Candolfi et al. 2000). For regulatory purposes, a tiered testing and assessment schedule is provided, and decision criteria and trigger values exist for both in-field and off-field assessment. When products are only applied in greenhouses, off-field assessment is not needed.

There are no specific testing guidelines for testing microbial plant protection products on beneficial organisms. Requirements for microbial pesticides are given in Directive 2001/36/EC. Information on non-target toxicity, infectiveness, and pathogenicity must be reported for the active ingredient as well as for the formulated product. This is required for bees and for arthropods which may be exposed to the plant protection product, with special attention for organisms used for biological control. The Directive does not specify which testing methods to use.

MPCPs have been incorporated in the IOBC side-effect testing programmes. Since 1980, standard guidelines for testing the side-effects of chemical pesticides on natural enemies and for rearing methods of beneficial arthropods have been developed. Results of laboratory, semi-field, and field experiments, and of the joint programmes to test the side-effects of pesticides on beneficial organisms have been published in the IOBC/WPRS Bulletin, the EPPO Bulletin, and various international scientific publications, see http://www.iobc-wprs.org/wg_sg/index.html. The IOBC Database on selectivity of pesticides lists all available information, including test results of biopesticides on natural enemies and bees (http://www.iobc.ch/news.html). The effects of chemical pesticides on microbial insecticides have been tested, mainly on entomopathogenic fungi (*B. bassiana, B. brongniartii, L. muscarium* and *M. anisopliae*), and on entomopathogenic nematodes (*S. carpocapsae* and *S. feltiae*). Effects on bacterial and viral products have not been tested. The effects of microbial insecticides (Bt and *L. muscarium*) on beneficial arthropods have been tested as well. Only a limited number of MPCPs has been part of the testing programme of the IOBC WG; baculoviruses have not been incorporated until now. The IOBC Working Group has also started to study the effects of fungal insecticides on beneficial organisms.

Generally, MPCPs appear to be harmless for natural enemies. Bt's and baculoviruses are safe to natural enemies while fungi and nematodes may affect some natural enemies. These effects are usually limited to some natural enemies and are minor. Many studies have been published on compatibility of entomopathogens and natural enemies, however, reports on tests on greenhouse beneficial arthropods are limited. For instance, Sterk et al. (2003) tested the effects of PreFeRal (*I. fumosorosea*), a whitefly pathogen, and Scutello (*Bt* subsp. *kurstaki*) and Xentari (*Bt* subsp. *aizawai*) on four greenhouse natural enemies. There were no adverse effects found on *Phytoseiulus persimilis, Macrolophus caliginosus, Aphidius* spp. and *Encarsia formosa*. I will briefly review the effects of each type of entomopathogen on greenhouse natural enemies, followed by overall recommendations for testing side-effects of MPCPs on these organisms. Side-effects of chemical pesticides and MPCPs on natural enemies, as well as side-effects of chemicals on MPCPs can be found on websites of the larger producers of natural enemies like Biobest and Koppert Biological Systems, and on the website of the IOBC/WPRS. Manufacturers and distributors of biopesticides study side-effects in order to know their product well and to give proper advice (Ravensberg et al. 1994; Sterk et al. 1995a).

3.2.1 Side-Effects of Entomopathogenic Bacteria on Natural Enemies

Subspecies and strains of Bt are highly specific to their respective hosts and have no effect through direct infection on natural enemies. Melin and Cozzi (1990) and Glare and O'Callaghan (2000, 2003) reviewed safety of Bt's to beneficial insects. The IOBC Working Group "Pesticides and Beneficial Organisms" has tested Bt on many natural enemies and found it to be harmless. Overall, Bt has rarely been found to be toxic to natural enemies at field dosages. Contradictory effects have been reported for some predators such as *Chrysopa* spp.: a brief review of these studies is given by Glare and O'Callaghan (2000). A direct toxic effect has been found on one predatory mite, *Metaseiulus occidentalis* from *Bt* subsp. *tenebrionis* (Chapman and Hoy 1991). Furthermore, various interactions between Bt and parasitoids and predators do occur with the same host. Egg parasitoids are fully compatible with Bt, but the interaction with larval parasitoids can be deleterious. The outcome of the competition is dependent on the interval between Bt infection, the ingested dose and the oviposition of the parasitoid. Indirect effects are reviewed by Navon (1993, 2000). In general, Bt has no or little toxic or pathogenic effects on natural enemies and Bt's can be safely integrated in an IPM programme.

3.2.2 Side-Effects of Entomopathogenic Fungi on Natural Enemies

These fungi can have a broad host spectrum and direct effects on natural enemies may occur. Particularly, strains of the species *M. anisopliae* and *B. bassiana* generally have a relatively broad host range, and testing on negative side-effects is indispensible for use within an IPM programme. Many papers report on non-target effects of fungi, including beneficial arthropods. Goettel et al. (1990) reviewed the non-target effects of entomopathogenic fungi and concluded that they "do pose inherent, albeit minimal risks". Indirect effects on parasitoids and predators may occur by reduction of host populations, but these are usually not lasting effects. A more recent overview of the safety of Hypocrealean (formerly Hyphomycetes) fungi on non-target organisms, including natural enemies used in greenhouses, is given by Vestergaard et al. (2003). The authors reviewed the information from laboratory studies and field studies of direct and persistent effects. They concluded that these fungi generally are remarkably safe to non-target organisms. This is exemplified by a detailed study on the combined use of *Amblyseius cucumeris* and the products Botanigard and Naturalis-L (both based on strains of *B. bassiana*) which illustrated the safety of this fungus to this commonly used predatory mite for control of *F. occidentalis* in cucumber (Jacobson et al. 2001). Other studies demonstrated negative effects for a number of insecticidal fungal species and strains, including registered mycoinsecticides, which are used in greenhouse crops. Effects were strongly dose dependent. The least harmful were strains of *L. muscarium* and *L. longisporum*, whereas strains of *B. bassiana, M. anisopliae* were harmful to non-target beneficial insects. *I. fumosorosea* had intermediate properties. Two kinds of studies were conducted: direct exposure and indirect exposure (i.e. via the host plant). As expected, direct exposure resulted in higher mortalities. Predatory mites were less susceptible than most of the other beneficial arthropods, they were even resistant to all the fungal species and strains tested (T. Butt, personal communication). Direct and sub-lethal effects are dose and species dependent as shown by Roy et al. (2008) with *B. bassiana* and three species of coccinellids. This study illustrated that grouping closely related natural enemies based on investigations of one species does not give reliable results. This calls for testing of each relevant species of natural enemies when dealing with broad-spectrum microbials. The study also showed that reduction in fecundity can be considerable. This warrants assessing sub-lethal effects besides mortality.

Regulators review all the tests carried out for determining the side-effects on non-target organisms and may come to more generic conclusions. This is illustrated by the product Botanigard (*B. bassiana*) which has been approved in the Netherlands for use in greenhouse crops, but has a warning on the label stating that: "it cannot be excluded that after application of the product at high humidity conditions beneficial insects can be infected by *Beauveria bassiana*. The effect of the product on beneficial insects should be monitored to prevent damage to the IPM system" (www.ctbg.agro.nl). The product Bio1020 (*M. anisopliae*) has been approved in the Netherlands for control of black vine weevil in strawberries, other berry crops, and ornamentals by means of a soil application. The label also has a warning stating: "This product is harmful to non-target arthropods. Avoid exposure". There are no such warnings on the products Mycotal (*Lecanicillium muscarium*) and PreFeRal (*Isaria fumosorosea*). The latter has been shown to be harmless for a range of natural enemies (Sterk et al. 1995a, 2003). In sum, the above illustrates that adverse effects of entomopathogenic fungi can not be excluded beforehand and that they need to be tested on beneficial insects used in an IPM programme.

3.2.3 Side-Effects of Baculoviruses on Natural Enemies

Baculoviruses are very host-specific and natural enemies are not susceptible to virus infections. Direct adverse effects of baculoviruses on beneficial insects have never been reported. An overview of tests and observations showing that deleterious effects do not occur is given by Gröner (1990). Hassan and Gröner (1977) tested direct effects of the *Mamestra brassicae* NPV on *C. carnea* and the egg parasitoid *Trichogramma cacoeciae* and found no direct harmful effects on both natural enemies. The nucleopolyhedrovirus of *Anticarsia gemmatalis* has been demonstrated to be safe to predators such as predatory bugs (Young and Yearian 1987) and pentatomid predators (Abbas and Boucias 1984). Carabid predators do not discriminate between healthy and diseased larvae of the cabbage moth *M. brassicae* and since virus infectivity was maintained after passage through the predator's gut, the beetles transfer the nucleopolyhedrovirus in the environment (Vasconcelos et al. 1996). The codling moth granulovirus showed no harmful effects on *Orius laevigatus*, an important predator used in greenhouses, after exposure to residue and by ingestion (Angeli et al. 2005). Baculoviruses have been used on many crops for many years and negative side-effects on non-targets have not been reported (Cory 2003). Testing for direct negative side-effects is therefore not relevant when baculoviruses are used as an inundative approach in an IPM system.

3.2.4 Side-Effects of Entomopathogenic Nematodes on Natural Enemies

When nematodes are applied to the soil, exposure to natural enemies is limited. Only for soil-dwelling biocontrol agents, or when natural enemies have a soil-dwelling life stage, is there a potential risk. When nematodes are applied to the foliage, many natural enemies are exposed and testing is necessary to establish a potential risk. Compatibility of nematodes with parasitoids and predators, and with bees has been reviewed by Akhurst (1990), Bathon (1996), Ehlers (2003b) and Koppenhöfer and Grewal (2005). Under laboratory conditions, nematodes appear to pose problems to many non-target insects; however, field evaluations showed that the use of nematodes had negligible impacts on natural non-target populations. Effects on larval stages of parasitoids depend on the time of infection by the nematode respectively the oviposition of the parasitoid of the host insect, and this could lead to reduced progeny of the parasitoids. Direct effects on the parasitoids have not been reported. Effects on predators are usually negligible in the field. Nematodes applied to foliage only survive for a short period of time, generally less than a few hours, and effects will be limited due to this fact and their limited mobility on leaves. Detailed studies on foliar applications and non-targets have not been performed. It can be safely assumed that direct negative effects of nematodes on beneficial insects in greenhouses will be limited and short-term.

3.2.5 Side-Effects of Entomopathogens on Pollinators

Honeybees and bumblebees are commonly used as pollinators in greenhouse crops, particularly in tomato, eggplant and strawberry. For the registration of pesticides it is obligatory to test contact and oral toxicity on honeybees. When a product is used in greenhouses, the effects on bumblebees are also required. This includes microbial pesticides. Given the pathogenicity to insects, harmful effects to bees cannot be excluded, particularly long-term effects on the development of the brood. The potential harmful effects of MPCPs on these beneficial insects need to be tested for registration and for integration of pollinators with MPCPs. Exposure of MPCPs is limited since these pollinators only visit flowers; they may be directly hit, however, during spraying. Bees may also collect entomopathogens by collecting contaminated pollen which is used as food for their brood. Indirect effects need to be considered too.

Bt's are reported to be safe for honeybees. Many studies have investigated direct effects on adult bees and on brood; a review is given by Glare and O'Callaghan (2000). Under field application conditions, the use of Bt is harmless to honeybees. Bt's have also been studied on bumblebees by Sterk et al. (2003). They found no harmful effects of Bt products on adult workers and on brood.

Entomopathogenic fungi can have a broad host range, depending on the species and the strain. Sterk et al. (2003) also tested fungal products on bumblebees. The fungi *I. fumosorosea* (PreFeRal) and *Trichoderma harzianum* (Trichodex) were completely safe to directly treated workers. In earlier work, Sterk et al. (1995b) tested PreFeRal on bumblebees through various ways of exposure: oral toxicity, *ad libitum* feeding, contact toxicity, direct spraying and inhalation toxicity, on workers and on brood, with three times the recommended field dose. There were no significant adverse effects found. But *B. bassiana* (Botanigard) showed detrimental effects on *Bombus terrestris* workers when the fungus was administered via sugar water or by direct topical contact. Another *B. bassiana* based product, Naturalis, was much less toxic to workers indicating that strain or product differences can be considerable, and testing each product is required to get reliable information. Sublethal effects on foraging behaviour were also observed. All these studies were laboratory studies and these indications suggest the need for further studies before conclusions about compatibility in the field can be drawn (Mommaerts et al. 2007,

2009). Hokkanen et al. (2003) investigated the effects of *B. bassiana* and *M. anisopliae* on several bumblebee species in laboratory as well as field studies and found that both fungi are potentially harmful to bumblebees. Honeybees are less susceptible to these fungi, although this depends on dose and strain (Butt et al. 1994). The product Botanigard (*B. bassiana*) has been approved in the Netherlands for use in greenhouse crops, but not in tomato. The label warning states: "Dangerous for bees. Remove or cover honeybee or bumblebee hives during the application of the product and during 16 h after the treatment" (www.ctgb.nl).

Baculoviruses are considered completely safe for honeybees and bumblebees. Few studies actually have been performed to test possible side-effects. Sterk et al. (2002) tested viral products on bumblebees. The *Cydia pomonella* GV (Granupom) and *Adoxophyes orana* GV (Capex) were completely safe when applied as a topical treatment and by feeding workers.

Beneficial nematodes are not applied in tomato, eggplant and strawberry as foliar treatments, so bees are not exposed to EPNs. Negative effects, however, can not be excluded, and in case foliar applications with EPNs take place, side-effect testing on workers and brood is recommended.

3.3 Integrated Use and Interactions Between Microbial Pest Control Products and Natural Enemies

In most cases, the use of MPCPs in IPM systems in greenhouse crops implies interactions with natural enemies, either direct or indirect. The outcome of these interactions will be negative, neutral or positive in terms of pest control. A MPCP could have direct negative effects on natural enemies that have been released to control the same pest as well as on natural enemies released for other pests. Pathogens with a broad host spectrum may infect natural enemies and kill them, or cause sublethal effects. These negative interactions are usually referred to as side-effects (see above). Indirect interactions among pathogens and natural enemies may impact populations of the natural enemy as well as the pathogen through competition-related effects. These intraguild interactions may also have consequences for the overall efficacy of biological control. For an extensive overview of trophic and guild interactions in relation to biological control, I refer to Brodeur and Boivin (2006). Intentional combinations of natural enemies and a MPCP to control the same pest will obviously be aimed at achieving enhanced efficacy. Another reason for using a MPCP may be the need as a corrective measure where a natural enemy is not able to suppress the pest sufficiently. Often, applications are temporally or spatially separated. Temporal is relative in this situation since most natural enemies are present in the crop for a large part of the season and interactions are bound to occur. But the release of a natural enemy and the application of a MPCP are usually separated in time. Next to spot treatments for corrective measures, MPCPs may be applied as a blanket treatment.

The purpose of intentional combined use is to enhance biological control. Wraight (2003) presented a detailed overview of synergistic interactions between insect pathogens and entomophagous insects. He also reviewed the terminology used describing these interactions and introduced clear definitions for synergistic, additive and negative interactions. I will use the definitions given by Wraight. A synergistic interaction results in a higher mortality rate than the level of combined independent uncorrelated joint actions (the combination of independent probabilities predicts a combined rate that is less than the arithmetic sum of the probabilities). An independent additive interaction is defined as when the effect equals the level of two independent uncorrelated joint actions. When the result is less than this, Wraight called this additive or sub-additive. When the effect is unchanged from the more effective agent, the interaction is called non-additive or neutral. When the effect is less than the effect of the more effective agent alone, it is called negative. Wraight considered the term antagonistic inappropriate to describe effects that are potentially beneficial even when they do not achieve a level of synergism. I agree with him and would use the term antagonism in the context of biological control when the purpose of the agent is to produce negative effects on a pest such as in biological disease control with biocontrol agents.

In the development of effective biological control programmes in which MPCPs are integrated with other natural enemies, an understanding of their interactions is imperative. Brooks (1993) reviewed interactions between hosts, parasitoids and pathogens. He elaborately described direct and indirect deleterious and beneficial aspects of interactions between baculoviruses, bacteria and fungi, and host insects and parasitoids. Parasitoids may be directly affected at the organismic level when they develop in infected hosts, as well by indirect effects at population level. A beneficial effect is when there is an increase of susceptibility for the pathogen in parasitized hosts, but this usually is at the expense of the parasitoid. Dispersal and transmission of pathogens may be increased by parasitoids. Wraight (2003) reviewed the various beneficial effects of interactions between control agents and hosts such as interactions that affect host immunity, host development, host behaviour, and host population dynamics. He also reviewed effects between natural enemies and pathogens such as interactions that affect natural enemy behaviour or development, pathogen dispersal and intraguild effects. Regarding combined use in glasshouse crops, Gillespie et al. (2006) reviewed three cases with predators and entomopathogenic fungi. In two cases, the predator *Dicyphus hesperus* was affected by the fungus (by *B. bassiana* and by *I. fumosorosea*) through sublethal effects and competition. In the case of *Aphidoletes aphidimyza,* a predatory gallmidge, in combination with *L. muscarium*, an additive mortality of aphids was observed.

3.3.1 Entomopathogenic Bacteria and Natural Enemies

Today, the only bacterium used in greenhouse crops is *B. thuringiensis,* either for control of caterpillars, the Colorado potato beetle or Sciaridae. Subspecies and strains of Bt are safe to natural enemies. Interactions with Bt do occur with larval parasitoids and with predators, however. These indirect effects are reviewed by Navon (1993, 2000). *Trichogramma* and other egg parasitoids are fully compatible with Bt, but the interaction with larval parasitoids can be deleterious. The outcome of the competition is dependent on the interval between Bt infection of the host and the oviposition of the parasitoid, and of the ingested Bt dose (Mohan et al. 2008). Sequential use is possible, and in cases where large caterpillars are not susceptible to Bt, parasitoids may be complementary. Pentatomid predators and the use of Bt showed a synergy in the control of the Colorado potato beetle in field trials (Cloutier and Jean 1998).

In general, Bt can be safely integrated. In greenhouse crops, however, there are currently no natural enemies used against caterpillars or against the Colorado potato beetle. Bti can be used against Sciaridae larvae in the soil, but this use is only approved in some countries. Natural enemies used against sciarids are *Hypoaspis* mites, but combined uses of Bti and natural enemies are rarely applied.

3.3.2 Entomopathogenic Fungi and Natural Enemies

Interactions between entomopathogenic fungi and natural enemies have been studied by many researchers. An overview of these studies is given by Roy and Pell (2000). The authors reviewed the possible effects of these interactions in relation to biological control, particularly between fungi and insect natural enemies. Enhancing effects can occur through an increased pathogen susceptibility, non-host transmission and dispersal. Foraging predators may spread propagules to new hosts or to new leaves or leaf parts where host insects may come in contact through secondary pick-up. Parasitoids and predators also may cause disturbance and movement of pest insects by which transmission of spores or contact with sporulating cadavers may be increased. These phenomena have been observed, for example, in aphids (Roy et al. 1998). Clearly these interactions are complex and unpredictable. Only few studies have been dedicated to this subject and most of them studied Entomophthoralean fungi which are not available as a biopesticide. Investigations with mycoinsecticides focused on these synergistic effects are lacking.

In contrast, negative effects have been studied for many mycoinsecticides. Direct infection could have an adverse effect on natural enemy populations. Furthermore, competition for hosts between a pathogen and a natural enemy may lead to decreasing numbers of either of them or both. Many mycoinsecticides are based on Hypocreales, and many species/strains have a wider host range than just the pest insect(s). Side-effects often occur and may have consequences for biological control efficacy. Natural enemies of the pest insect for which the fungal pathogen is applied too may be adversely affected. If the natural enemy is largely decimated this will have consequences for the longer term. It is much worse when natural enemies of another pest are affected and when this leads to a resurgence of the pest. In both cases new releases are needed. These unintentional effects must to be known as they have direct implications for the entire IPM system.

There are many investigations focussing on the side-effects of an entomopathogenic fungus on natural enemies of the same target insect. Fransen and van Lenteren (1993) studied the complementary activity of *Encarsia formosa* and *Aschersonia aleyrodis* on the greenhouse whitefly and the interaction between both beneficial organisms. They showed that the parasitoids were able to recognize and reject infected whitefly larvae in a certain life stage of the fungal infection. When whitefly larvae in a very early stage of infection were not rejected and parasitized, progeny would not develop and the fungus would develop and kill both the pest insect and the parasitoid. Whitefly larvae which had been parasitized by *E. formosa* and allowed to develop and where afterwards spores of *aleyrodis* were applied were less or not susceptible to fungal infection anymore, depending on the age of the developing parasitoid larvae (Fransen and van Lenteren 1994). This example shows that interactions can be complex and that by understanding the mutual effects, complementary use can be designed in an IPM system by using them with known intervals to minimize negative effects. Similar effects were found by Jazzar and Hammad (2004) in studies performed with *L. muscarium* and *E. formosa* and by Avery et al. (2008) who studied the compatibility of *I. fumosorosea* and *E. formosa*. The authors of both studies concluded that despite the negative effect, implementation of these biocontrol agents can succeed if precisely timed. Other examples are listed by Roy and Pell (2000).

Interactions between predators and fungal infection have also been reviewed by Roy and Pell (2000). In most cases intra-guild predation is caused by the pathogen through direct infection of the natural enemy. Predators, however, sometimes reduce the pathogen by consuming infected prey (Roy et al. 1998). Regarding greenhouse biocontrol, interactions with mirid predators have been investigated in detail. The predatory mirid bug *Dicyphus hesperus,* which preys amongst others on whitefly larvae, avoids feeding on larvae that have been infected by the fungus *B. bassiana,* particularly when the infection reached an advanced stage (Labbé et al. 2006). *D. hesperus* is only moderately susceptible to *I. fumosorosea* and is able to discriminate between infected and uninfected whitefly larvae (Alma 2005). Alma et al. (2007) studied the interaction between *D. hesperus* and the use of *I. fumosorosea,* with multiple applications, on control of the greenhouse whitefly on tomato in small greenhouse compartments. The combined use resulted in a mortality that was almost the same, but lower, than the independent additive mortality of both control agents, suggesting that interference was minimal. This study illustrated that with the combination of a generalist predator and entomopathogenic fungus increased pest mortality can be reached and that negative interference or competition has a minimal effect. Similar results were found with applications of *B. bassiana* and concurrent use of *D. hesperus* and *E. formosa* for control of the greenhouse whitefly on tomato (Labbé et al. 2009).

3.3.3 Baculoviruses and Natural Enemies

Baculoviruses are safe to natural enemies and therefore side-effects are rarely tested. Indirect interactions between baculoviruses and larval parasitoids and predators, however, do occur. Harper (1986) reviewed the interactions with parasitoids. With parasitoids, premature host death is the most common phenomena and competition for the host often results in death of the parasitoid. This interaction is usually unidirectional where the parasitoid "loses". The outcome of the "competition" is dependant on the sequence of and the interval between oviposition and virus infection. Some examples of investigations of these interactions are given by Matthews et al. (2004) and Nguyen et al. (2005). A case relevant for glasshouses was studied by Matthews and colleagues who investigated interactions between *Laconobia oleracea*, a pest in glasshouses, the *Laconobia oleracea granulovirus* and the parasitoid *Meteorus gyrator*. Combined use of the *Lo*GV and the parasitoid in a glasshouse trial did not result in a significant reduction in damage compared to the *Lo*GV alone.

The pentatomid bug *Podisus sagitta* readily feeds on virus-infected and virus-killed larvae of *Spodoptera exigua* and also aids in the dispersal of virus. Combined use of the predator and the *Se*MNPV gave better control of the beet army worm than the separate use of both control agents in chrysanthemums (Smits 1987). With these exceptions, few studies on the implications of combined use of baculoviruses and natural enemies in relation to enhancing biocontrol efficacy have been performed.

Natural enemies may increase dispersal and transmission; they are considered important players in the dispersal of baculoviruses in natural environments (Fuxa 2004). Matthews et al. (2004), however, only found 1.5% transmission by the parasitoid in a bio-assay test in the example mentioned above, and no transmission was found in the case studied by Nguyen et al. (2005). For commercial biocontrol, increased dispersal of pathogens is not relevant in terms of reliable pest control.

3.3.4 Entomopathogenic Nematodes and Natural Enemies

Few studies have investigated the use of EPNs in combination with natural enemies. EPNs have a broad host range and harmful effects to natural enemies cannot be excluded. On the other hand, most natural enemies are used in the plant canopy, while nematodes are generally applied against soil-inhabiting life stages of insects. Interactions occur in some cases where predators are soil-inhabiting and nematodes are applied to the soil. An example is control of the western flower thrips *Frankliniella occidentalis* with the predatory mites *Hypoaspis aculeifer* and *H. miles* and entomopathogenic nematodes (Premachandra et al. 2003). Both the mites and the nematodes exploit the pupal stages of thrips in the soil, which could lead to competition. Still, combined use of EPNs and *H. aculeifer* gave better control than use of either of the biocontrol agents alone. The combined effect, however, was less than the independent additive. Direct harmful effects of EPNs on predatory mites are highly unlikely given the size of the mites, and to my knowledge it has never been tested. A study in cut chrysanthemums showed that a combination of *Amblyseius cucumeris* with *S. feltiae* or with *L. muscarium* (Mycotal) can be safely used and that either combination resulted in a higher reduction of the western flower thrips *F. occidentalis* than any of the treatments alone (Beerling and van den Berg 2005).

Interactions between parasitoids and nematodes occur when life stages of the host of the parasitoid are found in the soil, or when nematodes are applied on the foliage. This may occur when leafminers are parasitized by larval parasitoids. Head et al. (2003) investigated the compatibility of nematodes and larval parasitoids of leafminers, the ectoparasitoid *Diglyphus isaea*, and the endoparasitoid *Dacnusa sibirica*, using foliar applications of *Steinernema feltiae* on lettuce. They found that the nematode infected all parasitized stages of the leafminer (from both parasitoids). The potential of wasps to survive to the adult stage was reduced by this interaction. When nematodes were applied first, *D. isaea* did not discriminate between infected and uninfected leafminer larvae for host-feeding, but it did discriminate for oviposition. The discriminatory behaviour of *D. sibirica* was not tested. The combined use of *S. feltiae* and *D. isaea* did not result in improved control compared to the use of the nematode alone. The nematodes had a detrimental effect on the parasitoids in this study. The authors concluded that this strategy would be uneconomical given the fact that *S. feltiae* alone causes a very high mortality of leafminer larvae. Similar intra-guild interactions were found for *D. begini* and *S. carpocapsae* in laboratory experiments with *L. trifolii* (Sher et al. 2000).

Interactions between parasitoids of lepidopteran caterpillars and nematodes are likely to occur too. The interaction between two parasitoids of the codling moth with *S. carpocapsae* has been investigated by Lacey et al. (2003). Developing larvae of the parasitoids are very susceptible to the nematode while full grown diapausing larvae are well protected by their cocoon. The adult parasitoids showed a strong preference for non-infected codling moth larvae over nematode-infected larvae. The authors concluded that the parasitoids can enhance the control of the codling moth larvae in combination with the entomopathogenic nematode. I have not found any study with regard to greenhouse caterpillar pests.

3.3.5 Integrated Use of Microbial Pest Control Products and Natural Enemies

In biocontrol, most studies initially consider the host-parasitoid or host-predator interaction. Integrated use of two or more biocontrol agents against one host insect or against multiple hosts increases the complexity of the interactions involved, sometimes including guild interactions. In an ecological setting, but also in an agricultural setting in which IPM is applied, these interactions occur in the short and long term. For a more elaborate review of these interactions involving pathogens, parasitoids and predators, and the consequences for biological control the reader is referred to Thomas et al. (2006). Understanding these multiple species interactions better could help in improving the efficacy of biological control using these organisms. On the other hand, the use of biopesticides in greenhouse crops aims for a rapid reduction of the host population and long term ecological effects play a minor role. In developing an IPM programme, a pragmatic approach is needed where direct negative and enhancing effects on the overall control of a pest need to be considered, rather than long-term ecological interactions. The grower can interfere in the system by releasing natural enemies again or by using a biopesticide again. This should, of course, be done with the goal of control of the pest in an effective and economical way.

Few studies have been performed on establishing the effect on the biocontrol efficacy of combinations of a MPCP and one or more natural enemies. The interactions have been studied in a number of cases in the laboratory or small greenhouse tests. Studies under commercial conditions are not available in the literature. When combinations are applied in practice, the goal is to enhance control of the pest, or to achieve a quicker control. When bioinsecticides are used in greenhouse vegetables, almost always multiple species effects are present. Jandricic et al. (2008) reviewed the effects of intraguild effects in greenhouse floriculture and concluded that, although the number of studies is small, the benefit of releasing two BCAs usually outweighs the risk of disruption of biocontrol. In practice, results of combined applications are assessed and monitored by the IPM advisors and the growers. With the help of the field expertise of the advisor and the experience of the grower they interpret the results and intervene again when necessary. The adverse effects of intraguild are usually difficult to detect in the field and I do not consider them relevant for the overall control in most cases. Combined use, however, is highly valuable where biocontrol agents complement each other, like in the case of *L. muscarium* and *E. formosa* where the fungus kills unparasitized larvae and does not affect parasitized larvae.

The application of a microbial pesticide in greenhouses as an inundative treatment instantaneously impacts the availability of prey and host insects for natural enemies. When an entomopathogen has direct harmful effects on a natural enemy, it should not be used. But if there is no other option to control the pest, natural enemies could be re-released. This solution is then preferred over the use of a chemical pesticide. Where host numbers have decreased and a part of the pest population is still present, the effects on the populations of natural enemies should be monitored in order to check whether the balance between those natural enemies and the pest re-establishes and control is maintained. If this is not happening, an extra release of the natural enemy must be considered, or another treatment with the microbial product. Using a combination is often carried out when control is insufficient or partial. It may be recommended when different life stages of the pest are controlled by the natural enemy respectively the pathogen, or when pest levels are too high and additive effects are quickly required. Understanding the consequences of combined use for the biological control of the pest is imperative for the grower or his consultant. Following the application, the crop must be monitored regularly to ensure a continuous effective pest control.

3.4 Combinations and Mixtures of Microbial Pest Control Products

MPCPs can be used together to achieve enhanced control of an insect pest, either by a combined application as a tank-mix or as alternate treatments. The goal is to achieve an additive effect when applied together or even a synergistic effect. When used alternatively, effects can also be additive or synergistically. Alternating applications may be chosen when tank-mixes are not compatible, or when simultaneous use results in reduced effects. Numerous studies have been carried out on interactions between insect pathogens. Many of those are laboratory studies investigating the mechanisms of the interactions. Only few of them are field studies which attempt to elucidate the combination effects in terms of an enhanced biocontrol. Below examples of combined pathogen applications are given to illustrate their potential.

3.4.1 Entomopathogenic Bacteria and Other Entomopathogens

Interactions and joint action of Bt with other microbial pest control agents has been studied by many researchers. A brief overview is given by Navon (1993) and by Glare and O'Callaghan (2000). Interactions with other pathogens were found to be variable and rarely synergistic, and Bt competed with other pathogens for the nutrition provided by the host insect.

A combination of Bt with EPNs did not give an additive effect. The nematode *S. feltiae* could not produce progeny in Bt diseased larvae of the silkworm *Bombyx mori* (Kaya and Burlando 1989). On the other hand, Koppenhöfer and Kaya (1997) demonstrated synergism and additive effects between *Bt* subsp. *japonensis* and *Heterorhabditis bacteriophora* and *S. glaseri* for control of white grubs. The effects of the interaction depended on the order of infection and the interval between the two pathogens.

Mixed applications of Bt and *B. bassiana* against the Colorado potato beetle gave low-level synergism in field trials (Wraight and Ramos 2005). The authors, however, could not exclude that the synergistic effects were caused by formulation materials in the commercial products.

Negative effects have been reported for the combination of a Bt mixed with a baculovirus. Ingestion of Bt causes a feeding arrestment which prevents larvae from ingesting sufficient viruses to enhance the lethal effect of Bt. The combined use of a Bt and a baculovirus with a time interval between applications may not have this reduced effect because of the loss of viability of propagules over time. Obviously, the intended enhanced effect is not achieved. Various combinations with sublethal dosages have been studied and this may give better results than with the normal dose of either pathogen alone (Navon 1993). In greenhouse crops, baculoviruses are used to control the beet armyworm *S. exigua* in chrysanthemum. The slow speed of kill still allows considerable plant damage after the application of the *Se*MNPV. A combination with Bt subsp. *aizawai* was investigated by Geervliet et al. (1991) in order to improve control and to reduce plant damage. Laboratory studies with mixtures of both pathogens revealed a negative interaction, even with sublethal Bt concentrations. Only low concentrations of both pathogens gave some synergism. This study showed that considerable research on fine-tuning is needed for successful use of mixed applications.

3.4.2 Entomopathogenic Fungi and Other Entomopathogens

The combined use of fungi with other entomopathogens has been studied in a few cases only (Inglis et al. 2001). Fungus-fungus interactions and the effects of mixed infections have rarely been studied (Thomas et al. 2003). In a study with *M. anisopliae* and *B. bassiana* and the desert locust *Schistocerca gregaria*, the complexity of such an interaction was shown. An avirulent pathogen may alter the insect's susceptibility to the virulent pathogen or alter the speed of kill. Co-infecting pathogens may act antagonistically, synergistically or independently, depending on the order of infection, the infective dose and the environmental conditions (Thomas et al. 2003). In terms of biocontrol, the result of the combination of fungal pathogens cannot be predicted and needs to be studied. Investigations concerning biocontrol of greenhouse pests with co-applications of mycoinsecticides are not available, but potentially enhance control.

3.4.3 Baculoviruses and Other Pathogens

Many naturally occurring multiple infections involving baculoviruses and other entomopathogens have been reported, and many studies on intentional multiple infections have been studied in the laboratory and the field (Harper 1986). Results varied considerably and were often additive, but no generalizations can be made in terms of insect control. The use of a combination of a baculovirus and entomopathogenic nematode has been investigated for the control of *S. exigua* on soybean. The combination of *S. carpocapsae* and *Se*MNPV resulted in significantly higher larval mortality than either pathogen used alone. The justification for a simultaneous application is that the virus is more effective on younger stages while the nematode is able to kill older stages. Another advantage is that when multiple caterpillar pests are present, these may be killed by the nematode too (Agra Gothama et al. 1996). This combination has rarely been tested. Sequential use of a baculovirus and EPNs has been studied for the control of codling moth in fruit trees. The virus is applied during the summer on the trees to achieve direct control, whereas EPNs are used to kill larvae on the trunk or in the soil and even in fruit bins. This complementary approach aims at a reduction of the overwintering population, resulting in enhanced overall control (Lacey et al. 2007).

3.4.4 Entomopathogenic Nematodes and Other Entomopathogens

Various combinations have been investigated on various insects. These combinations are reviewed by Koppenhöfer and Grewal (2005). The application of EPNs combined with other nematodes, fungi or viruses generally results in additive effects, while the nematode-bacteria combination effects range from negative to synergistic. Nematode-nematode interactions generally lead to competition without any additive effect. Combinations may be additive when two nematode species have a different searching behaviour or pathogenicity towards different life stages of a target insect. Or, obviously, when multiple pests are present which differ in susceptibility to any of the nematode species.

Nematode-bacterium interactions can result in negative effects. Nematodes did not produce progeny in hosts infected with Bt. Alternating applications of Bt and nematodes (*S. carpocapsae*) achieved better control results than a combined application in cabbage against the diamondback moth in Indonesia (Schroer et al. 2005; Yi and Ehlers 2006). Combined use of *S. carpocapsae* and Bti against early instars of *Tipula paludosa* did not give any synergistic effect in field trials (Oestergaard et al. 2006).

Nematode-fungus combinations generally are additive with regard to target mortality or at least give accelerated speed of kill. The use of *L. muscarium* and *S. feltiae* against *Thrips palmi* gave additive effects on the thrips population. *S. feltiae* caused a high mortality of juvenile stages of the thrips, whereas *L. muscarium* caused mortality in juveniles and adults (North et al. 2006). This offers the possibility to develop an IPM programme targeting all life stages of thrips on foliage as well in soil. This strategy is also exploited with the same pathogens in chrysanthemum for the control of western flower thrips, *Frankliniella occidentalis* (Beerling 2008). Sequential applications as well as tank-mixes may be used depending on the structure of the thrips population. When all life stages are prevalent, tank-mixes may be the best. Preferably synergistic effects are obtained, but this has not yet been documented. The interaction between the fungus *M. anisopliae* and various entomopathogenic nematode species, and the combined use was investigated for the control of the larvae of the Scarabaeid *Hoplia philantus* (Ansari et al. 2004, 2006) and larvae of the black vine weevil *Otiorhynchus sulcatus* (Ansari et al. 2008b, 2010). The researchers concluded that the effects ranged from additive to strong synergistic effects depending on the interval between the applications of the fungus and either of the nematodes. Effects differed also between nematode species.

The beet armyworm *S. exigua* is a difficult pest to control and combinations of pathogens have been studied in order to improve its management. The use of *S. carpocapsae* combined with the *Se*MNPV in soybean gave additive mortality on the larvae of beet armyworm in a field trial, indicating the potential of a combined scenario (Agra Gothama et al. 1996). Nematodes successfully produced progeny in moribund hosts infected with an NPV (Kaya and Burlando 1989). Progeny from such hosts carried enough viruses to infect healthy insect larvae.

Negative effects between nematodes and other MPCAs may occur. This was demonstrated in laboratory tests by Ansari et al. (2005) who exposed entomopathogenic fungi to bacterial symbionts of EPNs. *Photorhabdus luminescens* bacteria inhibited growth and sporulation of several fungi, but *Xenorhabdus poinarii* bacteria did not. Crude extract of *M. anisopliae* inhibited growth of the bacteria, but had no effect on the dispersal of *S. glaseri* and *H. megidis*. This study highlighted the possible negative interaction between pathogens. In field experiments, however, the combined use in sequential applications of *M. anisopliae* and four weeks later of *S. glaseri* on *Hoplia philanthus* larvae gave synergistic effects resulting in high mortality rates (Ansari et al. 2006). This demonstrates that laboratory results should be considered as a worst-case scenario, and that ultimately applications in the field should be investigated on combination effects.

The result of an application consisting of a combination of a nematode and a pathogen can be additive or negative depending on the order of the infections and the interval between applications of them. Combined infections generally are a competition for resources where only one is able to successfully produce progeny.

3.4.5 Considerations for the Combination of Entomopathogens

Any combination of entomopathogens is theoretically possible. The result of a two-agent system can be negative, neutral, additive or synergistic. In most cases mortality is increased (Koppenhöfer and Kaya 1997 and references therein);

however, often the effect is additive, *e.g.* less than the sum of the agents alone. I have not found any study of a three agent system with pathogens. The aim of a two agent system is to enhance biocontrol, or to reduce costs. Reduced dosages of one or of both of the agents may be possible and still give good control of the pest. This could be done for cost reasons. I have, however, found very few studies investigating this with pathogens (Ansari et al. 2008b; Wraight and Ramos 2005). Simultaneous use (but also sequential use) of two or more entomopathogens can be considered in different situations:

- when several larval stages of a pest are present with a different susceptibility. For instance, small stages may be susceptible to a baculovirus or to Bt, whereas large larvae are no longer susceptible. In this case a nematode or a fungus may induce mortality of larger stages;
- when the speed of kill is slow, combined use may hasten death of the pest. In most fungi and baculoviruses it takes considerable time before the insect dies, while an enhanced activity may be reached with EPNs, or a feeding arrestment with Bt, or there may be reduced crop damage;
- when larval stages and pupae are located in different parts of the plant or in the soil. Some pathogens are used on the foliar plant parts while others may be more effective in the growing medium. Simultaneous use may be targeted at these various life stages and accelerate reduction of the population;
- two pathogens may increase the susceptibility of a pest and cause an accelerated death;
- a combination can be targeted at multiple pest insects which each have a different susceptibility to a pathogen, so only one application is needed.

In practice, few combinations of pathogens are used in greenhouse systems to control pests, and even fewer tank-mixes. An example that seems successful is the sequential or alternate use of nematodes and fungi for the control of thrips in chrysanthemums and some other ornamental crops in the Netherlands, UK, Denmark and some other countries.

3.5 *Combined Use of Microbial Pest Control Products and Other Pesticides*

The combination of MPCPs with chemicals, preferably at low dosages, has been frequently suggested to improve the overall effect. This idea has been based on observations that the pest target is weakened by the chemical insecticide and becomes more susceptible to pathogens. Another possibility is that the insect's behaviour has been modified resulting in more movement and an increased exposure to the pathogens. Even low dosages of either, or both control products may give additive or synergistic effects. On the other hand, use of low dosages of chemicals may be difficult to register due to efficacy and resistance issues. The agrochemical industry may also be reluctant to do this for commercial reasons. Combinations may be advantageous when two control products have different effects on different life stages of the target. The costs of control may also be reduced when combinations are successful, particularly application costs. This is an important reason to investigate combined use. A prerequisite of this approach is that the chemical have no side-effects on the microbial agent. The same approach can be followed with natural pesticides such as neem based products or products based on plant extracts, fatty acids, etc., although the availability of this kind of products is limited. I will briefly review examples from the literature with the four groups of pathogens. This is focussing on combined use where the aim is to enhance control of the pest through a direct interaction between both applied control means. These can be tank-mixes or sequential applications within a short interval, i.e. more or less simultaneously applied control means. Alternating applications of products with longer intervals as commonly done in IPM are not considered here.

3.5.1 Entomopathogenic Bacteria and Chemical Pesticides

Various combinations of Bt's and chemical insecticides have been tested and synergistic as well as negative effects have been reported. Brief overviews have been given by Navon (1993) and by Glare and O'Callaghan (2000). In general, chemicals enhanced Bt activity. When the activity of the chemical pesticides depended on the uptake of the material, the feeding arrestment caused by Bt resulted in a reduced effect. This was the case with IGRs and neem. The combination with contact insecticides such as pyrethroids and organophosphates generally resulted in synergistic effects, even with low dosages of the chemicals (Navon 1993).

3.5.2 Entomopathogenic Fungi and Chemical Pesticides

The combination of entomopathogenic fungi and chemicals has been investigated for many pests. Brief overviews have been given by Zimmermann (1994) and by Inglis et al. (2001). Some studies reported additive and synergistic interaction between fungi of the genera *Beauveria* and *Metarhizium* and sublethal doses of imidacloprid and other chemical insecticides for various insects (Quintela and McCoy 1998; Jaramillo et al. 2005; Shah et al. 2007a). Sequential use of a chemical whitefly insecticide, followed after 24 h by an application with *L. muscarium* did not increase mortality of whitefly larvae (Cuthbertson et al. 2005b). Although these treatments were compatible, it did not give any additive effect. Using sublethal doses of chemicals can lead to a change in behaviour of the pest insect resulting in an increased pick-up of spores and in a higher mortality. By using *L. longisporum* (Vertalec) with 1% of the recommended dose of imidacloprid, applied systemically, aphid movement increased dramatically and the number of mycosed insects was significantly higher than without the use of the insecticide (Roditakis et al. 2000). Caution is warranted since some chemicals could repel insects and reduce the exposure to fungal spores (Kepler and Bruck 2006). Most of these studies were laboratory or small scale studies. Few studies have reported the use of such combinations under commercial conditions. Mycotal (*L. muscarium*) has been used in combination with

IGR's (buprofezin, pyriproxifen) for control of whitefly in greenhouse vegetables with good results (J. Rodenrijs, personal communication).

3.5.3 Baculoviruses and Chemical Pesticides

Many studies have investigated the interaction and the combined use of baculoviruses with chemical pesticides. Often the slow kill of the virus was the incentive to study potential additive effects, including use of low doses of chemicals. An overview of these studies is given by Harper (1986) and McCutchen and Flexner (1999). Results can go either way and each combination needs to be studied. Synergistic effects have been obtained, even with low concentrations of chemicals, indicating the interesting potential of these combinations.

3.5.4 Entomopathogenic Nematodes and Chemical Pesticides

The combined use of nematodes with chemicals has been investigated for control of soil pests. An overview of these studies and the results are given by Koppenhöfer and Grewal (2005). Combined use of EPNs with chemical insecticides and the results on insect control can be additive or synergistic, but the authors warned that lab results are not always checked in the field. Simultaneous applications as well sequential applications of imidacloprid and *H. bacteriophora* resulted in strong synergistic effects on white grubs (Koppenhöfer and Kaya 1998). Further research indicated that the level of synergism varied with the nematode species and that results in the field were more variable than in greenhouse trials (Koppenhöfer et al. 2000). Combined uses are useful when target pests become more susceptible to a pathogen due to the effect of the chemical. The combined use of nematodes and chemicals was investigated by Cuthbertson et al. (2003, 2008b) on foliage of tomato and verbena. When nematodes were exposed for 24 h to the chemical spray solution, there was a large reduction of infectivity. The nematodes were not affected by the one day old residue of the chemicals on the foliage. Sequential applications of a chemical insecticide and *S. feltiae* or *S. carpocapsae* increased whitefly larvae mortality in some cases.

3.5.5 Microbial Pest Control Products and Natural Pesticides

Natural products have also been investigated in combination with MPCPs. For instance, neem enhanced the efficacy of *M. anisopliae* against the black vine weevil. Even a 100-fold lower dose of the fungus in combination with neem produced similar mortality to the recommended fungal dose without neem, and protected plants from black vine weevil damage (Shah et al. 2008). This offers a potential benefit to growers with regard to efficacy and costs. Neem also improved the efficacy of the nematicidal fungus *Paecilomyces lilacinus* against plant parasitic nematodes (Nagesh et al. 2003). Since neem has anti-feeding effects, usage in combination with pathogens that need to be ingested, like Bt and baculoviruses, is not recommended. Nevertheless, laboratory studies showed additive and synergistic effects of combined ingestion of Bt and neem in the larvae of the Colorado potato beetle, even with reduced doses of both (Trisyono and Whalon 1999). Additive effects of the combination were found in *Helicoverpa armigera* larvae (Singh et al. 2007). The IGR mode of action of neem makes it potentially more interesting for combinations with entomopathogenic fungi (Walter 1999).

The combination of nematodes with insecticidal soap has been investigated for simultaneous control of soil and foliage pests when sprayed as a tank-mix. Immediate application after preparing the mix was harmless to the nematodes, but longer exposure periods killed nematodes (Kaya et al. 1995). The combination of both control means to control different pests occupying different habitats is feasible, but there was no additive or synergistic effect on either of the pests. Investigations on combinations of other natural products, apart from neem and soap, with MPCPs are hardly available.

3.5.6 Microbial Pest Control Products Combined with Other Control Techniques

Few studies report the use of a pathogen combined with other non-pesticide tools. In a study of Ludwig and Oetting (2002) the efficacy of *B. bassiana* in combination with insect attractants was investigated. Attractants were tank-mixed with the fungus and applied three times in chrysanthemum at bud break to control *F. occidentalis*. The attractants were either behaviour-influencing compounds or feeding-stimulants/stickers and had no adverse effect on the fungus. Thrips populations were not reduced in any combination of the fungus with the attractants compared to the treatment with *B. bassiana* alone. The use of attractants and baits, possibly in combination with autodissemination devices, has been advocated as a way to improve biocontrol with entomopathogens (Butt and Brownbridge 2001; Roy et al. 2007). Similarly, using trap plants and treating those to control the pests has been suggested by many authors. This is often called the push-pull strategy. However, these technologies are not yet used with entomopathogens in protected crops (Cook et al. 2007).

3.6 Use of Insects as Vectors to Disseminate Microbial Pest Control Products

Pollinators such as honeybees and bumblebees have been studied as vectors to disseminate MPCAs. Bees can spread small particles of powder formulations of bacteria, fungi and baculoviruses (which they can accumulate by means of a dispenser) to flowers. The prerequisite is safety of the material to the foraging bees as well as to the bee colony. Bt's are safe for honeybees; a review is given by Glare and O'Callaghan (2000). Bt's and baculoviruses are also safe for bumblebees (Sterk et al. 2002). Entomopathogenic fungi, however, can have detrimental effects on bees (Butt et al. 1994) and caution is required when using them with bees.

Kovach et al. (2000) demonstrated that the use of honeybees and bumblebees for disseminating spores of *T. harzianum* to strawberry flowers reduced *Botrytis* disease incidence in the field. Honeybees were successfully used in sunflower to disseminate

Bt to flower heads to control the banded sunflower moth (Jyoti and Brewer 1999), and to disseminate *Heliothis* NPV to crimson clover flowers (Gross et al. 1994). Carreck et al. (2007) showed that honeybees can be used to deliver inoculum of *M. anisopliae* to flowers of oilseed rape for control of the pollen beetle and the cabbage seed weevil. There was no evidence of any adverse effect on the honeybee colonies.

More recently bumblebees have been studied as a vector of fungal agents in greenhouse tomato and sweet pepper crops (Kapongo et al. 2008; Kevan et al. 2007; Shipp et al. 2008). Dissemination of the combination of an entomopathogenic fungus with an antagonistic fungal agent was also investigated by Shipp et al. (2008). Results indicated that this method is potentially appropriate for dispersal of inoculum to flowers and plants for pests and diseases such as whitefly, *Lygus* bugs, and grey mould. Mortality of bees is a critical factor when using entomopathogenic agents with a broad host spectrum such as *B. bassiana* (Shipp et al. 2008) and, requires detailed side-effect testing. This dissemination method is still in its infancy and commercial applications have not yet been successfully applied in greenhouse crops, nor have they been approved by regulatory authorities. Approval of this method may be a challenge due to greenhouse personnel exposure to spores on the plants, and due to direct non-target effects on bees.

The predatory bug *Orius laevigatus* has been studied for use as a vector for fungal entomopathogens in greenhouse crops. Bugs artificially surface-dosed with conidia of *L. longisporum* or *L. muscarium* successfully disseminated conidia around a plant, resulting in infected pest insects and a higher reduction in aphids respectively thrips than with the fungal pathogen alone. There was no additional effect in the case of whitefly larvae. The benefits of using bugs contaminated with the pathogens were not significant due to the predatory behaviour of the vector (Down et al. 2009). I doubt whether this method can be developed into a practical application resulting in enhanced control of pests.

4 Resistance and Microbial Pest Control Products

4.1 Resistance Against Entomopathogens

For a long time, resistance to entomopathogens had been considered improbable due to the complex and multi-site targeting mode of action that has evolved with the host insects. Pathogens do not represent a new selection pressure in nature, and novel resistant mutants were not likely to occur. This is in contrast to chemicals, which offer a new selection pressure on often single-sited mode of actions, and thus new mutants were to be expected (Homan 1981). Nevertheless, resistance to bacteria and to viruses was recognized by Homan (1981) as a possibility. This hypothesis was based on theoretical considerations about possible mutations inducing increased resistance by the insect to infections. This possibility was confirmed when resistance to Bt was found in the field in the diamond back moth *Plutella xylostella* in the late 1980s (Tabashnik et al. 1990). Resistance had developed because of repeated applications against this major pest in continuous cropping of crucifers in large areas in subtropical and tropical regions. A review on the occurrence of resistance to Bt in various insects is given by Tabashnik (1994) and by Glare and O'Callaghan (2000). For details on the biochemical and genetic mechanisms of this phenomenon I refer to Van Rie and Ferré (2000). Even cross-resistance between Bt strains has been reported, and for Bt resistance management programmes have been designed to avoid this problem. Resistance to Bt has also been found in greenhouse crops in Canada in the cabbage looper, *Trichoplusia ni*, due to overuse and using rates above the recommended ones (Janmaat and Myers 2003; Janmaat 2007).

Factors influencing resistance to baculoviruses by insects have been elaborately discussed by Briese (1986); he first recognized the risks of reduced susceptibility to virus infections. Recently, resistance in a baculovirus in the field has been found. The use of *Cydia pomonella* GV in apple orchards in Europe as the main component for control of the codling moth for many years has lead to a high degree of resistance in some populations (Sauphanor et al. 2006; Frisch et al. 2007). This is the first example where field resistance to a commercially applied baculovirus has been documented (Eberle and Jehle 2006). These cases illustrate that also with entomopathogens there is a risk for the development of resistance when products are overused, or used as a single or predominant control method. I have, however, not found any examples of field development of resistance to entomopathogenic fungi or nematodes. This is confirmed by Shelton et al. (2007) who gave an overview of cases of resistance to insect pathogens. Studies indicate that the mechanisms of natural resistance to fungi (Wilson et al. 2001, 2002) as well as to nematodes (Kunkel et al. 2004) are present in insects, indicating that resistance to these pathogens is not impossible.

The above illustrates that with the use of MPCPs a resistance strategy also needs to be developed. These cases of resistance concerned lepidopteran pests, but most pest insects and mites in greenhouses have a history of developing resistance to chemical insecticides. Resistance concerns should be raised when insects have many generations in one cropping season and when a pathogen is frequently applied, or applied as a single control method. General recommendations for the use of MPCPs should include recommendations to avoid resistance. These resistance management recommendations are similar to the ones for chemical pesticides: alternate insecticides with a different mode of action, reduce the number of applications and do not apply dosages higher than the recommended ones (Shelton et al. 2007).

Resistance management tactics for biopesticides, particularly for Bt's, in agriculture are given by Roush (1999, 2000). In general, the solution is to develop an IPM programme with different control methods and to use MPCPs with care. This approach should start at the launch of every new product and should be proactive rather than reactive. The monitoring of resistance is important and as soon as products seem to fail to give sufficient control, any suspicious populations should be tested for susceptibility. In my experience, resistance to biopesticides has not occurred in greenhouse crops in Europe to date. Biopesticides have not been used to a large extent, except Bt. Decreased susceptibility to Bt in the tomato looper, *Chrysodeixis chalcites*, has been suspected, but resistance has not been documented.

In Canada, resistance to Bt developed in *T. ni* in greenhouses due to heavy reliance on Bt (Janmaat and Myers 2003). The relative containment of the moth populations in greenhouses may be highly conducive to resistance development (Janmaat and Myers 2003). In the laboratory the high level of resistance to Bt rapidly declined in some populations once the selection pressure was no longer present (Janmaat and Myers 2003). This example shows that when biopesticides are commonly used in greenhouse crops, resistance management must become part of the application strategy.

4.2 Use of Microbial Pest Control Products in Resistance Management Programmes

In pest control programmes where pest insects have developed resistance to commonly used pesticides, or in cases where a new pesticide is likely to develop resistance due to its mode of action, MPCPs are an appropriate tool to prevent resistance. Resistance management programmes should be designed that incorporate rotation schedules of control products with different modes of action. Many insects have developed resistance to chemicals and many insect groups are notorious for doing so. Due to the particularly favourable conditions for population growth of pests in greenhouses, treatment must take place frequently, and as a result, the development of resistance in greenhouses will take place more quickly than outside. Numerous examples have been documented, and all the key pests such as whitefly, spider mites, thrips, aphids, leafminers, sciarids and caterpillars are capable of quickly developing resistance (see http://www.pesticideresistance.org/DB). In addition to the use of natural enemies, which often are the backbone of an IPM programme in greenhouse vegetable crops, MPCPs can be used as corrective tools for pest control or as additional products besides the beneficials. Chemical insecticides should only be used in IPM when no other methods are available anymore. This also helps to prevent resistance to these chemicals and keeps them efficacious for a longer period. Selective chemicals can be essential in IPM programmes when used as corrective measures and the development of resistance is unfavourable for the growers as well as for the manufacturer.

5 Practical Aspects of Side-Effect Testing

5.1 Side-Effects of Chemical Pesticides on Entomopathogens

Several standard procedures have been developed to test side-effects of chemicals on pathogens by members of the IOBC Working Group "Pesticides and Beneficial Organisms". They differ per class of pathogen. Methods have been developed for EPN (Rovesti 1991; Vainio 1992; Peters 2003), and for fungi (Hokkanen and Kotiluoto 1992; Coremans-Pelseneer 1994; Keller 1994; Tuset 1985). The Working Group has not developed methods for testing effects of chemicals on bacteria and baculoviruses, and consequently has not tested side-effects on these pathogens. The most likely reason for this is the lack of susceptibility of bacteria (in the case of Bt both spores and proteins) and baculoviruses towards chemical pesticides. Furthermore, there are no chemical bactericides and virucides available for the control of plant diseases that could have adverse effects on beneficial bacteria and viruses. The developed methods generally describe laboratory tests. Testing methods with plant material or with whole plants or with soil are often not available. There is a need to go beyond such laboratory tests to generate data that reflect "real world" conditions. Given that side-effects of chemicals, particularly fungicides, are the most critical group for entomopathogenic fungi, I will illustrate this need by discussing the case of fungi.

Testing of side-effects of chemicals on entomopathogenic fungi is usually carried out by in vitro Petri-dish agar tests and/or by in vivo bio-assays. The latter are preferred since they simulate more closely the interaction in the field. In the case of *L. muscarium*, many authors have reported side-effects of fungicides, and most of them refer to germination and mycelial growth inhibition on agar, incorporated with the chemical test compound. For an overview of results, see Schuler et al. (1991). Recently, Cuthbertson et al. (2005b) tested the side-effects of four chemical insecticides for whitefly control on *L. muscarium* (Mycotal). They tested the effect of 24 h exposure of spores in the insecticide solution on germination, a worst-case scenario. Only one chemical did not influence germination. The others reduced it greatly. Tests that were conducted on plants (tomato and verbena) which had been sprayed with an insecticide and followed 24 h later by the application of the fungal product showed no difference in either whitefly mortality or in mycelial growth for any of the insecticides. This indicated that tests on plants showed compatibility of the insecticides with *L. muscarium*. This example illustrates that caution needs to be taken when translating laboratory data into a compatibility profile for practical use. The negative results from the 24 h exposure of spores to a fungicide are not very relevant either because the producer's recommendation for Mycotal precludes tank-mixing. Even if a grower wants to spray a combination, it would normally not take more than 8–10 h before the tank will be emptied. Leaving a spray-solution overnight is strongly discouraged in all cases.

Testing on plants is required to get relevant data for commercial conditions, particularly when laboratory results show harmful effects. Hall (1981) and Ravensberg et al. (1994) tested side-effects directly with host insects in bio-assays and on whole plants. Direct effects as well as persistent effects can be tested in this way and the results are easier to compare to the field situation. Testing compatibility of pathogens with chemicals should follow this approach rather than just conducting agar tests.

For fungi applied to the soil, Keller (1994) proposed a semi-field and a field test including a bio-assay with bait insects in the presence of the pesticide, because results then directly indicate effects on virulence and efficacy. This was based on earlier work where Keller et al. (1993) had found that in vitro results are difficult to extrapolate to the field. Despite complete inhibition of mycelial growth of the fungus *B. brongniartii* by some fungicides in in-vitro tests, soil sprayed with the fungicides was much less inhibitory to the growth of the fungus.

EPNs are often tank-mixed with other products, thus compatibility testing is required. Peters (2003) discussed the methodology of testing harmful effects of pesticides on nematodes. He suggested a short exposure of nematodes to the chemical in a water solution, followed by an assessment of mortality. This is the high level exposure. The low level exposure is encountered in the substrate and can be assessed in a bio-assay by infectivity and fecundity of the nematodes on lesser mealworm larvae. According to Peters (2003), the methods used earlier tend to overestimate the pesticides' effects on nematode survival and underestimate the effect on infectivity. The proposed new method supposedly gives more realistic results. Nematode propagation is measured also as this may be important in some cases for prolonged pest control by the following generation of nematodes. This is seen in control of white grubs in turf.

Since there are no specific methods developed for bacteria and baculoviruses, the method for fungi can be used and adapted where necessary. The laboratory in vitro test can be performed with bacteria (viability), but not with baculoviruses. Bio-assays and field tests used for fungi can be used for both these types of pathogens.

For testing compatibility a tiered approach should be chosen. For fungi and bacteria, this starts with agar tests to assess germination and growth. When inhibitory effects are noticed a bio-assay with plant material, or where relevant with soil, and insects is recommended. For baculoviruses and EPNs, the first step is a bio-assay. When harmful effects are seen in the bio-assay, a semi-field greenhouse test should follow. To test persistent effects, I recommend testing this under commercial conditions in the crop and region for which it is relevant. In laboratory in vitro tests a concentration range can be easily tested. The slope of the graph will give information on the dose-effect relationship and help estimate how serious the effects can be. In greenhouse or field tests I recommend to test the label rates only because of time and costs. Side-effect testing should also be done with the formulated products because formulation compounds may influence the results.

5.2 Side-Effects of Entomopathogens on Beneficial Organisms

Natural enemies and bees can be adversely affected by various entomopathogens. This concern needs to be addressed by testing putative negative side-effects of MPCPs on these non-target organisms. Relevant considerations and guidelines for testing are given by Hajek and Goettel (2008). The most significant one concerns the method of testing the pathogens' host range. For the IPM practioner, the ecological host range is relevant, rather then the physiological host range. Often, laboratory studies are carried out under optimal conditions for the pathogen such as high humidity conditions for entomopathogenic fungi, high dosages and/or direct exposure routes. In these "worst-case" scenario tests, "physiological susceptibility" is shown, but this does not necessarily represent the "ecological susceptibility", *e.g.* effects under field conditions. Differences between laboratory and field studies with entomopathogenic fungi, reviewed by Roy and Pell (2000), illustrate that side-effect testing methods should reflect the conditions of use.

Besides direct effects, indirect effects may also occur such as reduced longevity or fecundity. Test methods should also take this into consideration. There are no standardized testing methods for microbials. Methods developed for chemical pesticides (OECD and ESCORT guidelines, see above) can be used and they should be adapted where necessary. For registration purposes official OECD guidelines are required. Testing is particularly difficult when the host or the prey is directly affected by the pathogen. Several examples are discussed above. For predators, additional prey animals can be added during the test. For parasitoids, the relative timing of the pathogen application and the moment of release of the parasitoids is often decisive, and the trial design needs to take this into consideration. Direct effects of the pathogens as well as residual effects must be investigated, the latter only when directs effects are present. A tiered approach (Table 6.3) for testing side-effects is also suggested for this type of studies.

Results from these tests need to be seen as indicative for the side-effects, and monitoring of the interactions in commercial greenhouses should be carried out in order to validate the data. The indicative character is also reflected by the IOBC classification of a side-effect, and expressed as mortality or reduction in beneficial capacity caused by a pesticide. The side-effect level is given in the following categories: harmless or slightly harmful: reduction 0–50%; moderately harmful: 51–75% reduction, and harmful: > 75% reduction. Based on experiences in the field, the effect level may need to be adapted. The large variation of relevant parameters

Table 6.3 Tiered approach for side-effect testing. The subsequent test is only required in case the previous test shows harmful effects

Test location	Type of test	Effect	Testing method
Laboratory tests	(1) Petri dish agar test	Direct effects	Worst case testing on germination and growth
	(2) Bio-assay	Direct effects	(2.1) Topical spray on insects and detached leaf
		Indirect effects	(2.2) First the leaf is sprayed and insects are transferred once residue is dry
		Indirect residual effects	(2.3) As 2.2: after x days on aged residue
Field tests	(3) Experimental greenhouse	Direct effects	(3.1) On whole plants: spray on plants and insects
		Indirect effects	(3.2) On whole plants: spray on plants, insects introduced once residue is dry
		Persistent effect	(3.3) On whole plants: spray on plants, insects introduced after x days
	(4) Commercial conditions	Field persistence	Crop treatment on plants and insects, under slow degradation conditions (North Europe, winter)
	(5) Commercial conditions, various regions	Field persistence	As 4; under different climates, testing different degradation conditions

(environmental conditions, application strategy, cultural practices, pest and disease control measures, etc.) influence the interaction, and gathering information from commercial usages needs to be ongoing to ensure proper advice over time.

6 Implementation of Microbial Pest Control Products and Knowledge Transfer

6.1 Adoption of a Microbial Pest Control Product

Adoption of a new product requires growers to change their current IPM practices. In general, changes are perceived as undesired, unless the benefits, technical and/or economical, are clear. Adoption of innovations is reviewed and described for conservation biological control for farmers by Pannell et al. (2006). He distinguished three main elements that play a role: the process of learning, the relative advantage of the innovation over existing practices, and the "trialability" of the new product, which refers to the ease of testing and learning before adoption. Cullen et al. (2008) argued that demonstrating the economic benefits of conservation biocontrol to farmers is crucial in order to change their practices. Much of this discussion is true for the adoption of biopesticides. The biocontrol industry needs to show the benefits of the change to the user such as cost savings, better yield or better market access. It is crucial to motivate a grower to start with something new. This should be the focus of marketers and salespeople involved in biopesticides. An interesting example of this approach has been described by Kirkpatrick and Georgis (2007) who demonstrated in field trials the cost-benefit of using Bt products in IPM programmes to cabbage and tomato farmers in Florida. They showed that it is worth changing to a new technology since it delivered a higher percentage of marketable produce with reduced pesticide costs.

6.2 Implementation of a Microbial Pest Control Product in an IPM System

Biopesticides act differently than chemical pesticides. Growers' expectations of control agents are still often based on the chemical paradigm where a single product gives a quick knockdown and control of the pest(s). Biocontrol products are often new to the users and the users need to be properly informed about these products. Growers need to understand the strengths and weaknesses of each biopesticide and they have to have a realistic view of what to expect and what not to expect (Straus and Knight 1997; Warrior 2000). MPCPs are knowledge-intensive and implementation in an IPM system requires understanding the product as well the biology of the pests in much more detail than with chemicals. Growers must be provided with effective use strategies to ensure that the product is used in an optimal way. If the grower's perspective of the product is incorrect, the product is bound to fail and the grower will be disappointed and not adopt the new product in his IPM system. Many authors have discussed this critical element of biological control, and of biopesticides (Waage 1996; Straus and Knight 1997; Williamson 1998; Whalon and Norris 1999).

It is well understood that proper training of the practical aspects of a biological product for all stakeholders in the chain, from manufacturer to end-user, should be part of the development of a biopesticide and an IPM programme. The role of the distributors should not be underestimated, since they often have direct contact with the growers. The various steps in the process from research to implementation in the field and the roles of stakeholders are described for natural enemies by Onillon and Gullino (1999); this is basically the same for microbial BCAs. An extended review on the IPM implementation process is presented by Wearing (1988). He focussed on insect and mite pests in various cropping systems from all over the world. Despite more than 20 years of IPM developments, this process is still rather slow and many of the obstacles described by Wearing that hamper adoption of IPM, are still valid. He concluded that IPM must be designed in such a way that it fits the farmer. And he recommended that the primary objective be that "the farmer's confidence be nurtured and maintained". There are very few papers treating this subject in such an elaborate manner and therefore it is worth taking notice of it for the IPM worker. Biopesticides have not been reviewed is such a way, but implementation into an IPM system needs similar considerations as those mentioned by Wearing. Vital aspects are that the product fit the farmer, is simple to implement, and has a cost advantage.

6.3 Stakeholders' Involvement, with Emphasis on Growers

Key to the success of a biopesticide is growers' involvement in the learning process of using and integrating a biopesticide in their pest and disease control programme. De Buck and Beerling (2006) presented an interesting case of implementation of biological control in the Netherlands in which the key role of growers' networks is exemplified. This process is complex and involves many different stakeholders. Another example, in this case from Finland, showed that knowledge transfer and participative learning by growers are essential elements in the process of implementing IPM (Vanninen et al. 2008). But what can an individual MPCP company do? The role for manufacturers of biopesticides is to participate in these processes and to present all information required to make growers understand the benefits of a biopesticide. Demonstration trials at commercial holdings are a convincing tool for all stakeholders. These should be done at the launch of the product, but also over time to maintain interest, and to adapt implementation strategies where necessary with new practices, cultivars, pesticides, etc. The presentation of data should be accessible by growers in the form of technical booklets (see Box 6.1), websites, articles in growers' magazines, presentations at grower meetings, etc.

Box 6.1 Mycotal Manual: The Contents of the Booklet with Vital Users Information

PRODUCT INFORMATION
Mycotal – Microbial insecticide
Active ingredient – *Verticillium lecanii*
Formulation

Mode of action
Quality control
Storage
Disposal

DIRECTIONS FOR USE
Crops
Timing and number of application
Dosage
Application
Environmental conditions

EFFECTS
Appearance of control
Effects in various crops
Testing new applications

COMPATIBILITY
Biological control organisms
Pollination by bumblebees
Chemical pesticides

TROUBLE SHOOTING

ENVIRONMENT AND SAFETY

REFERENCES

The above refers to interactions with individual growers. Furthermore, producers should communicate and interact with distributors, extension services, and research institutes concerning the product in order to promote it and to gain a better understanding of the product and its role in the ever-changing practices of an integrated pest control programme. Biocontrol with natural enemies in greenhouses used this approach and was very successful (Van Lenteren 2007). For biopesticides a similar approach is required, and we can use the lessons learned with natural enemies. Without effective transfer of knowledge there will be no successful innovation of technology, e.g. no successful adoption of new biopesticides.

6.4 Ecological Effects Versus Direct Effects

Many authors highlighted the fact that biopesticides are different from chemicals. In addition to killing the pest by the "numerical response" there are sublethal effects and the potential to reproduce and to continue the control effect over time on the pest population by the "functional response" (Waage 1997; Gelernter and Trumble 1999; Thomas 1999; Inglis et al. 2001). They argued that this extra benefit is not exploited enough, and therefore biopesticides cannot escape from the pesticide paradigm. This ecological approach may be valid in field crops, but in greenhouse crops it is not a relevant phenomenon. I have seen epizootics with entomopathogenic fungi, but they are unpredictable with regard to time and scope. Naturally-occurring epizootics have been reported from Entomophthoralean fungi and baculoviruses in agro-ecosystems, and in some cases they provided significant control of pests

(Steinkraus 2007). There are, however, no reports of epizootics giving a predictable and considerable contribution to control of the pest following the application of a MPCP.

The same can be said about sub-lethal effects caused by pathogens. Various parameters of the pests' biology may be affected by the pathogen, such as fecundity, feeding patterns and developmental rate. Many of these effects are not obvious and consequently are not appreciated as part of the product's effectiveness. Only in the case of Bt where feeding ceases quickly after oral uptake, and thus the damage to the plants does not become larger, is this sub-lethal effect appreciated.

I do not want to neglect or underestimate the potential of the "functional response" of a MPCA, but in seasonal and annual greenhouse crops the reality is that MPCPs are used as the inundative approach. Microbials can only be advised based on their known direct effectiveness, and they need to be repeated when the pests re-appear.

6.5 Successful Implementation Requires Continued Efforts

No matter how much research has been conducted with the product in the developmental phase, only when it has been used on a large scale in commercial greenhouses will its strengths and weaknesses become apparent. Advisers need to communicate with each other and the developers of the product in order to find the best way of using the product and the best ways of integrating it successfully in each cropping system. For a grower, the initial use of a microbial product is not an easy job and first-hand advice from a specialist is required. When an adviser visits a grower, not only does the grower learn how to use this new microbial, but also the adviser comes across many different situations from which he again learns more about the product and how best to integrate it in various cropping systems and pest situations. This ultimately leads to a well-balanced use of the microbial. Nevertheless, this should be an ongoing process since many changes occur over time in commercial horticulture and feedback should lead to a cycle of continuous improvement of the use of the microbial product. This is an essential element in the commercial development of any new product, but particularly so with a knowledge-intensive microbial pesticide. This requires considerable effort of the sales and marketing team of a producer. It can be called "technical marketing" of a biopesticide. This element is often forgotten or neglected in the commercial marketing of a new biopesticide, but it is indispensible for the successful uptake of a product.

7 Conclusions and Recommendations

7.1 Application and Implementation Strategy

A microbial pest control product and its proper and effective application are two elements that cannot be seen separately. Therefore, application and implementation of the product needs considerable research within the total development programme of

Table 6.4 Interaction matrix between four types of entomopathogens and various system aspects: importance with regard to the knowledge needed for successful implementation of a microbial insecticide within an IPM programme

System factor/pathogen	Bacteria	Fungi	Baculo-viruses	EPN
Target pest(s)	5	5	5	5
Application strategy	3	5	4	3
Equipment	3	4	3	4
Host plant	3	4	2	1
Cropping system	2	4	2	1
Growing medium[a]	5	5	–	5
Environmental factors				
– Temperature	3	4	3	4
– Humidity	3	5	2	5
– UV radiation	2	2	4	3
Chemical insecticides	2	2	1	2
Chemical fungicides	1	5	2	1
Natural pesticides	1	3	1	3
Parasitoids				
– Direct effects	1	3	1	1
– Indirect effects	4	4	4	4
Predators				
– Direct effects	1	2	1	1
– Indirect effects	4	3	3	2
Pollinators	1	3	1	1
Resistance	4	2	3	1
Knowledge-intensive	2	5	2	4
Technology transfer	2	5	2	4
Costs				
– Product	2	4	3	4
– Application	2	3	2	4
Combination potential	2	4	3	4

1: not important; 2: slightly important; 3: moderately important; 4: very important; 5: critical.
[a]In case of application to the root system.

a MPCP (Table 6.4). The introduction of a new tool in a complex multi-component IPM system is likely to impact many other components. This calls for identification of the relevant interactions, and for studying the effects, and makes the system highly knowledge-based. Elements that are imperative in the development of an IPM system with a new product are listed in Table 6.5.

7.2 Compatibility Profile of Microbial Pest Control Products

A "two-sided" compatibility profile must be established for products based on entomopathogens. On one side, they are susceptible to various chemical and natural pesticides, and knowledge of these effects needs to be generated. On the other hand, microbial products may affect natural enemies depending on their mode of action,

Table 6.5 Essential elements of the implementation process of microbial pest control products in an IPM programme

Requirements for the successful use of a microbial insecticide
• Develop an efficient application strategy for the product • Develop a two-sided compatibility profile of the product • Identify relevant interactions with other control agents • Study the impact of these interactions and develop a strategy to optimize using the MPCP: exploit positive ones, avoid negative ones • Develop simple and manageable recommendations for use • Keep the costs of the product and the applications acceptable • Ensure that transfer of knowledge is sufficient and efficient • Involve all stakeholders in biocontrol, emphasize growers' involvement • Detect problems and concerns in practice and find solutions • Treat implementation and optimization of the IPM system as an ongoing process and as imperative for sustainable use of the MPCP • Position the product as part of a season-long IPM system, not as a single temporary solution for a pest problem

their host range, and the conditions of use. Side-effect testing should be performed when there are indications for harmful effects on any natural enemy or pollinator.

In general, negative effects of biopesticides are of minor importance to natural enemies and pollinators. Nevertheless, for each new MPCP a compatibility profile must be developed since the formulation, the dose, the application strategy, and the conditions of use may lead to effects that are different from those of other strains or products. Often, laboratory studies are carried out under optimal conditions for the pathogen, but side-effect testing methods should reflect the conditions of use. Recommendations on testing methods and conditions are given the paragraph "practical aspects of side-effect testing" above and in Table 6.3. Only testing under relevant conditions generates reliable information that can be used to give advice on the side-effects under commercial conditions.

7.3 *Microbial Pest Control Products and Interactions with Other Control Components*

Combinations of pathogens and natural enemies are often used to enhance biological control or to reduce high pest levels, but this generates interactions in various forms. Many interactions have been studied in the laboratory or in small scale field trials, often in a two agent-one host system. But many of these interactions have not been demonstrated to result in synergism in terms of pest or damage control in commercial settings. Wraight (2003) concluded that: "any attempt to exploit laboratory-discovered synergism or other beneficial interactions among natural enemies certainly warrants caution". At the same time, negative effects and interference are more likely to occur than synergistic effects since control agents target the same pest. Natural enemies of the pest insect for which the pathogen is applied may be adversely affected. This, however, is of relative importance if control by the pathogen is effective. It is much worse when natural enemies of another pest are affected by the pathogen and when this leads to a resurgence of that other pest. These unintentional effects must be known because they have direct implications for the whole IPM system.

IPM systems are much more complex than a two-agent/single host system and understanding all those interactions in detail is impossible. It is therefore pivotal to identify those interactions that hamper or impede effective control, or those that enhance control significantly. Only those require further investigation in order to optimize the integration of various IPM components. Over time, concerns or promising effects may be encountered in practice and those need to be studied to see if they can be avoided or exploited.

7.4 Advantages and Disadvantages of Microbial Pest Control Products in IPM Programmes

The use of a MPCP has the following advantages: they are easy to apply; they act relatively quickly; are generally compatible; have a limited persistence, with few if any lasting adverse effects; are selective; and are relatively cheap compared with natural enemies. They are also generally safe to workers, to the crop, to natural enemies and pollinators. In addition, there is no MRL (maximum residue level); a short or no re-entry period; no pre-harvest period; and they can be tank-mixed. Furthermore, they are a valuable tool in resistance management programmes and they may give the grower a position as a preferred partner for a supermarket.

They have a few disadvantages as they are knowledge-intensive; relatively expensive compared with chemicals; have a narrow spectrum; are slow-acting, and it can be difficult to evaluate their effect. These disadvantageous are considered negative when one compares biopesticides with chemicals which have their own disadvantages. In a biocontrol focussed crop protection system, MPCPs are a good fit as a complementary or corrective component and in some cases as a stand-alone control means.

7.5 Costs of IPM Implementation

A biopesticide producer needs to communicate the cost of the implementation of a MPCP to the grower. IPM programmes have been become quite complex in greenhouse crops. To adopt a new product into this system, the monetary aspect is a significant factor for a grower's decision. It is not the decisive factor, but the costs need to be in the range of products that the grower typically uses. This means not just the product end-user price, but also the labour involved in the application and the potential advantage in the market. Demonstrating the costs and benefits to the grower is vital for the adoption of the product and the IPM programme.

The implementation of a biopesticide means costs for a biopesticide producer. For a successful uptake of a new MPCP in IPM systems, considerable efforts are needed. Education and training is required for its sales force, its distributors and the end-users. Training materials, leaflets, website information, promotional talks, demonstration trials for growers and regularly visiting growers are all part of this. Knowledge acquisition and marketing activities cost money and within the development of a biopesticide these should be budgeted. Since it is an ongoing exercise, these costs must be considered as part of the product price (see Chapter 3, Section 4.1). It is difficult to estimate the extent of these costs. It will depend on the type of pathogen and the targets.

7.6 The Potential of Combinations

Combinations of MPCPs with other control agents have been investigated in numerous studies for their putative enhancing effects in control of the target (Wraight and Ramos 2005 and references discussed by them). Few of these studies have demonstrated improved control in field tests, and few of these ideas are practiced by growers. Moreover, biopesticides are generally not cheap, and mixtures must enhance control in order to justify the higher costs for the grower. Still, the potential of enhanced effects is worth investigating.

The potential of low dosage effects should not be overlooked either. The focus should be on reducing costs of products and labour, on simultaneous as well as sequential use, on avoiding negative interactions, and exploiting positive interactions. I believe that the value of these areas of study is still underestimated. It is much easier and cheaper to develop better biocontrol in this way then to develop new products. More research is required, particularly to study the effects under commercial conditions. Results should be followed by efficient knowledge transfer and "technical marketing". This element of IPM implementation could offer new solutions in the short term to growers and could increase the use of microbial pest control products in general.

References

Abbas, M.S.T. and D.C. Boucias, 1984. Interaction between nuclear polyhedrosis virus-infected *Anticarsia gemmatalis* (Lepidoptera: Noctuidae) larvae and predator *Podisus maculiventris* (Say) (Hemiptera: Pentatomidae). *Environ. Entomol.* 13: 599–602.

Agra Gothama, A.A., G.W. Lawrence, and P.P. Sikorowskii, 1996. Activity and persistence of *Steinernema carpocapsae* and *Spodoptera exigua* nuclear polyhedrosis virus against *S. exigua* larvae on soybean. *J. Nematol.* 28(1): 68–74.

Akhurst, R.J., 1990. Safety to nontarget invertebrates of nematodes of economically important pests. In: M. Laird, L.A. Lacey, E.W. Davidson (eds), *Safety of microbial insecticides*. CRC Press, Boca Raton. pp. 233–240.

Ali, M.I., G.W. Felton, T. Meade, and S.Y. Young, 1998. Influence of interspecific and intraspecific host plant variation on the susceptibility of Heliothines to a baculovirus. *Biol. Control* 12(1): 42–49.

Alma, R.C., 2005. Intraguild interactions between two natural enemies of *Trialeurodes vaporariorum* (Homoptera: Aleyrodidae), the predator, *Dicyphus hesperus* (Heteroptera: Miridae), and the entomopathogenic fungus, *Paecilomyces fumosoroseus* Apopka-97 (Deuteromycotina: Hyphomycetes) (PFR-97). Master Thesis, Department of Biological Sciences, Simon Fraser University. 156 pp.

Alma, R.C., M.S. Goettel, B.D. Roitberg, and D.R. Gillespie, 2007. Combined effects of the entomopathogenic fungus, *Paecilomyces fumosoroseus* Apopka-97, and the generalist predator, *Dicyphus hesperus*, on whitefly populations. *BioControl* 52(5): 669–681.

Angeli, G., M. Baldessari, R. Maines, and C. Duso, 2005. Side-effects on the predatory bug *Orius laevigatus* (Heteroptera: Anthocoridae) in the laboratory. *Biocontro. Sci. Technol.* 15(7): 745–754.

Ansari, M.A., M. Brownbridge, F.A. Shah, and T.M. Butt, 2008a. Efficacy of entomopathogenic fungi against soil-dwelling life stages of western flower thrips, *Frankliniella occidentalis*, in plant-growing media. *Entomol. Exp. Appl.* 127(2): 80–87.

Ansari, M.A., F.A. Shah, and T.M. Butt, 2008b. Combined use of entomopathogenic nematodes and *Metarhizium anisopliae* as a new approach for black vine weevil, *Otiorhynchus sulcatus*, control. *Entomol. Exp. Appl.* 129: 340–347.

Ansari, M.A., F.A. Shah, and T.M. Butt, 2010. The entomopathogenic nematode *Steinernema kraussei* and *Metarhizium anisopliae* work synergistically in controlling overwintering larvae of the black vine weevil, *Otiorhynchus sulcatus*, in strawberry growbags. *Biocontrol Sci. Technol.* 20(1): 99–105.

Ansari, M.A., F.A. Shah, L. Tirry, and M. Moens, 2006. Field trials against *Hoplia philanthus* (Coleoptera: Scarabaeidae) with a combination of an entomopathogenic nematode and the fungus *Metarhizium anisopliae* CLO 53. *Biol. Control* 39: 453–459.

Ansari, M.A., F.A. Shah, M. Whittaker, M. Prasad, and T.M. Butt, 2007. Control of western flower thrips (*Frankliniella occidentalis*) pupae with *Metarhizium anisopliae* in peat and peat alternative growing media. *Biol. Control* 40: 293–297.

Ansari, M.A., L. Tirry, and M. Moens, 2004. Interaction between *Metarhizium anisopliae* CLO 53 and entomopathogenic nematodes for control of *Hoplia philanthus*. *Biol. Control* 31: 172–180.

Ansari, M.A., L. Tirry, and M. Moens, 2005. Antagonism between entomopathogenic fungi and bacterial symbionts of entomopathogenic nematodes. *BioControl* 50: 465–475.

Avery, P.B., J. Faull, and M.S.J. Simmonds, 2008. Effects of *Paecilomyces fumosoroseus* and *Encarsia formosa* on the control of the greenhouse whitefly: preliminary assessment of a compatibility study. *BioControl* 53: 303–316.

Bardin, M., J. Fargues, L. Couston, C. Troulet, G. Philippe, and P.C. Nicot, 2004. Compatibility of intervention to control grey mould, powdery mildew and whitefly on tomato, using three biological methods. *IOBC/WPRS Bull.* 27(8): 5–9.

Barrett, K., N. Grandy, E.G. Harrison, S. Hassan, and P. Oomen, 1994. Guidance document on regulatory testing procedures for pesticides with non-target arthropods. *ESCORT Workgroup*, Wageningen, The Netherlands, 1994. Society of Environmental Toxicology and Chemistry – Europe (SETAC).

Bateman, R.P., G.A. Matthews, and F.R. Hall, 2007. Ground-based application equipment. In: L.A. Lacey, H.K. Kaya (eds), *Field manual of techniques in invertebrate pathology*, 2nd edition. Springer, Dordrecht. pp. 73–98.

Bathon, H., 1996. Impact of entomopathogenic nematodes on non-target hosts. *Biocontrol Sci. Technol.* 6: 421–434.

Beerling, E., 2008. The switch to IPM in cut-chrysanthemum in the Netherlands. *IOBC/WPRS Bull.* 32: 17–20.

Beerling, E. and D. van den Berg, 2005. Evaluation of two microbial products and an insecticide for integrated control in greenhouse chrysanthemums. *IOBC/WPRS Bull.* 28(3): 179–184.

Blümel, S., G.A. Matthews, A. Grinstein, and Y. Elad, 1999. Pesticides in IPM: selectivity, side-effects, application and resistance problems. In: R. Albajes, M.L. Gullino, J.C. van Lenteren,

Y. Elad (eds), *Integrated pest and disease management in greenhouse crops*. Kluwer, Dordrecht. pp. 150–167.

Briese, D.T., 1986. Insect resistance to baculoviruses. In: R.R. Granados, B.A. Federici (eds), *The biology of baculoviruses. Vol. II. Practical application for insect control*. CRC Press, Boca Raton. pp. 237–263.

Brodeur, J. and G. Boivin (eds), 2006. *Trophic and guild interactions in biological control*. Springer, Dordrecht.

Brooks, W.M., 1993. Host-parasitoid-pathogen interactions. In: N.E. Beckage, S. Thompson, B.A. Federici (eds), *Parasites and pathogens of insects*, Vol. 2. Academic Press, New York. pp. 231–272.

Bruck, D.J., 2009. Impact of fungicides on *Metarhizium anisopliae*in the rhizosphere, bulk soil and in vitro. *BioControl* 54: 597–606.

Bruck, D.J. and K.M. Donahue, 2007. Persistence of *Metarhizium anisopliae* incorporated into soilless potting media for control of the black vine weevil, *Otiorhynchus sulcatus* in container-grown ornamentals. *J. Invertebr. Pathol.* 95: 146–150.

Brusselman, E., W. Steurbaut, and B. Sonck, 2007. Optimizing the application of entomopathogenic nematodes: experimental set-up. *Commun. Agric. Appl. Biol. Sci.* 72(2): 81–86.

Butt, T.M., 2002. Use of entomogenous fungi for the control of insect pests. In: F. Kempken (ed), *The Mycota XI, Agricultural applications*. Springer, Berlin. pp. 111–134.

Butt, T.M. and M. Brownbridge, 2001. Increasing the efficacy of entomogenous fungi. In: M. Vurro, J. Gressel (eds), *Novel biotechnologies for biocontrol agents enhancement and management*. Springer, Dordrecht. pp. 52–63.

Butt, T.M., L. Ibrahim, B.V. Ball, and S.J. Clark, 1994. Pathogenicity of the entomogenous fungi *Metarhizium anisopliae* and *Beauveria bassiana* against crucifer pests and the honey bee. *Biocontrol Sci. Technol.* 4: 207–214.

Candolfi, M., K. Barrett, P. Campbell, R. Forster, N. Grandy, M.-C. Huet, G. Lewis, P. Oomen, R. Schmuck, and H. Vogt, 2000. Guidance document on regulatory testing procedures for pesticides with non-target arthropods. *ESCORT Workgroup,* Wageningen, The Netherlands 2000. Society of Environmental Toxicology and Chemistry – Europe (SETAC).

Carreck, N.L., T.M. Butt, S.J. Clark, L. Ibrahim, E.A. Isger, J.K. Pell, and I.H. Williams, 2007. Honey bees can disseminate a microbial control agent to more than one inflorescence pest of oilseed rape. *Biocontrol Sci. Technol.* 17(2): 179–191.

Chapman, M.H. and M.A. Hoy, 1991. Relative toxicity of *Bacillus thuringiensis* var. *tenebrionis* to the spotted spider mite (*Tetranychus urticae* Koch) and its predator *Metaseiulus occidentalis* (Nesbitt) (Acari, Tetranychidae and Phytoseiidae). *J. Appl. Entomol.* 111(2): 147–154.

Chapple, A.C., R.A. Downer, and R.P. Bateman, 2007. Theory and practice of microbial insecticide application. In: L.A. Lacey, H.K. Kaya (eds), *Field manual of techniques in invertebrate pathology*, 2nd edition. Springer, Dordrecht. pp. 9–34.

Cloutier, C. and C. Jean, 1998. Synergism between natural enemies and biopesticides: a test case using the stinkbug *Perillus bioculatus* (Hemiptera: Pentatomidae) and *Bacillus thuringiensis tenebrionis* against the Colorado potato beetle (Coleoptera: Chrysomelidae). *J. Econ. Entomol.* 91(5): 1096–1108.

Cook, S.M., R.K. Zeyaur, and J.A. Pickett, 2007. The use of push-pull strategies in integrated pest management. *Ann. Rev. Entomol.* 52: 375–400.

Coremans-Pelseneer, J., 1994. Laboratory tests on the entomopathogenic fungus *Beauveria*. *IOBC/WPRS Bull.* 17(10): 147–154.

Cory, J.S., 2003. Ecological impacts of virus insecticides: host range and non-target organisms. In: H.M.T. Hokkanen, A.E. Hajek (eds), *Environmental impacts of microbial insecticides*. Kluwer, Dordrecht. pp. 73–91.

Cory, J.S. and J.D. Ericsson, 2010. Fungal entomopathogens in a tritrophic context. *BioControl* 55: 75–88.

Cory, J.S. and K. Hoover, 2006. Plant-mediated effects in insect-pathogen interactions. *Trends Ecol. Evol.* 21(5): 278–286.

Cory, J.S. and J.H. Myers, 2003. The ecology and evolution of insect baculoviruses. *Annu. Rev. Ecol. Evol. Syst.* 34: 239–272.

Costa, H.S., K.L. Robb, and C.A. Wilen, 2001. Increased persistence of *Beauveria bassiana* spore viability under high ultraviolet-blocking greenhouse plastic. *Hortscience* 36(6): 1082–1084.

Cullen, R., K.D. Warner, M. Jonsson, and S.D. Wratten, 2008. Economics and adoption of conservation biological control. *Biol. Control* 45: 272–280.

Cuthbertson, A.G.S., L.F. Blackburn, P. Northing, W. Luo, R.J.C. Cannon, and K.F.A. Walters, 2008a. Further compatibility tests of the entomopathogenic fungus *Lecanicillium muscarium* with conventional insecticide products for control of sweetpotato whitefly, *Bemisia tabaci* on poinsettia plants. *Insect. Sci.* 15: 355–360.

Cuthbertson, A.G.S., J. Head, K.F.A. Walters, and A.W.A. Murray, 2003. The integrated use of chemical insecticides and the entomopathogenic nematode, *Steinernema feltiae*, for the control of sweetpotato whitefly. *Bemisia tabaci. Nematol.* 5(5): 713–720.

Cuthbertson, A.G.S., J.J. Mathers, P. Northing, A.J. Prickett, and K.F.A. Walters, 2008b. The integrated use of chemical insecticides and the entomopathogenic nematode, *Steinernema carpocapsae* (Nematoda: Steinernematidae), for the control of sweetpotato whitefly, *Bemisia tabaci* (Hemiptera: Aleyrodidae). *Insect Sci.* 15: 447–453.

Cuthbertson, A.G.S., J.P. North, and K.F.A. Walters, 2005a. Effect of temperature and host plant leaf morphology on the efficacy of two entomopathogenic biocontrol agents of *Thrips palmi* (Thysanoptera: Thripidae). *Bull. Ent. Res.* 95: 321–327.

Cuthbertson, A.G.S., K.F.A. Walters, and C. Deppe, 2005b. Compatibility of the entomopathogenic fungus *Lecanicillium muscarium* and insecticides for eradication of sweetpotato whitefly. *Bemisia tabaci. Mycopathologia* 160: 35–41.

De Buck, A.J. and E.A.M. Beerling, 2006. Implementation of biocontrol and IPM in Dutch horticulture. A socio-economic and technical approach. In: J. Eilenberg, H. Hokkanen (eds), *An ecological and societal approach to biological control.* Kluwer, Dordrecht. pp. 73–90.

Dent, D.R., 1997. Integrated pest management and microbial insecticides. In: H.F. Evans (ed), *Microbial insecticides: novelty or necessity? BCPC Symposium Proceedings* No. 68, Coventry, April 16–18, 1997. BCPC, Farnham. pp. 127–138.

Down, R.A., A.G.S. Cuthbertson, J.J. Mathers, and K.F.A. Walters, 2009. Dissemination of the entomopathogenic fungi, *Lecanicillium longisporum* and *L. muscarium*, by the predatory bug, *Orius laevigatus*, to provide concurrent control of *Myzus persicae, Frankliniella occidentalis* and *Bemisia tabaci. Biol. Control* 50(2): 172–178.

Eberle, K.E. and J.A. Jehle, 2006. Field resistance of codling moth against *Cydia pomonella* granulovirus (CpGV) is autosomal and incompletely dominant inherited. *J. Invertebr. Pathol.* 93(3): 201–206.

Ehlers, R.-U., 2003a. Entomopathogenic nematodes in the European biocontrol market. *Commun. Appl. Biol. Sci. Ghent Univ.* 68(4a): 3–15.

Ehlers, R.-U., 2003b. Biocontrol nematodes. In: H.M.T. Hokkanen, A.E. Hajek (eds), *Environmental impacts of microbial insecticides.* Kluwer, Dordrecht. pp. 177–225.

Evans, H.F., 1999. Principles of dose acquisition for biopesticides. In: F.R. Hall, J.J. Menn (eds), *Biopesticides: use and delivery.* Humana Press, Totowa. pp. 553–573.

Fargues, J., N. Smits, M. Rougier, T. Boulard, G. Ridray, J. Lagier, B. Jeannequin, M. Mermier, C. Vidal, P. Reich, and P. Nicot, 2003b. Biological control of the greenhouse whitefly (*Trialeurodes vaporariorum*) with entomopathogenic Hyphomycetes: how does it work in Mediterranean greenhouses on tomato crops? *Colloque international tomate sous abri, protection intégrée-agriculture biologique*, Avignon, 17–19 septembre 2003. pp. 107–112.

Fargues, J., C. Vidal, N. Smits, M. Rougier, T. Boulard, M. Mermier, P. Nicot, P. Reich, B. Jeannequin, G. Ridray, and J. Lagier, 2003a. Climatic factors on entomopathogenic hyphomycetes infection of *Trialeurodes vaporariorum* (Homoptera: Aleyrodidae) in Mediterranean glasshouse tomato. *Biol. Control* 28: 320–331.

Fife, J.P., R.C. Derksen, H.E. Ozkan, and P.S. Grewal, 2003. Effects of pressure differentials on the viability and infectivity of entomopathogenic nematodes. *Biol. Control* 27: 65–72.

Fransen, J.J. and J. van Lenteren, 1993. Host selection and survival of the parasitoid *Encarsia formosa* on greenhouse whitefly, *Trialeurodes vaporariorum*, in the presence of hosts infected with the fungus *Aschersonia aleyrodis*. *Entomol. Exp. Appl.* 69: 239–249.

Fransen, J.J. and J. van Lenteren, 1994. Survival of the parasitoid *Encarsia formosa* after treatment of parasitized whitefly larvae with fungal spores of *Aschersonia aleyrodis*. *Entomol. Exp. Appl.* 71: 235–243.

Frisch, E., K. Undorf-Spahn, J. Kienzle, C.P.W. Zebitz, and J. Huber, 2007. Codling moth granulosisvirus: first indications of variations in the susceptibility of local codling moth populations. *IOBC/WPRS Bull.* 30(1): 181–186.

Fuxa, J.R., 2004. Ecology of insect nucleopolyhedroviruses. *Agric. Ecosyst. Environ.* 103: 27–43.

Gan-Mor, S. and G.A. Matthews, 2003. Recent developments in sprayers for application of biopesticides-an overview. *Biosyst. Eng.* 84(2): 119–125.

Geervliet, J.B.F., J.M. Vlak, and P.H. Smits, 1991. Effects of *Spodoptera exigua* nuclear polyhedrosis virus and *Bacillus thuringiensis* subsp. *aizawai* mixtures on mortality of beet armyworm. *Spodoptera exigua. Med. Fac. Landbouww. Rijksuniv. Gent.* 56(2a): 305–311.

Gelernter, W.D. and J.T. Trumble, 1999. Factors in the success and failure of microbial insecticides in vegetable crops. *Integr. Pest Manag. Rev.* 4: 301–306.

Gillespie, D., J. Brodeur, C. Cloutier, M. Goettel, P. Jamarillo, R. Labbé, B. Roitberg, C. Thompson, and S. van Laerhoven, 2006. Combining pathogens and predators of insects in biological control. *IOBC/WPRS Bull.* 29(4): 3–8.

Glare, T.R. and M. O'Callaghan, 2000. *Bacillus thuringiensis: biology, ecology and safety*. Wiley & Sons, Chichester.

Glare, T.R. and M. O'Callaghan, 2003. Environmental impacts of bacterial biopesticides. In: H.M.T. Hokkanen, A.E. Hajek (eds), *Environmental impacts of microbial insecticides*. Kluwer, Dordrecht. pp. 119–149.

Goettel, M.S., T.J. Poprawski, J.D. Vandenberg, Z. Li, and D.W. Roberts, 1990. Safety to nontargets of fungal biocontrol agents. In: M. Laird, L.A. Lacey, E.W. Davidson (eds), *Safety of microbial insecticides*. CRC Press, Boca Raton. pp. 209–231.

Gross, H.R., J.J. Hamm, and J.E. Carpenter, 1994. Design and application of a hive-mounted device that uses honey bees (Hymenoptera: Apidae) to disseminate *Heliothis* nuclear polyhedrosis virus. *Environ. Entomol.* 23: 492–501.

Gröner, A., 1990. Safety to nontarget invertebrates of baculoviruses. In: M. Laird, L.A. Lacey, E.W. Davidson (eds), *Safety of microbial insecticides*. CRC Press, Boca Raton. pp. 135–147.

Gutiérrez, C., R. Campos-Herrera, and J. Jiménez, 2008. Comparative study of the effect of selected agrochemical products on *Steinernema feltiae* (Rhabditidae: Steinernematidae). *Biocontrol Sci. Technol.* 18(1): 101–108.

Hajek, A.E. and M.S. Goettel, 2008. Guidelines for evaluating effects of entomopathogens on non-target organisms. In: L.A. Lacey, H.K. Kaya (eds), *Field manual of techniques in invertebrate pathology*, 2nd edition. Springer, Dordrecht. pp. 815–833.

Hall, R.A., 1981. Laboratory studies on the effects of fungicides, acaracides and insecticides on the entomopathogenic fungus, *Verticillium lecanii*. *Entomol. Exp. Appl.* 29: 39–48.

Harper, J.D., 1986. Interactions between baculoviruses and other entomopathogens, chemical pesticides, and parasitoids. In: R.R. Granados, B.A. Federici (eds), *The biology of baculoviruses. Vol. II. Practical application for insect control*. CRC Press, Boca Raton. pp. 133–155.

Hassan, S.A. and A. Gröner, 1977. Die Wirkung von Kernpolyedern (*Baculovirus* spec.) aus *Mamestra brassicae* auf *Trichogramma cacoeciae* (Hym: Trichogrammatidae) und *Chrysopa carnea* (Neur.: Chrysopidae). *Entomophaga* 22: 281–288.

Hassan, S.A. and M. van de Veire, 2004. Compatibility of pesticides with biological control agents. In: K.M. Heinz, R.G. van Driesche, M.P. Parrella (eds), *Biocontrol in protected crops*. Ball Publishing, Batavia. pp. 129–147.

Head, J., L.F. Palmer, and K.F.A. Walters, 2003. Compatibility of control agents used for the control of the south American leafminer, *Liriomyza huidobrensis*. *Biocontrol Sci. Technol.* 13: 77–86.

Hofstein, R. and B. Fridlender, 1994. Development of production, formulation and delivery systems. *Proceedings BCPC Conference* 1994 – *Pests and diseases.* 3(9B): 1273–1280. BCPC Publications, Farnham.
Hokkanen, H.M.T. and R. Kotiluoto, 1992. Bioassay of the side-effects of pesticides on *Beauveria bassiana* and *Metarhizium anisopliae*: standardized sequential testing procedure. *IOBC/WPRS Bull.* XV(3): 148–151.
Hokkanen, H.M.T., Q.-Q. Zeng, and I. Menzler-Hokkanen, 2003. Assessing the impacts of Metarhizium and Beauveria on bumblebees. In: H.M.T. Hokkanen, A.E. Hajek (eds), *Environmental impacts of microbial insecticides.* Kluwer, Dordrecht. pp. 63–71.
Homan, H.G., 1981. Insect responses to microbial infections. In: H.D. Burges (ed), *Microbial control of pests and plant diseases 1970–1980.* Academic Press, London. pp. 769–784.
Hoover, K., J.L. Yee, C.M. Schultz, D.M. Rocke, B.D. Hammock, and S.S. Duffey, 1998. Effects of plant identity and chemical constituents on the efficacy of a baculovirus against *Heliothis virescens*. *J. Chem. Ecol.* 24(2): 221–252.
Inbar, M. and D. Gerling, 2008. Plant-mediated interactions between whiteflies, herbivores, and natural enemies. *Ann. Rev. Entomol.* 53: 431–448.
Inglis, G.D., M.S. Goettel, T.M. Butt, and H. Strasser, 2001. Use of hyphomycetous fungi for managing insect pests. In: T.M. Butt, C. Jackson, N. Magan (eds), *Fungi as biocontrol agents: progress, problems and potential.* CAB International, Wallingford. pp. 23–69.
Jacobson, R.J., D. Chandler, J. Fenlon, and K.M. Russell, 2001. Compatibility of *Beauveria bassiana* (Balsamo) Vuillemin with *Amblyseius cucumeris* Oudemans (Acarina: Phytoseiidae) to control *Frankliniella occidentalis* Pergande (Thysanoptera: Thripidae) on cucumber plants. *Biocontrol Sci. Technol.* 11: 391–400.
Jacques, R.P. and O.N. Morris, 1981. Compatibility of pathogens with other methods of pest control and with different crops. In: H.D. Burges (ed), *Microbial control of pests and plant diseases 1970–1980.* Academic Press, London. pp. 695–715.
Jagdale, G.B., M.L. Casey, P.S. Grewal, and R.K. Lindquist, 2004. Application rate and timing, potting medium, and host plant effects on the efficacy of *Steinernema feltiae* against the fungus gnat, *Bradysia coprophila*, in floriculture. *Biol. Control* 29: 296–305.
Jandricic, S., J. Sanderson, and S. Wraight, 2008. Intraguild predation among biological control agents used in greenhouse floriculture crops: a preliminary review. *IOBC/WPRS Bull.* 32: 91–94.
Janmaat, A.F., 2007. Development of resistance to the pesticide Bacillus thuringiensis kurstaki. In: C. Vincent, M.S. Goettel, G. Lazarovits (eds), *Biological control – a global perspective.* CAB International, Wallingford. pp. 179–184.
Janmaat, A.F. and J. Myers, 2003. Rapid evolution and the cost of resistance to *Bacillus thuringiensis* in greenhouse populations of cabbage loopers, Trichoplusia ni. *Proc. R. Soc. London Ser. B. Biol. Sci.* 270: 2263–2270.
Jaramillo, J., C. Borgemeister, L. Ebssa, A. Gaigl, R. Tobón, and G. Zimmermann, 2005. Effect of combined applications of *Metarhizium anisopliae* (Metsch.) Sorokin (Deuteromycotina: Hyphomycetes) strain CIAT 224 and different dosage of imidacloprid on the subterranean burrower bug *Cyrtomenus bergi* Froeschner (Hemiptera: Cydnidae). *Biol. Control* 34: 12–20.
Jaronski, S.T., 2010. Ecological factors in the inundative use of fungal entomopathogens. *BioControl* 55: 159–185.
Jazzar, C. and E.A.-F. Hammad, 2004. Efficacy of multiple biocontrol agents against the sweet potato whitefly *Bemisia tabaci* (Gennadius) (Homoptera: Aleyrodidae) on tomato. *J. Appl. Entomol.* 128: 188–194.
Jyoti, J.L. and G.J. Brewer, 1999. Honey bees (Hymenoptera: Apidae) as vectors of *Bacillus thuringiensis* for control of banded sunflower moth (Lepidoptera: Tortricidae). *Environ. Entomol.* 28(6): 1172–1176.
Kapongo, J.P., L. Shipp, P. Kevan, and B. Broadbent, 2008. Optimal concentration of *Beauveria bassiana* vectored by bumble bees in relation to pest and bee mortality in greenhouse tomato and sweet pepper. *BioControl* 53: 797–812.

Kaya, H.K. and T.M. Burlando, 1989. Development of *Steinernema feltiae* (Rhabditidae: Steinernematidae) in diseased insect hosts. *J. Invertebr. Pathol.* 53: 164–168.

Kaya, H.K., T.M. Burlando, H.Y. Choo, and G.S. Thurston, 1995. Integration of entomopathogenic nematodes with *Bacillus thuringiensis* or pesticidal soap for control of insect pests. *Biol. Control* 5: 432–441.

Keller, S., 1994. Side effects of pesticides on insect pathogenic fungi: some remarks and a proposition. *IOBC/WPRS Bull.* 17(3): 193–196.

Keller, S., B. Pärli, M. Lujan, and C. Schweizer, 1993. Der Einfluss von Fungiziden auf den insektenpathogenen Pilz *Beauveria brongniartii* (Sacc.) Petch. *Anz. Schädlingskunde Pflanzensch. Umweltsch* 66: 108–114.

Kepler, R.M. and D.J. Bruck, 2006. Examination of the interaction between the black vine weevil (Coleoptera: Curculionidae) and an entomopathogenic fungus reveals a new tritrophic interaction. *Environ. Entomol.* 35(4): 1021–1029.

Kevan, P.G., J. Sutton, and L. Shipp, 2007. Pollinators as vectors of biocontrol agents- the B52 story. In: C. Vincent, M.S. Goettel, G. Lazarovits (eds), *Biological control – a global perspective*. CAB International, Wallingford. pp. 319–327.

Kirkpatrick, B. and R.Georgis, 2007. Cost benefit of using Bt-based products in IPM programs. *Paper presented at the 2nd Annual Biocontrol Industry Meeting (ABIM) Lucerne*, Switzerland, October 22–23, 2007. http://www.abim-lucerne.ch/archive/conference2007. Cited June 29, 2008.

Koppenhöfer, A.M., 2007. Nematodes. In: L.A. Lacey, H.K. Kaya (eds), *Field manual of techniques in invertebrate pathology*, 2nd edition. Springer, Dordrecht. pp. 249–264.

Koppenhöfer, A.M., I.M. Brown, R. Gaugler, P.S. Grewal, H.K. Kaya, and M.G. Klein, 2000. Synergism of entomopathogenic nematodes and imidacloprid against white grubs: greenhouse and field evaluation. *Biol. Control* 19: 245–251.

Koppenhöfer, A.M. and P.S. Grewal, 2005. Compatibility and interactions with agrochemicals and other biocontrol agents. In: P.S. Grewal, R.-U. Ehlers, D.I. Shapiro-Ilan (eds), *Nematodes as biocontrol agents*. CAB International, Wallingford. pp. 363–381.

Koppenhöfer, A.M. and H.K. Kaya, 1997. Additive and synergistic interaction between entomopathogenic nematodes and *Bacillus thuringiensis* for scarab grub control. *Biol. Control* 8: 131–137.

Koppenhöfer, A.M. and H.K. Kaya, 1998. Synergism of imidacloprid and an entomopathogenic nematode: a novel approach to white grub control in turfgrass. *J. Econ. Entomol.* 91: 618–623.

Kovach, J., R. Petzoldt, and G.E. Harman, 2000. Use of honeybees and bumble bees to disseminate *Trichoderma harzianum* 1295-22 to strawberries for *Botrytis* control. *Biol. Control* 18: 235–242.

Krishnayya, P.V. and P.S. Grewal, 2002. Effect of neem and selected fungicides on viability and virulence of the entomopathogenic nematode *Steinernema feltiae*. *Biocontrol Sci. Technol.* 12: 259–266.

Kunkel, B.A., P.S. Grewal, and M.F. Quigley, 2004. A mechanism of acquired resistance against an entomopathogenic nematode by *Agrotis ipsilon* feeding on perennial ryegrass harboring a fungal endophyte. *Biol. Control* 29(1): 100–108.

Labbé, R., C. Cloutier, and J. Brodeur, 2006. Prey selection by *Dicyphus hesperus* of infected or parasitized greenhouse whitefly. *Biocontrol Sci. Technol.* 16(5): 485–494.

Labbé, R.M., D.R. Gillespie, C. Cloutier, and J. Brodeur, 2009. Compatibility of an entomopathogenic fungus with a predator and a parasitoid in the biological control of greenhouse whitefly. *Biocontrol Sci. Technol.* 19(4): 429–446.

Lacey, L.A., S.P. Arthurs, A.L. Knight, and J. Huber 2007. Microbial control of lepidopteran pests in apple orchards. In: L.A. Lacey, H.K. Kaya (eds), *Field manual of techniques in invertebrate pathology*, 2nd edition. Springer, Dordrecht. pp. 527–546.

Lacey, L.A., T.R. Unruh, and H.L. Headrick, 2003. Interactions of two idiobiont parasitoids (Hymenoptera: Ichneumonidae) of codling moth (Lepidoptera: Tortricidae) with the entomopathogenic nematode *Steinernema carpocapsae* (Rhabditidae: Steinernematidae). *J. Invertebr. Pathol.* 83: 230–239.

Laczyński, A., W. Dierickx, and A. de Moor, 2007. The effect of agitation system, temperature of the spray liquid, nematode concentration, and air injection on the viability of *Heterorhabditis bacteriophora. Biocontrol Sci. Technol.* 17(8): 841–851.

Laczyński, A., A. de Moor, W. Dierickx, M. Moens, P. Darius, B. Sonck, and H. Ramon, 2006. The effect of hydraulic agitation on the viability of the nematode *Heterorhabditis bacteriophora. Crop Protect.* 25: 1135–1141.

Lindquist, R.K. and T.L. Short, 2004. Effects of greenhouse structure and function on biological control. In: K. Heinz, R.G. van Driesche, M.P. Parrella (eds), *Biocontrol in protected culture*. Ball Publishing, Batavia. pp. 37–53.

Lomer, C.J., R.P. Bateman, H. de Groote, D. Dent, C. Kooyman, J. Langewald, D. Johnson, R. Peveling, and M.B. Thomas, 1999. Development of strategies for the incorporation of microbial pesticides into the integrated management of migratory pests. *Agric. Forest Entomol.* 1: 71–88.

Ludwig, S.W. and R.D. Oetting, 2002. Efficacy of *Beauveria bassiana* plus insect attractants for enhanced control of *Frankliniella occidentalis* (Thysanoptera: Thripidae). *Florida Entomol.* 85(1): 270–272.

Mason, J.M., G.A. Matthews, and D.J. Wright, 1999. Evaluation of spinning disc technology for the application of entomopathogenic nematodes against a foliar pest. *J. Invertebr. Pathol.* 73: 282–288.

Matthews, G.A., 2000. Applying biological agents: needs and new developments for controlling foliar pests. *Proceedings of the BCPC Conference – Pests and diseases 2000*, 2(6B): 681–686. BCPC Publications, Farnham, UK.

Matthews, H.J., I. Smith, H.A. Bell, and J.P. Edwards, 2004. Interactions between the parasitoid *Meteorus gyrator* (Hymenoptera: Braconidae) and a granulovirus in *Lacanobia oleracea* (Lepidoptera: Noctuidae). *Environ. Entomol.* 33(4): 949–957.

McCutchen, W.F. and L. Flexner, 1999. Joint actions of baculoviruses and other control agents. In: F.R. Hall, J.J. Menn (eds), *Biopesticides: use and delivery*. Humana Press, Totowa. pp. 341–355.

Meekes, E.T.M., S. van Voorst, N.N. Joosten, J.J. Fransen, and J.C. van Lenteren, 2000. Persistence of the fungal whitefly pathogen, *Aschersonia aleyrodis*, on three different plant species. *Mycol. Res.* 104: 1234–1240.

Melin, B.E. and E.M. Cozzi, 1990. Safety to nontarget invertebrates of lepidopteran strains of Bacillus thuringiensis and their β-exotoxins. In: M. Laird, L.A. Lacey, E.W. Davidson (eds), *Safety of microbial insecticides*. CRC Press, Boca Raton. pp. 149–167.

Miles, M., 2006. The effects of spinosad on beneficial insects and mites used in integrated pest management in greenhouses. *IOBC/WPRS Bull.* 29(10): 53–59.

Mohan, M., S.N. Sushil, J.C. Bhatt, G.T. Gujar, and H.S. Gupta, 2008. Synergistic interaction between sublethal doses of *Bacillus thuringiensis* and *Campoletis chlorideae* in managing *Heliothis armigera. BioControl* 53: 375–386.

Mommaerts, V., J. Boulet, G. Sterk, and G. Smagghe, 2007. Effects of biological control agents (BCAs) on the pollinator, *Bombus terrestris. Proceedings of the BCPC XVI International Plant Protection Congress 2007*, 15–18 October 2007, Glasgow. pp. 862–863.

Mommaerts, V., G. Sterk, L. Hoffmann, and G. Smagghe, 2009. A laboratory evaluation to determine the compatibility of microbiological control agents with the pollinator *Bombus terrestris. Pest Manag. Sci.* 65(9): 949–955.

Nagesh, M., S.S. Hussaini, S.P. Singh, and S.R. Biswas, 2003. Management of root-knot nematode, *Meloidogyne incognita* (Kofoid & White) Chitwood in chrysanthemum using *Paecilomyces lilacinus* (Thom) Samson in combination with neem cake. *J. Biol. Control* 17(2): 125–131.

Navon, A., 1993. Control of lepidopteran pests with Bacillus thuringiensis. In: P.F. Entwistle, J.S. Cory, M.J. Bailey, S. Higgs (eds), *Bacillus thuringiensis, an environmental biopesticide: theory and practice*. Wiley & Sons, New York. pp. 125–146.

Navon, A., 2000. *Bacillus thuringiensis* insecticides in crop protection – reality and prospects. *Crop Protect.* 19: 669–676.

Nguyen, D.H., M. Nakai, J. Takatsuka, S. Okuno, T. Ishii, and Y. Kunimi, 2005. Interaction between a nucleopolyhedrovirus and the braconid parasitoid *Meteorus pulchricornis* (Hymenoptera: Braconidae) in the larvae of *Spodoptera litura* (Lepidoptera: Noctuidae). *Appl. Entomol. Zool.* 40(2): 325–334.

Nilsson, U. and E. Gripwall, 1999. Influence of application technique on the viability of the biological control agents V*erticillium lecanii* and *Steinernema feltiae*. *Crop Protect.* 18: 53–59.

North, J.P., A.G.S. Cuthbertson, and K.F.A. Walters, 2006. The efficacy of two entomopathogenic biocontrol agents against adult *Thrips palmi* (Thysanoptera: Thripidae). *J. Invertebr. Pathol.* 92(2): 89–92.

Oestergaard, J., C. Belau, O. Strauch, A. Ester, K. van Rozen, and R.-U. Ehlers, 2006. Biological control of *Tipula paludosa* (Diptera: Nematocera) using entomopathogenic nematodes (*Steinernema* spp.) and *Bacillus thuringiensis* subsp. *israelensis*. *Biol. Control* 39: 525–531.

Olleka, A., N. Mandour, and S. Ren, 2009. Effect of host plant on susceptibility of whitefly *Bemisia tabaci* (Homoptera: Aleyrodidae) to the entomopathogenic fungus *Beauveria bassiana* (Ascomycota: Hypocreales). *Biocontrol Sci. Technol.* 19(7): 717–727.

Onillon, J.-C. and M.L. Gullino, 1999. Implementation of IPM: from research to the consumer. In: R. Albajes, M.L. Gullino, J.C. van Lenteren, Y. Elad (eds), *Integrated pest and disease management in greenhouse crops*. Kluwer, Dordrecht. pp. 411–419.

Osborne, L.S., Z. Landa, A. Bohata, and C. McKenzie, 2008. Potential of entomopathogenic fungus *Isaria fumosoroseus* to protect potted ornamental plants against *Bemisia tabaci* during shipping. *IOBC/WPRS Bull.* 32: 159–165.

Pannell, D.J., G.R. Marshall, N. Barr, A. Curtis, F. Vanclay, and R. Wilkinson, 2006. Understanding and promoting adoption of conservation practices by rural landholders. *Austrlas. J. Exp. Agric.* 46: 1407–1424.

Peters, A., 2003. Pesticides and entomopathogenic nematodes – current status and future work. *IOBC/WPRS Bull.* 26(5): 107–110.

Peters, A. and D. Poullot, 2004. Side effects of surfactants and pesticides on entomopathogenic nematodes assessed using advanced IOBC guidelines. *IOBC/WPRS Bull.* 27(6): 67–72.

Poprawski, T.J., S.M. Greenberg, and M.A. Ciomperlik, 2000. Effect of host plant on *Beauveria bassiana*- and *Paecilomyces fumosoroseus*-induced mortality of *Trialeurodes vaporariorum* (Homoptera: Aleyrodidae). *Environ. Entomol.* 29: 1048–1053.

Poprawski, T.J. and W.J. Jones, 2001. Host plant effects on activity of the mitosporic fungi *Beauveria bassiana* and *Paecilomyces fumosoroseus* against two populations of *Bemisia* whiteflies (Homoptera: Aleyrodidae). *Mycopathologia* 151: 11–20.

Premachandra, W.T., C. Borgemeister, O. Berndt, R.-U. Ehlers, and H.-M. Poehling, 2003. Combined releases of entomopathogenic nematodes and the predatory mite *Hypoaspis aculeifer* to control soil-dwelling stages of western flower thrips *Frankliniella occidentalis*. *BioControl* 48(5): 529–541.

Quinlan, R.J., 1988. Use of fungi to control insects in glasshouses. In: M.N. Burge (ed), *Fungi in biological control systems*. Manchester University Press, Manchester. pp. 19–36.

Quintela, E.D. and C.W. McCoy, 1998. Conidial attachment of *Metarhizium anisopliae* and *Beauveria bassiana* to the larval cuticle of *Diaprepes abbreviatus* (Coleoptera: Curculionidae) treated with imidacloprid. *J. Invertebr. Pathol.* 72: 220–230.

Ravensberg, W.J., M. Malais, and D.A. van der Schaaf, 1990. *Verticillium lecanii* as a microbial insecticide against glasshouse whitefly. *Proceedings BCPC Conference – Pest and diseases 2000* (3): 265–268.

Ravensberg, W.J., A.C. van Buysen, and R. Berns, 1994. Side-effects of pesticides on *Verticillium lecanii*: in vivo tests with whitefly and aphids. *IOBC/WPRS Bull.* 17(3): 234–238.

Roditakis, E., I.D. Couzin, K. Balrow, N.R. Franks, and A.K. Charnley, 2000. Improving secondary pick up of insect fungal pathogen conidia by manipulating host behaviour. *Ann. Appl. Biol.* 137: 329–335.

Roush, R.T., 1999. Strategies for resistance management. In: F.R. Hall, J.J. Menn (eds), *Biopesticides: use and delivery*. Humana Press, Totowa. pp. 575–593.

Roush, R.T., 2000. Resistance management for agricultural pests. In: J.-F. Charles, A. Delécluse, C. Nielsen-LeRoux (eds), *Entomopathogenic bacteria: from laboratory to field application*. Kluwer, Dordrecht. pp. 399–417.

Rovesti, L., 1991. Compatibility of heterorhabditid and steinernematid nematodes with pesticides. *IOBC/WPRS Bull.* 14(1): 60–61.

Roy, H.E., J. Baverstock, and J.K. Pell, 2007. Manipulating behaviour: A strategy for pest control? In: S. Ekesi, N.K. Maniania (eds), *Use of entomopathogenic fungi in biological pest management*. Research Signpost, Kerala, India. pp. 179–195.

Roy, H.E., P.M.J. Brown, P. Rothery, R.L. Ware, and M.E.N. Majerus, 2008. Interactions between the fungal pathogen *Beauveria bassiana* and three species of coccinellid: *Harmonia axyridis, Coccinella septempunctata* and *Adalia punctata*. *BioControl* 53: 265–276.

Roy, H.E. and J.K. Pell, 2000. Interactions between entomopathogenic fungi and other natural enemies: implications for biological control. *Biocontrol Sci. Technol.* 10: 737–752.

Roy, H.E., J.K. Pell, S.J. Clark, and P.G. Alderson, 1998. Implications of predator foraging on aphid pathogen dynamics. *J. Invertebr. Pathol.* 71: 236–247.

Sauphanor, B., M. Berling, J.-F. Toubon, M. Reyes, J. Delnatte, and P. Allemoz, 2006. Carpocapse des pommes. Cas de résistance au virus de la granulose en vergers biologique. *Phytoma – La Défense des Végétaux* 590: 24–27.

Schnorbach, H.-J., 2006. Considerations of side-effects on beneficial organisms during development in the agrochemicals industry. *IOBC/WPRS Bull.* 29(10): 113.

Schroer, S., D. Sulistyanto, and R.-U. Ehlers, 2005. Control of *Plutella xylostella* using polymer-formulated *Steinernema carpocapsae* and *Bacillus thuringiensis* in cabbage fields. *J. Appl. Entomol.* 129(4): 198–204.

Schuler, T., M. Hommes, H.-P. Plate, and G. Zimmermann, 1991. Verticillium lecanii (Z.) Viégas (Hyphomycetales: Moniliaceae): Geschichte, Systematik, Verbreitung, Biologie und Anwendung im Pflanzenschutz. *Mitt. Biol. Bundesanstalt fur Land und Forstwirtschaft Berlin_Dahlem, heft 269*, März 1991. pp. 1–154.

Shah, F.A., M.A. Ansari, M. Prasad, and T.M. Butt, 2007a. Evaluation of black vine weevil (*Otiorhynchus sulcatus*) control strategies using *Metarhizium anisopliae* with sublethal doses of insecticides in disparate horticultural growing media. *Biol. Control* 40(2): 246–252.

Shah, F.A., M.A. Ansari, J. Watkins, Z. Phelps, J. Cross, and T.M. Butt, 2009. Influence of commercial fungicides on the germination, growth and virulence of four species of entomopathogenic fungi. *Biocontrol Sci. Technol.* 19(7): 743–753.

Shah, F.A., M. Gaffney, M.A. Ansari, M. Prasad, and T.M. Butt, 2008. Neem seed cake enhances the efficacy of the insect pathogenic fungus *Metarhizium anisopliae* for the control of black vine weevil, *Otiorhynchus sulcatus* (Coleoptera: Curculionidae). *Biol. Control* 44: 111–115.

Shah, F.A., M. Prasad, and T.M. Butt, 2007b. A novel method for the quantitative assessment of the percolation of *Metarhizium anisopliae* conidia through horticultural growing media. *BioControl* 52: 889–893.

Shapiro-Ilan, D.I., D.H. Gouge, S.J. Piggott, and J.P. Fife, 2006. Application technology and environmental considerations for use of entomopathogenic nematodes in biological control. *Biol. Control* 38: 124–133.

Shelton, A.M., P. Wang, J.-Z. Zhao, and R.T. Roush, 2007. Resistance to insect pathogens and strategies to manage resistance: an update. In: L.A. Lacey, H.K. Kaya (eds), *Field manual of techniques in invertebrate pathology*, 2nd edition. Springer, Dordrecht. pp. 793–811.

Sher, R.B., M.P. Parrella, and H.K. Kaya, 2000. Biological control of the leafminer *Liriomyza trifolii* (Burgess): implications for intraguild predation between *Diglyphus begini* Ashmead and *Steinernema carpocapsae* (Weiser). *Biol. Control* 17(2): 155–163.

Shipp, L., J.P. Kapongo, P. Kevan, J. Sutton, and B. Broadbent, 2008. Using bees to disseminate multiple fungal agents for insect pest control and plant disease suppression in greenhouse vegetables. *IOBC/WPRS Bull.* 32: 201–204.

Shipp, L., Y. Zhang, D. Hunt, and G. Ferguson, 2002. Influence of greenhouse microclimate on the efficacy of *Beauveria bassiana* (Balsamo) Vuillemin for control of greenhouse pests. *IOBC/WPRS Bull.* 25(1): 237–240.

Singh, G., P.J. Rup, and O. Koul, 2007. Acute, sublethal and combination effects of azadirachtin and *Bacillus thuringiensis* toxins on *Helicoverpa armigera* (Lepidoptera: Noctuidae) larvae. *Bull. Entomol. Res.* 97: 351–357.

Smits, P.H., 1987. Nuclear polyhedrosis virus as biological control agent of *Spodoptera exigua.* PhD Thesis, Laboratory of Virology, University Wageningen, Wageningen. 127pp.

Smits, P.H., M. van de Vrie, and J.M. Vlak 1987. Nuclear polyhedrosis virus for control of *Spodoptera exigua* larvae on glasshouse crops. *Entomol. Exp. Appl.* 43: 73–80.

Steinkraus, D.C., 2007. Documentation of naturally occurring pathogens and their impact in agroecosystems. In: L.A. Lacey, H.K. Kaya (eds), *Field manual of techniques in invertebrate pathology*, 2nd edition. Springer, Dordrecht. pp. 267–281.

Sterk, G., K. Bolckmans, R. de Jonghe, L. de Wael, and J. Vermeulen, 1995b. Side-effects of the microbial insecticide Preferal WG (*Paecilomyces fumosoroseus* strain Apopka 97) on *Bombus terrestris. Med. Fac. Landbouww. Univ. Gent.* 60: 713–717.

Sterk, G., K. Bolckmans, M. van de Veire, B. Sels, and W. Stepman, 1995a. Side-effects of the microbial insecticide PreFeRal (*Paecilomyces fumosoroseus* (Wize) Brown and Smith, strain Apopka 97) on different species of beneficial arthropods. *Med. Fac. Landbouww. Univ. Gent.* 60: 719–723.

Sterk, G., F. Heuts, N. Merck, and J. Bock, 2002. Sensitivity of non-target arthropods and beneficial fungal species to chemical and biological plant protection products: results of laboratory and semi-field trials. *Proceedings of the 1st International Symposium on Biological control of arthropods*, Honolulu, Hawaii, January 14–18, 2002. pp. 308–313.

Sterk, G., K. Jans, K. Put, O.V. Wulandari, and M. Uyttebroek, 2003. Toxicity of chemical and biological plant protection products to beneficial arthropods. *Colloque international tomate sous abri, protection intégrée – agriculture biologique,* Avignon, France, 17–18 et 19 septembre 2003. pp. 113–118.

Straus, E.E. and J.D. Knight, 1997. Microbial insecticides: can farmers be persuaded to use them? In: H.F. Evans (ed), *Microbial insecticides: novelty or necessity? BCPC Symposium Proceedings* No. 68, Coventry, April 16–18, 1997. BCPC, Farnham. pp. 141–150.

Tabashnik, B.E., 1994. Evolution of resistance to *Bacillus thuringiensis. Ann. Rev. Entomol.* 39: 47–79.

Tabashnik, B.E., N.L. Cushing, N. Finson, and M. Johnson, 1990. Field development of resistance to *Bacillus thuringiensis* in diamondback moth. *J. Econ. Entomol.* 83: 1671–1676.

Thomas, M.B., 1999. Ecological approaches and the development of "truly integrated pest management". *Proc. Nat. Acad. Sci. USA.* 96: 5944–5951.

Thomas, M.B., S.P. Arthurs, and E.L. Watson, 2006. Trophic and guild interactions and the influence of multiple species on disease. In: J. Brodeur, G. Boivin (eds), *Trophic and guild interactions in biological control.* Springer, Dordrecht. pp. 101–122.

Thomas, M.B., E.L. Watson, and P. Valverde-Garcia, 2003. Mixed infections and insect-pathogen interactions. *Ecol. Lett.* 6: 183–188.

Tomalak, M., S. Piggott, and G.B. Jagdale, 2005. Glasshouse applications. In: P.S. Grewal, R.-U. Ehlers, D.I. Shapiro-Ilan (eds), *Nematodes as biocontrol agents.* CAB International, Wallingford. pp. 147–166.

Trisyono, A. and M.E. Whalon, 1999. Toxicity of neem applied alone and in combinations with *Bacillus thuringiensis* to Colorado potato beetle (Coleoptera: Chrysomelidae). *J. Econ. Entomol.* 92(6): 1281–1288.

Tuset, J.J., 1985. *Verticillium lecanii* (Zimm.) Viegas (Moniliaceae, Hyphomycetes). *IOBC/WPRS Bull.* 15: 236–237.

Ugine, T.A., S.P. Wraight, and J.P. Sanderson, 2005. Differential susceptibility of western flower thrips (*Frankliniella occidentalis*) to *Beauveria bassiana*, as a function of host plant species. *IOBC/WPRS Bull.* 28(1): 271–274.

Ugine, T.A., S.P. Wraight, and J.P. Sanderson, 2007a. Effects of manipulating spray-application parameters on efficacy of the entomopathogenic fungus *Beauveria bassiana* against western flower thrips, *Frankliniella occidentalis*, infesting greenhouse impatiens crops. *Biocontrol Sci. Technol.* 17(1/2): 193–219.

Ugine, T.A., S.P. Wraight, and J.P. Sanderson, 2007b. A tritrophic effect of host plant on susceptibility of western flower thrips to the entomopathogenic fungus *Beauveria bassiana*. *J. Invertebr. Pathol.* 96: 162–172.

Vainio, A., 1992. Guideline for laboratory testing of the side-effects of pesticides on entomophagous nematodes *Steinernema* spp. *IOBC/WPRS Bull.* XV(3): 145–147.

Vainio, A., 1994. Effects of pesticides on long-term survival of *Steinernema feltiae* in the field. *IOBC/WPRS Bull.* 17(3): 70–76.

Van Lenteren, J.C., 2007. Biological control for insect pests in greenhouses: an unexpected success. In: C. Vincent, M.S. Goettel, G. Lazarovits (eds), *Biological control. A global perspective.* CAB International, Wallingford. pp. 105–117.

Van Rie, J. and J. Ferré, 2000. Insect resistance to Bacillus thuringiensis insecticidal crystal proteins. In: J.-F. Charles, A. Delécluse, C. Nielsen-LeRoux (eds), *Entomopathogenic bacteria: from laboratory to field application.* Kluwer, Dordrecht. pp. 219–236.

Vanninen, I., M. Linnamäki, and P. Laitinen, 2008. Knowledge transfer of IPM to Finnish ornamental growers in 2004–2007. *IOBC/WPRS Bull.* 32: 249–252.

Vasconselos, S.D., T. Williams, S. Hails, and J.S. Cory, 1996. Prey selection and baculovirus dissemination by carabid predators of Lepidoptera. *Ecol. Entomol.* 21(1): 98–104.

Vestergaard, S., A. Cherry, S. Keller, and M. Goettel, 2003. Safety of hyphomycete fungi as microbial control agents. In: H.M.T. Hokkanen, A.E. Hajek (eds), *Environmental impacts of microbial insecticides.* Kluwer, Dordrecht. pp. 35–62.

Vidal, C., L.S. Osborne, L.A. Lacey, and J. Fargues, 1998. Effect of host plant on the potential of *Paecilomyces fumosoroseus* (Deuteromycotina: Hyphomycetes) for controlling the silverleaf whitefly, *Bemisia argentifolii* (Homoptera: Aleyrodidae), in greenhouses. *Biol. Control* 12: 191–199.

Waage, J., 1996. "Yes, but does it work in the field?" The challenge of technology transfer in biological control. *Entomophaga* 41(3/4): 315–332.

Waage, J.K., 1997. *Biopesticides at the crossroads: IPM products or chemical clones?* In: H.F. Evans (ed), *Microbial insecticides: novelty or necessity? BCPC Symposium Proceedings* No. 68, Coventry, April 16–18, 1997. BCPC, Farnham. pp. 11–19.

Walter, J.F., 1999. Commercial experience with neem products. In: F.R. Hall, J.J. Menn (eds), *Biopesticide use and delivery.* Humana Press, Totowa. pp. 155–163.

Warrior, P., 2000. Living systems as natural crop-protection agents. *Pest Manag. Sci.* 56: 681–687.

Wearing, C.H., 1988. Evaluating the IPM implementation process. *Ann. Rev. Entomol.* 33: 17–38.

Whalon, M.E. and D.L. Norris, 1999. Field management. Delivery of new technologies to growers. In: F.R. Hall, J.J. Menn (eds), *Biopesticides: use and delivery.* Humana Press, Totowa. pp. 595–608.

Williamson, S., 1998. Understanding natural enemies; a review of training and information in the practical use of biological control. *Biocont. News Inform.* 19(4): 117 N–126 N.

Wilson, K., S.C. Cotter, A.F. Reeson, and J.K. Pell, 2001. Melanism and disease resistance in insects. *Ecol. Lett.* 4: 637–649.

Wilson, K., M.B. Thomas, S. Blanford, M. Doggett, S.J. Simpson, and S.L. Moore., 2002. Coping with crowds: density-dependent disease resistance in desert locusts. *Proc. Nat. Acad. Sci. USA.* 99: 5471–5475.

Wraight, S.P., 2003. Synergism between insect pathogens and entomophagous insects, and its potential to enhance biocontrol efficacy. In: O. Koul, G.S. Dhaliwal (eds), *Predators and parasitoids.* Taylor and Francis, New York. pp. 139–161.

Wraight, S.P. and M.E. Ramos, 2005. Synergistic interaction between *Beauveria bassiana*- and *Bacillus thuringiensis tenebrionis*-based biopesticides applied against field populations of Colorado potato beetle larvae. *J. Invertebr. Pathol.* 90: 139–150.

Wright, D.J., A. Peters, S. Schroer, and J.P. Fife, 2005. Application technology. In: P.S. Grewal, R.-U. Ehlers, D.I. Shapiro-Ilan (eds), *Nematodes as biocontrol agents*. CAB International, Wallingford. pp. 91–106.

Yi, X. and R.-U. Ehlers, 2006. Combining *Steinernema carpocapsae* and *Bacillus thuringiensis* strains for control of diamondback moth (*Plutella xylostella*). *Comm. Appl. Biol. Sci. Ghent Univ.* 71(3a): 633–636.

Young, S.Y. and W.C. Yearian, 1987. Nuclear polyhedrosis virus-infected and healthy *Anticarsia gemmatalis* larvae as prey for *Nabis roseipennis* adults in the laboratory. *J. Invertebr. Pathol.* 54: 139–143.

Zimmermann, G., 1994. Strategies for the utilization of entomopathogenic fungi. *Proceedings of the VIth International colloquium of invertebrate pathology and microbial control*, Montpellier, France, 28 August–2 September. pp. 67–73.

Chapter 7
Critical Factors in the Successful Commercialization of Microbial Pest Control Products

Abstract Commercialization is the final and most difficult step in the development and the introduction to the market of a microbial pest control product. The critical factors that determine success or failure are identified for a company as well as for a product, and recommendations are provided that will facilitate success. The history of the biopesticide industry will be highlighted. Currently available products will be presented, as well as the biopesticide markets in Europe and the Netherlands. The main crops in which biopesticides are used are presented as well as the most successful products. Profitability of products in relation to their market size will be analyzed. Critical success and failure factors for a biopesticide company and for a microbial pest control product are analyzed, and recommendations will be provided on essential factors that need to be considered. Tools will be provided to facilitate decision-making in the commercialization process. Company profiles will be reviewed, and a business model will be identified that currently performs best. Total developmental costs and time to market are essential for a company's success. Costs amount to €10–15 million for a company that still needs to be built; in an existing company, costs reach €5–10 million. Time to market including registration is 5–7 years. The distribution and sale strategy is reviewed, and various salient options are discussed. Crucial factors for a successful commercialization will be identified, and recommendations will be provided that will increase the chances to develop a successful biopesticide and to become a profitable and sustainable biopesticide company.

1 Introduction

The development and commercialization of biopesticides has attracted many companies during the last 40 years. Many products have reached the market, but the majority has not really been successful. Overall, biopesticides only comprise a small share of the world market for crop protection products, 1–2%. Many companies have failed and left this business; some remain, but the profitability of the

W.J. Ravensberg, *A Roadmap to the Successful Development and Commercialization of Microbial Pest Control Products for Control of Arthropods*, Progress in Biological Control 10, DOI 10.1007/978-94-007-0437-4_7, © Springer Science+Business Media B.V. 2011

survivors is doubtful. Nevertheless, the biopesticide market is steadily growing, by approximately 10% annually. Many people consider the future for these products to be positive, due to environmental awareness and the negative perception of chemical pesticides in general. Microbial control agents offer an array of opportunities for the development of crop protection products. Academic scientists as well as industrial researchers have made enormous progress over the years in improving all kinds of biological and technical features. Entrepreneurs who are contemplating the development of microbial pest control products, however, face a complex matrix of biological, technical, regulatory and commercial challenges. The first two have extensively been addressed in previous chapters, as is registration which is still a major hurdle before products can be introduced in the market. In this chapter I will address the critical factors in the process of commercialization that determine success or failure. I consider a company successful when the production and sale of a biopesticide provides a sustainable, growing, and profitable business which is able to develop new markets and products. Both product and company determine whether success can be achieved.

The concept commercialization is frequently mentioned in papers on the development and use of biopesticides. In general, authors refer to commercialization as the whole process of product development including all its stages, and further, registration and market introduction (Törmälä 1995; Carlton 1996; Cross and Polonenko 1996; Stewart 2001; Montesinos 2003; Fravel 2005). This broad concept is confusing and meaningless, and in no way assists in understanding what it takes to commercialize a new product in a business environment. I separate new product development, as described in Chapters 2, 3, 4 and 6, from the commercialization process. To distinguish these developmental and operational aspects from the task of bringing a new product to the market I will use a narrower definition here: commercialization is the management process that provides structure in developing and bringing a new product to the market. Effective implementation of this step-by-step process is needed to coordinate the gathering of information, the establishment a project plan, the taking of decisions, and the translating of these into resources, operational requirements and functions, and the coordination of these activities to bring the product to the market in an economical and profitable way. In this chapter I will focus on the business and management aspects of this commercialization process. Further, the history of biocontrol companies and entomopathogenic products is presented with an emphasis on the European market. The currently active companies as well as the available products based on bacteria, fungi, baculoviruses, and nematodes will be briefly presented.

Reliable data on the size of the biopesticide market are hard to find. Only direct industry surveys have been able to report reliable data (CPL 2006a). Recent data will be reported, particularly for the most important protected crop markets, the Netherlands and Spain. Microbial pest control products are predominantly used in niche markets which will be reported. Use in agricultural crops is limited, but it is slowly increasing. This demonstrates that these products have a role to play in any crop. For a long time, the Bt products have been the most successful products. The importance of products based on other entomopathogens is briefly discussed.

An analysis is made on the profitability of the industry based on the total sales and the number of available products in the European market. This analysis leads to the question why some companies are successful and others not. The most critical factors from a commercial point of view will be highlighted.

Companies that are contemplating the development of a microbial product need to make a final decision to start. Determinative factors related to the new product and the potential market will be identified and discussed. A decision process that facilitates making the right decision will be introduced. Various types of enterprises from small to large multinational companies have engaged in developing biopesticides. Gelernter (2005) reviewed the history of this field of business where many failed. I will present an analysis on which business model is most suited to perform well in the challenging area of biological crop protection.

Microbial pesticides are often criticized on many of their characteristics which limit their adoption in the market. Their strong and weak aspects will be reviewed, as well as new developments and technological breakthroughs that could lead to better products.

The development and registration are expensive and lengthy procedures that need to be conducted before any sales will generate income. A realistic estimate of the expenses and the time to market is critical for a company in the business plan and will influence the decision process. Every product is unique and forecasts provided in the literature vary greatly. I will provide an estimate for various situations and types of products.

The development and commercialization of each microbial pest control product is different and unique. Five decisive success factors are identified and discussed in the conclusions. When these five factors are carefully observed, the likelihood that a company will perform well, increases. Finally, I will provide ten recommendations that facilitate the development of a successful product, and company requirements that are necessary to accomplish a profitable and sustainable business.

2 The Historical Development of Biopesticide Companies and Microbial Pest Control Products

The development of a biopesticide company and a microbial pest control product are inextricably linked. The development of a novel product to solve a pest problem, or a fortuitous discovery that potentially leads to an innovative product could lead to the establishment of a new company, or a new department in an existing company. Potential products that can be commercialized will be the incentive. In the early days of this industry, research ideas from public research generally motivated committed persons to initiate product development and commercialization, and often lead to the establishment of a company. The early steps in applied insect pathology leading to the development of microbial insecticides have been reviewed by Lord (2005). He highlighted the discovery and the initial use of entomopathogenic bacteria, fungi, nematodes, and viruses, and the commercialization of the first products around the world. His review predominantly focussed on the research side of microbial insect control. Reviews on commercialization of microbials and the companies involved have been presented by many authors. Comprehensive overviews are provided by Burges and Hussey (1971), Burges (1981a), Payne (1987), Rodgers (1993), Starnes et al. (1993), Hall and Menn (1999), Butt (2002), and Szewczyk et al. (2006). The commercial development of nematode-based products and the producers are reviewed by Georgis (2002), Ehlers (2003, 2007) and Kaya et al. (2006). Important historical discoveries and events in the biopesticide industry are listed by Thakore (2006). Success stories in biocontrol, including microbial biocontrol, are reported by Gurr and Wratten (2000). Interesting and instructive cases that involve companies' stories are provided by Vincent et al. (2007).

2.1 History of the Biopesticide Industry

Numerous enterprises have been active in the field of biopesticide production and marketing. Lisansky (CPL 2006a) listed nearly four hundred companies worldwide which manufactured and/or marketed biopesticides (broad definition) between 1980 and 2006. The companies involved in the early period of biopesticides and some of their histories have been presented in papers of Lisansky (1997), Lisansky and Coombs (1994), Rodgers (1993), Starnes et al. (1993), Gelernter (2005), and for nematodes, in papers of Georgis (2002) and Ehlers (2007). More than 200 companies have been active in the last two decades in this field (Harwood et al. 2007); a considerable number remains active to date. In Europe eighty-two manufacturers were identified, and their activities and products have been described in The Biopesticide Companies of Europe (CPL 2006b). The earlier edition of this report (2002) listed just twenty companies. Many different types of companies such as companies primarily active in the field of agrochemicals, pharmaceuticals, fertilizers, and several other areas, have engaged in the production of microbial pest control products. Others just specialised in biocontrol. Interest in the development of biopesticides increased in the 1980s and early 1990s, when large (agrochemical) companies and venture capital companies became active in this area, predominantly in the USA. Business interest in biocontrol was peaking and research resources seemed inexhaustible (Gelernter 2005). Large agricultural markets were envisaged, but projections of sales never met expectations (Lisansky 1997). Many companies overestimated their insight in crop protection markets, and most of these companies left the biocontrol field or went bankrupt. Interest waned in the large agrochemical companies; they changed their focus to transgenic crops. Small biocontrol companies continued the effort.

Gelernter (2005) reviewed 50 years of biocontrol products and companies. She divided this period in three eras: the pioneering era; the era in which large agrochemicals and venture capital companies explored this field without success; and the period since 1995 where primarily small companies try to be successful. She concluded that the incorrectly perceived size of the market has been leading these developments. With a more realistic approach, a small group of "feet on the ground" entrepreneurs continue to develop the biocontrol market. The biopesticide industry, if one could call it this, consisted and still consists of a diverse group of players. It has gone through periods of turmoil and many changes. This illustrates that this sector is a young sector which faces many challenges in a field where survival and sustainable business is far from easy to achieve. However, the agrochemical world has gone through many of the same challenges and many companies have disappeared. Only a handful is still active and developing new products (Pallett 2005).

2.2 Currently Active Biopesticide Manufacturers

An accurate picture of the currently active biopesticide manufacturers is very difficult to provide. Every year, new ones enter this field and others leave. Acquisitions of products and whole companies occur frequently. The status of companies in North America and Western Europe is well known, but for the rest of the world it is impossible to provide reliable information. There are two industrial associations of the biopesticide industry. The global International Biocontrol Manufacturers Association (IBMA), founded in 1995, has about hundred sixty members from all over the world. In Europe, approximately thirty members are producers of microbial pesticides. The Biopesticide Industry Alliance (BPIA) in the USA has more than fifty members, of which about ten produce microbial pesticides. Other members of these associations have activities in the field of macrobial pest control products, pheromones, and natural or biochemical plant protection products.

Detailed information (history, products, and markets) on individual biopesticide manufacturers has been provided in a small number of market reports such as Frost and Sullivan (2001) and The 2007 Biopesticides Study of CPL (CPL 2006a). The Frost and Sullivan Study (2001) presented an elaborate picture of the main companies with background information, products, and market shares. The CPL study described history, structure, products and markets of a company in a detailed manner in the volumes on biopesticides companies (CPL 2006a). BCC Research (BCC 2006) provided mainly information on American companies including transgenic crop producers. These reports are very expensive and not widely available in libraries. A worldwide list of manufacturers of microbial pesticides is provided in the Directory of Microbial Pesticides for Agricultural Crops in OECD Countries (Tabaluk and Gaznik 2007). In general, these reports focussed on the biocontrol industry in the Western world. The latest report of CPL (2006a) provided volumes with company information for Europe, for North America, for Latin America, for Africa and the Middle East, and for Asia and Oceania. Most companies nowadays provide a website with elaborate information on their history and on their current products and activities.

I will focus on manufacturers of microbial insecticidal products in Europe, and on non-European manufacturers that market microbial insecticides in Europe. It is, however, not always clear whether companies manufacture products themselves, or whether production and/or formulation are out-contracted, or whether products are purchased and marketed. I present here the companies involved in bioinsecticide production in Tables 7.1 and 7.2. The types of bioinsecticides are mentioned in the order of importance within a certain company. I only list those who are active in Europe with officially registered products (see also Chapter 5, Table 5.1), and the producers of entomopathogenic nematodes. EPNs are produced by three large producers, Becker Underwood, Enema, and Koppert. All three market their products on the European and American market. Three small producers, Andermatt, Bionema, and Owiplant, mainly sell in their domestic market. The first three produce three to nine species, and use submerged fermentation for the manufacturing process, while the small ones only produce one species, using the solid medium technology.

A number of companies produce other microbial pesticides such as products for control of foliar and soil-borne diseases, for post-harvest diseases, or for control of

Table 7.1 European manufacturers of microbial pest control products

Company	Production	Company	Production
Andermatt (CH)	BV, Fungi (I), EPN	Koppert (Nl)	EPN, Fungi (I, F)
Agrifutur (It)	Fungi (I, F)	NPP-Arysta (Fr)	BV, Fungi (I)
Biocolor (Sp)	BV	Owiplant (Po)	EPN
Bionema (Sw)	EPN	Probelte (Sp)	Bt
Becker Underwood (UK)	EPN	Probis (Ge)	BV
Enema (Ge)	EPN, Bact (I), Fungi (F)	Schweizer (CH)	Fungi (I)
FuturEco (Sp)	Fungi (I)	Sipcam (It)	BV
Intrachem (It)	Bt, Fungi (I, F)	Sourcon Padena	Bacteria (F)

BV: Baculoviruses.
EPN: Entomopathogenic nematodes.
I: Bioinsecticide.
F: Biofungicide.

Table 7.2 Non-European manufacturers of microbial pest control products, active on the European market

Company	Production
Becker Underwood (USA)	EPN, Fungi (I)
Certis USA	Baculoviruses, Bt, Fungi (I, F)
Laverlam (Colombia)	Fungi (I, F)
Novozymes (USA)	Fungi (I)
Valent BioSciences (USA)	Bt

EPN: Entomopathogenic nematodes.
I: Bioinsecticide
F: Biofungicide.

nematodes. The following European companies are active with this type of biocontrol products: Agrauxine (Fr), Arcadis (Nl), Binab (Sw), BioAgri (Sw), Bio-ferm (At), Bionext (Be), Biopreparaty (Cz), Isagro (It), NewBiotech (Sp), Prophyta (Ge), Sourcon-Padena (Ge), and Verdera (Fi). Two American companies are active in the European market with biofungicides: Agraquest and BioWorks. Bioherbicides are not manufactured in Europe.

A small number of companies are active in various biocontrol activities, while others specialize in one type of products. For instance, Koppert produces macrobial, microbial and natural products for plant protection. Andermatt Biocontrol produces mainly baculoviruses; Prophyta only fungal products, and Verdera microbial fungicides. A specialization in production technology occurs as well as a specialization in market approaches with bioinsecticides or biofungicides. Many combinations of activities are found in manufacturing activities as well as in the marketing of products. Certis, Isagro, Natural Plant Protection-Arysta LifeScience and Sipcam also manufacture and market chemical pesticides. Some producers present themselves as toll manufacturers only, such as Vitalin (Ge) and ELPE (It). Others like Enema and Prophyta produce their own products as well as products or unformulated active ingredients for third parties. Remarkably, more new companies are starting in the field of disease control than in insect control. Examples are Agrauxine, Bio-ferm, BioNext, NewBiotech, and Sourcon-Padena. Occasionally, companies from other parts of the world try to market their products in Europe. Companies from India, South-America, Israel, Kenya, and South Africa have tried to market their products in Europe, but the difficult and expensive registration procedure withholds them still. When this will become easier in the future, these companies will surely become active on the European market. I do not expect that in the next decade.

2.3 Overview of Currently Available Products

A comprehensive overview of available products worldwide is provided by Copping in subsequent editions of the Biopesticide Manual/Manual of Biocontrol Agents (Copping 1998, 2001, 2004, 2009). These editions listed 59, 96, 110 respectively 149 microorganisms on which biopesticides are based. Product availability per country is provided in the market study in the Biopesticide series of CPL (CPL 2006a). A list of products and producers is provided in the Directory of Microbial Control Products and Services of the Society of Insect Pathology (Shah and Goettel 1999) and in the Directory of Microbial Pesticides for Agricultural Crops in OECD Countries (Tabaluk and Gaznik 2007), but both lists are not very accurate with regard to currently registered and available products. The situation is continuously changing, however, and a correct picture of the producers and distributors and the registered products is difficult to provide. Producers may stop, may be taken over or merge; products may fail, may be withdrawn, but are still listed as registered. At any time, a report on this subject may be outdated soon after publication. There are no official data banks that provide regularly updated information. Data bases of regulatory authorities provide the number of registered products, but products are not necessary "alive" as is the case for instance with Bt products in the USA: only about 230 out of 469 registered labels are marketed actively (Braverman 2007). Nor do the IBMA and BPIA provide any commercial information due to confidentiality.

Many authors described the commercial developments with microorganisms as MPCPs and listed the available products at a certain time. I will only refer to papers that describe the situation over the last 10–15 years since the registration status and commercial availability of products change frequently. Authors that reported a general overview are Butt et al. (1999, 2001), Copping and Menn (2000), Wraight et al. (2001), Hynes and Boyetchko (2006), Rosell et al. (2008); and for nematodes Ehlers (2003) and Kaya et al. (2006). For an overview of products based on entomopathogenic bacteria, I refer to Glare and O'Callaghan (2000) and Navon (2000). A comprehensive overview of the status of fungal products for control of insects and mites is presented by De Faria and Wraight (2007). They listed 171 products and 80 manufacturers and updated availability and the registration status of these products worldwide. Approximately 43% of the products were developed in South America. Szewczyk et al. (2006) presented the worldwide historical development and the current situation of insecticides based on baculoviruses. Kaya et al. (2006) outlined the status of EPNs and their uses for most regions of the world.

Below I will describe the history of microbial insecticides in Europe, and the situation at present. Biofungicides seem to be the largest group of products on the market currently. For an overview of microbial disease control products I refer to Whipps and Davies (2000), Stewart (2001) and Fravel (2005). The status of microbial products for control of foliar and soil-borne diseases in greenhouse systems is reported by Paulitz and Bélanger (2001) and Ravensberg and Elad (2001). Bioherbicides are conspicuous by their absence in both the EU and elsewhere (Hallett 2005). One bionematicide, BioAct, is registered in the EU, but sales are small and limited to greenhouse crops (www.prophyta.de). Two microbial products for post-harvest disease control are available on the market. A few have disappeared and some are under development. The use of those products is limited and their commercialization faces many obstacles and constraints similar to the ones for other biopesticides (Droby et al. 2009).

The development and commercial availability of microbial products has also been described per region and per country. The situation for the USA is described by Carlton (1996). The use of entomopathogenic microbials in greenhouse crops has been presented by Lipa and Smits (1999) for Europe, USA and Japan. In Canadian greenhouses, use of biopesticides is limited to Bt and nematodes for control of fungus gnats (Shipp et al. 2007). An overview of all available microbial biopesticides in Canada is provided by Bailey et al. (2009). The current status of microbial control in South-East Asia is provided by Gelernter (2007) and Skovmand (2007). The status of microbial insecticides for Japan is reported by Kunimi (2007), for China by Huang et al. (2007) and Li et al. (2010), for India by Vasantharaj David (2008), for South America by Alves et al. (2003), for Latin America by Rodríguez and Niemeyer (2005), for Brazil by Li et al. (2010), and the use and perspectives for Africa are provided by Cherry and Gwynn (2007). Some information on the use of

Bt and baculoviruses in Spanish greenhouses is provided by Van der Blom (2009). The availability and use of biopesticides in the UK is provided by Gwynn (2009). The status and use of entomopathogenic nematodes is provided by Kaya et al. (2006) for many regions in the world. For many countries and regions recent overviews on the availability and use of microbial pesticides are lacking.

2.3.1 Bacterial Insecticides

The first bacterial insecticide was based on *Bacillus popilliae* (=*Paenibacillus popilliae*) and developed for control of larvae of the Japanese beetle. It was developed by the USDA in the mid-1940s, and the first registered biopesticide in the USA in 1948. The product "Milky Spore Powder" was produced by Fairfax Laboratories until recently (Lord 2005), and is still available through St. Gabriel Laboratories (www.epa.gov). The first bioinsecticides on the European market were *Bacillus thuringiensis* (Bt) products. The history of the development of Bt as a bioinsecticide has been reviewed by numerous authors (e.g. Burges 1981a, 2001; Beegle and Yamamoto 1992; Navon 2000; Federici 2005; Coté 2007). Lord (2005) summarized the highlights with regard to applied aspects and reviewed the development of the first products. The first commercial product, Sporeine, was developed in France in 1938, but it was only briefly available just before World War II. Later, in the 1960s, a number of companies around the world commercialized Bt products with various degrees of success (Lambert and Peferoen 1992; Lisansky and Coombs 1994). Thuricide was the first registered product for control of caterpillars in the USA in 1961. In Europe, the development and production was started by large agrochemical companies, such as Solvay (Belgium) and Sandoz (Switzerland). Thuricide was registered in 1964. In the 1970s Dipel and Bactospeine reached the market. Solvay marketed their Bt through their affiliates Biochem Products and Duphar, but sold the business to Novo Nordisk (Denmark). Novo Nordisk sold its Bt business in 1995 to Abbott Laboratories. Valent BioSciences (USA), which is part of Sumitomo Chemical (Japan), took over the Bt business from Abbott Laboratories in 1999, and became the largest Bt producer worldwide. Sandoz merged with Ciba Geigy and later became Novartis, which merged with Zeneca into Syngenta. The Sandoz/Syngenta product line was acquired by Thermo Trilogy in 1997. A few years later, the biopesticide business of Thermo Trilogy was acquired by Certis USA, a subsidiary of Mitsui & Co, Japan. Certis also purchased Ecogen's product line in 2002, which also contained products based on transconjugant and recombinant Btk strains.

A number of companies currently market Bt products in the European market. Valent BioSciences markets their products, mainly Dipel and XenTari, through distributors. Certis uses its subsidiaries and distributors for their product Turex. Intrachem, Isagro and Probelte are marketing their own Bt products. Intrachem developed three formulations based on one strain EG 2348: Lepinox Plus, Rapax and Wormox. The strain is included on Annex I and product registrations have been obtained in a number of European countries (Ladurner et al. 2008).

2.3.2 Fungal Insecticides

The first fungal insecticide, actually an acaracide, was Mycar, a product based on the fungus *Hirsutella thompsonii*, for control of rust mites in citrus in Florida, USA. It was manufactured by Abbott Laboratories, and was registered in 1981. The product was, however never commercially successful and was withdrawn in 1985 (McCoy et al. 2009). At the same time, similar developments started in Europe. Tate and Lyle, a large food ingredient and sugar manufacturer, started with the development of entomopathogenic fungi and Bt's in the mid 1970s in the United Kingdom. Two fungal products were registered in 1981 in the United Kingdom and marketed in the greenhouse sector: Vertalec for the control of aphids, and Mycotal for the control of whitefly. A third product for the control of thrips, Thriptal, never made it to the market. All products were based on strains of *Verticillium lecanii*. In 1984 the biopesticide department was turned into an independent small venture capital company called Microbial Resources Limited (MRL), because the interest of the parent company in this business faded due to lack of revenue. MRL did not make it and all the registrations and technical know-how were sold in 1986 to Novo Industria (S. Lisansky, personal communication). This Danish company did not develop the products any further and in 1988 Koppert bought the intellectual property rights of the fungal products and initiated the production and marketing of Mycotal and Vertalec. Tate and Lyle as well as Koppert investigated the production of *Aschersonia aleyrodis* for whitefly control, but a product was never brought to the market. Production was complicated and expensive, and the efficacy was not greater then that of Mycotal. A Danish company, Christian Hansen's Biocontrol, also developed products based on *V. lecanii*. The product MicroGermin consisted of an aphid and whitefly strain and was produced and marketed in Denmark for some time, but production was later discontinued (Butt et al. 1999). Currently, there are two products on the market: Vertalec and Mycotal, both produced and marketed by Koppert. Vertalec will have to be withdrawn from the EU since the strain has not been defended; the Mycotal strain is included on Annex I.

The large German agrochemical company Bayer showed an interest in biopesticides and developed a product based on *Metarhizium anisopliae* strain F52. Their first product was Bio1020 for control of black vine weevil larvae (Reinecke et al. 1990). The intention was to develop more products based on this fungus for control of whitefly and thrips, but this did not succeed. The management lost interest in this kind of products and a new company was formed, Taensa (USA), that took over the business. This small company failed and was bought by EarthBiosciences (USA). After a few years, all IP was again sold to Novozymes (USA). In 2008, this company re-launched Bio1020 in Germany and the Netherlands and seems committed to continuing with these products. They defended the strain in the EU re-registration process successfully. There are two formulations: a granular and an emulsion concentrate. The product was launched as Met52 in the USA, and they also market a product Tick-Ex against ticks in the USA. Agrifutur, Italy, produces a granular formulation called Granmet for control of white grubs. Currently, there are two products on the EU market: Bio1020 and Granmet, based on the same strain, Bipesco5/F52, which is included on Annex I. In Switzerland, a third product, Metarhizium Schweizer produced by Schweizer Seeds, is available for grub control.

In 1999, Biobest received approval for Certis' product PreFeRal, a mycoinsecticide based on *Isaria fumusorosea* (=*Paecilomyces fumosoroseus*) strain "Apopka 97" for control of whitefly in greenhouse crops in Belgium (Sterk et al. 1996). This was the first microorganism to obtain Annex I inclusion in 2001 according to the new EU Directive 91/414/EC. Futureco (Spain) submitted a dossier for strain "Fe 9901" of *I. fumusorosea* (*P. fumusoroseus*) in 2005 for Annex I inclusion, and applied for provisional approval for the product Nofly for whitefly control in several countries. Approval has been obtained in Belgium. Currently, there are two products on the European market: Preferal and Nofly.

The first product based on the fungus *Beauveria bassiana* was developed in France by Natural Plant Protection, a subsidiary of Calliope, now part of Arysta LifeScience, Japan. The product Ostrinil was a granular formulation for control of the corn borer in corn. It was registered in France in 1993. Production was discontinued for sometime, but the product will be re-launched for control of the palm borer *Paysandisia archon* (Besse 2008). Boverol was produced and registered by Fytovita in the Czech Republic for use against the Colorado potato beetle and the spruce bark beetle, but this strain was not defended in the re-registration programme of the EU and therefore will have to be withdrawn from the market. Botanigard is another product based on *B. bassiana* and developed by Mycotech, USA, which was acquired in 2004 by Laverlam, Colombia. It was brought to the European market by Certis Europe for the greenhouse market. There is a liquid formulation and a wettable powder, both for whitefly and thrips control in vegetables and ornamentals. These were first registered in the Netherlands and are now available in a few more countries. Naturalis, another product based on a different strain of this fungus, was developed in the USA by Troy Biosciences and was marketed in Spain and Italy by subsidiaries of Intrachem, Switzerland. Troy Bioscience no longer exists and the IP is now in the hands of Intrachem. Currently, there are two products on the European market: Botanigard and Naturalis based on Annex I included strains.

Some products have been developed and commercialized based on *B. brongniartii*. These were Melocont, produced by Agrifutur, Italy, and marketed by Kwizda, Austria. This strain was not defended and the product will have to be withdrawn. In Switzerland, two products were developed, Beaupro produced by Andermatt Biocontrol, and Beauveria Schweizer from Schweizer Seeds. All these products were applied to control white grubs of the May beetle. Currently, there are two products on the market, but only in Switzerland. The one that is marketed in the EU will have to be withdrawn by 2014 at the latest.

2.3.3 Baculovirus-Based Insecticides

The first product, Elcar, was developed by Sandoz in the USA and marketed from 1975 to 1982. It was later re-launched in 1996 by Biosys as Gemstar. Currently, it is sold by Certis USA. It is applied against *Heliothis* and *Helicoverpa* species in field crops. The past and present of baculovirus products have been reviewed by Arif

(2005) and by Szewczyk et al. (2006). Many baculovirus products have been developed for agricultural crops around the world (Hunter-Fujita et al. 1998; Moscardi 1999; Copping 2004, 2009). In Europe, the main use of viruses is with products based on *Cydia pomonella* GV against of the codling moth, in apple and pear orchards. Baculovirus products have been developed for control of Lepidopteran larvae and sawfly larvae in forestry too. Production and application were often conducted by public forestry departments, some products were developed by companies. Most of these are no longer produced and sold (Evans 1997; Moscardi 1999).

For protected crops, the first baculovirus product, Spod-X, was registered in 1994 in the Netherlands for control of the beet army worm (*Spodoptera exigua*) in greenhouse vegetables and flowers. It was first produced by Crop Genetics, USA; currently it is produced and marketed by Certis USA. The registration is pending for the product in Spain for use in greenhouse vegetables. Other *Spodoptera exigua* NPV products developed are Virex and Spexit, produced by Biocolor, Spain, respectively Andermatt Biocontrol, Switzerland. Today, there are nine products on the market in the EU, the three mentioned SeNPV products, and five CpGV-based products: Madex (Andermatt), Cyd-X (Certis Europe), Carpovirusine (Natural Plant Protection (NPP)), Granupom (Probis) and Virgo (Sipcam). Furthermore, Andermatt produces Capex, which is based on the *Adoxophyes orana* GV, for control of summer fruit tortrix in apples. It is available in five European countries. Two other virus products, Littovir based on the *Spodoptera littoralis* NPV, for control of *S. littoralis* and, Helicovex based on the *Helicoverpa armigera* NPV, for control of cotton bollworm, are being developed by Andermatt (Kessler et al. 2008).

2.3.4 Entomopathogenic Nematode-Based Products

The first nematode-based product contained *Steinernema glaseri* and was produced by the USDA in the USA. It was released on large areas for grub control in the 1930s (Lord 2005). The commercialization of nematodes started in the 1980s after Bedding had developed a semi-artificial medium (Bedding 1984). The company Bioenterprises in Tasmania was the first to sell nematode-based products with *H. bacteriophora*. Simultaneously, commercial production started in Europe in the Netherlands. Koppert and De Groene Vlieg started production of *H. megidis* and *S. feltiae* on "the Bedding-system" in the mid 1980s and sold nematodes for control of black vine weevil and Sciarid larvae. Koppert produces three species today: *S. feltiae, S. carpocapsae* and *H. bacteriophora.* De Groene Vlieg stopped the production of nematodes in the nineties. Research in the UK also led to commercial interest, and the company AGC MicroBio started to produce and sell nematodes (*S. feltiae*) for horticulture and mushroom production in 1989. Later, they developed a nematode product for slug control. MicroBio was taken over by Becker Underwood, USA, in 2000. At present they produce eight different species of entomopathogenic nematodes and one slug-parasitic nematode. Bionema, a small Swedish company, produced nematodes from 1988 until today. They produce *S. feltiae* and *S. carpocapsae*, and sell to the non-professional market. Andermatt

Biocontrol started to produce nematodes on solid medium in 1989 and still produces *H. megidis*, a species that is difficult to produce in fermentors. In Poland, Owiplant started producing *S. feltiae* in 1994 for use in horticulture and mushrooms. The three latter mentioned companies still produce nematodes on a small scale on solid medium. Enema, Germany, was founded 1997 (Ehlers 2007), and currently produces three species: *S. feltiae, S. carpocapsae* and *H. bacteriophora.*

In the USA, two venture capital enterprises were established to produce nematodes, among others. Both Biosys and Ecogen were founded in 1983. Biosys was the first to produce nematodes by large-scale liquid fermentation, and launched a *S. carpocapsae*-based product in 1988. Later, four other species were developed: *S. feltiae, S. glaseri, S. riobrave*, and *H. bacteriophora* (Georgis 2002). Ecogen took over Bioenterprises in 1992 and set up subsidiaries in Germany, Israel, and Italy in order to develop nematode-based products, in addition to other biopesticides. Ecogen also started production in submerged fermentors and developed several products. Both companies failed eventually. Biosys went bankrupt in 1996 (Georgis 2002), and was bought by Thermo Trilogy. This company was acquired by Mitsui & Co, Japan, in 2001 and was renamed Certis USA. Ecogen closed its subsidiaries in 1996 and scaled down its research and activities on nematodes. The company sold its remaining nematode business in 2002 to Certis. In 2005, Becker Underwood acquired Certis's nematode business.

Some small producers in the USA produce nematodes on solid medium such as BioLogic Company, or even on insects, such as Hydro-Gardens (Kaya et al. 2006). In Europe, three companies produce nematodes in large-scale submerged fermentors. These are Enema, Koppert, and Becker Underwood, USA, which has its production facility in the UK. They dominate the world market today. Ten species of entomopathogenic species (five *Steinernema* and five *Heterorhabditis* species) are being produced at present (Kaya et al. 2006). The various uses of these nematodes are provided by Georgis et al. (2006). Companies generally produce a single strain per species; various companies use the same strain. Also, local strains have been developed and are produced. Strain identification is usually not mentioned by companies nor is it clear which strain is present in a product. Species that are commercially available in Europe are *H. bacteriophora, H. megidis, S. carpocapsae, S. feltiae,* and *S. kraussei.* A new species, *H. downesi*, is under development for control of the pine weevil (Dillon et al. 2007) by several companies.

3 Overview of Current Use of Microbial Pest Control Products

3.1 Global Microbial Pesticide Market

Reliable figures for the global use and markets for biopesticides are not available. At best, fragmented information is available. Information on sales in Europe and North America is probably the most reliable. Gelernter (2007) estimated these markets together to value $200 million in 2003, and sales in the rest of the world to be equivalent or more. Sales of biopesticides in South America and Asia are probably considerable, but figures are lacking. There are limited sales in Oceania, while estimates of sales in Africa are low. Industry organizations such as IBMA and BPIA do not collect market data. Only scattered information is available from some governmental authorities. The EU (Eurostat) tried to collect data on biological plant protection systems, but the survey did not deliver the required information. Companies did not want to share their information due to competition and confidentiality in a difficult market. Business consultants such as Frost and Sullivan (2001), Agrow (Evans 2004b), BCC Research (2006) and CPL (2006a) which work with direct industry surveys, probably gather the most reliable data.

In 2000, the world market for biopesticides neared $300 million, less than 1% of all pesticides. In this study on biopesticides, Frost and Sullivan (2001) included all non-synthetic chemical crop protection products such as microbial pesticides, beneficial arthropods, natural pesticides, and pheromones. Bt for caterpillar and mosquito control accounted for the majority of the microbial sales. Jarvis (2001) estimated the global biopesticide (true microbials) market at $160 million in 2000; over 90% were Bt sales. This meant about $16 million for microbials world-wide, excluding Bt sales. This estimate seemed lower than that of Frost and Sullivan (2001). Guillon (2004) reported a worldwide market for biopesticides (bacteria, fungi and viruses) of $235 million, with $31 million in Europe and $109 million in the NAFTA region. In a global market study on biopesticides, Thakore (2006) reported a turnover of biopesticides of $672 million in 2005. It is, however, not clear which product categories have been included in this figure. It seems to be a broad definition including natural enemies, but the table with formulation types is confusing since these refer to true microbials. According to the author, the share of biopesticides has risen from 0.2% in 2000 to 2.5% of the total pesticide market in 2006. This share is expected to increase to 4.3% in 2010, reaching greater than a billion dollar. Bt had a market share of 70% of the microbials; bacterial biopesticides represented 74% of all biopesticides, followed by 10% fungal biopesticides, 5% viral, 8% natural enemies, and 3% "others", including nematodes. Harwood et al. (2007) reported the global biopesticide (true microbials) market to be about $280 million in 2007. In the past, Bt dominated this market with a 90% share, but this has gone down to 60%. This is partly accounted for by the rise of transgenic Bt-crops, and partly because the use of other biopesticides has increased. Evans (2008) reported that annual sales of microbial pesticides are $750 million globally, amounting to 2.5% of the chemical market. The crop protection market is increasing considerably after years of declining sales. The total of pesticide sales in 2008 was over $40 billion with a growth rate of 21% (CPM 2009a).

Few of the authors that report data on the use of biopesticides clearly presented how they defined biopesticides. Nor is it indicated whether amounts mentioned refer to manufacturer level or end-user level, and this makes quite a difference. Generally, figures of end-use markets are provided. Obviously, there are no reliable data on the biopesticide markets. Nevertheless, trends can be observed and sales are steadily increasing from about $150 million in 2000 to more than $500 million in 2008 (Table 7.3).

Table 7.3 Overview of reported worldwide sales per annum of biopesticides

Year of sales	Sales (in million $)	Definition of biopesticides	Source
2000	300	Broad	Frost and Sullivan (2001)
2001	160	Microbials	Jarvis (2001)
2003	235	Microbials	Guillon (2004)
2005	672	Broad (?)	Thakore (2006)
2006	280	Microbials	Harwood et al. (2007)
2006	400	Microbials	Gelernter (2007)
2008	750	Microbials (?)	Evans (2008)

3.2 The European Microbial Pesticide Market

In Europe, sales of biopesticides including beneficials, microbial pesticides and pheromone products were $97 million in 2000 – about 2% of the total European pesticide market. Sales of microbials were $25 million. In the category of microbials "soft" pesticides were also included such as fatty acids (Frost and Sullivan 2001). The authors expected annual growth of 11.7% leading to about $210 million in 2007 for all biopesticides in Europe. Guillon (2004) reported that in Europe in 2004 the market for biopesticides (bacteria, fungi and viruses) amounted to $31 million. According to Thakore (2006), the European market is estimated to represent $135 million in 2005; this market is expected to grow the fastest to $270 million by 2010. This figure is most likely based on a broad definition of biopesticides.

Lisansky (CPL 2006a) reported that in Europe annual sales of microbial- and nematode-based pesticides were estimated to be around $43 million at user-level in 2005. The share of Bt-based products has declined from an estimated 90% in 2000 to 72.4% in 2005. The fastest growing products have been nematodes and baculoviruses; where sales of both may have doubled since 2000 to reach $6 million and $5.45 million respectively. By comparison, the North American market for microbial- and nematode-based biopesticides is estimated to be worth US$110 million at user-level in 2004, with a Bt share of 67%. The total pesticide market in Europe was $12.8 billion in 2008 as a reference (CPM 2009a). Thus the microbial pesticide market is approximately 0.5%. The largest individual European biopesticide market is Spain, followed by France and Italy. Although overall growth in the biopesticide market has not met expectations of the past, Lisansky (CPL 2006a) expected the potential to remain high and that the total market could rise to $200 million by 2015. Table 7.4 provided the most recent estimates for the European market. In the project ENDURE (the European Network for the Durable Exploitation of Crop Protection Strategies, a network of excellence funded by the EU) IBMA made an estimate of the biopesticide market in Europe through a farmers survey, figures are presented in Table 7.4 (B. Blum, personal communication).

Literature provides little data on the total nematode market. From the report of Frost and Sullivan (2001) a $7 million market in 2000 for Europe can be deducted. CPL (2006a) reported a worldwide market of $14.65 million for 2005: $8.25 million for the USA and $6 million for Europe, while almost none for the rest of the world.

Table 7.4 Estimated sales per annum of microbial and nematode-based pesticides in Europe

Organism	Sales estimate[a] in 2005 (in US $) (CPL 2006b)	Sales estimate[b] in 2008 (in €) (IBMA)	Sales estimate[a] in 2008 (in €) (this author)
Bt products	26.7	52[c]	30
Other bacteria	2.6		3.5
Fungi	2.2		8
Baculoviruses	5.5		7.5
Nematodes	6.0	12	8
Total	42.9	64	57

[a]Market level.
[b]Manufacturer level.
[c]Total sales of bacteria, fungi and baculoviruses.

IBMA estimated the market to be around €12 million at the manufacturer level, while my own estimate of €8 million is much lower, even at market level. All these estimates vary greatly, indicating the difficulty of obtaining reliable data on the sales of nematode products. Apparently, the competition in the market causes producers to keep their information very confidential.

3.3 Microbial Pesticide Market in the Netherlands

Nefyto, the association of agrochemical companies in the Netherlands, provides figures on the use of crop protection products in terms of kg active ingredient. In 2007, biological insecticides (in kg a.i.) were reported to be < 0.7% (biofungicide use is negligible) of the total use of insecticides/acaracides (Nefyto 2008). If microbial pesticide sales from non-members were included, the total might be between 1 and 2%. Annual sales of pesticides account for approximately €330 million, biological pesticides only comprise €5–6 million (at end-user level). The Netherlands is probably number four or five in the European biopesticide market. The use of Bt has decreased tremendously the last few years due to new chemicals against caterpillars (J. de Hoog, personal communication). Table 7.5 provides sales estimates per pathogen group.

3.4 Microbial Pesticide Market for Protected Crops

The area under protection for vegetables, ornamentals and soft fruit is estimated to be approximately 2.4 million ha (Van Lenteren 2006a). Data from China alone reveal that about over 2 million ha of protected crops are grown (Zheng et al. 2005). The worldwide acreage of protected crops is still increasing. Biological control with natural enemies is successfully used on about 32,000 ha out of the estimated 2.4 million ha in 2006 (van Lenteren 2006a). This has increased recently since the

Table 7.5 Estimated sales of microbial pest control products and their main uses in the Netherlands in 2008

Type of pathogen	Sales	Main pests	Products	Main uses
Bacteria (Bt)	€1 million	Caterpillars	Delphin, Dipel, Turex, Xentari, Scutello	Protected vegetables and ornamentals
Nematodes	€1 million	Thrips, Sciaridae, black vine weevil, slugs	Bionem, Entonem, Larvanem, Nemasys, Nematop, Nemaplus, Nemagreen Nemaslug	Protected ornamentals, nursery, field vegetables
Baculoviruses	€1,5 million	Codling moth	Carpovirusine Plus, Cyd-X, Madex	Apple, pear
Entomopathogenic fungi	€300,000	Whitefly, thrips	Botanigard, Mycotal, Preferal	Protected crops
Antagonistic fungi	€1,5 million	Root diseases	Contans, Mycostop, Trianum	Protected crops, field vegetables
Various	€300,000	Various	Bio 1020, Cerall, Dutch Trig, Spod-X	Various crops

adoption of biocontrol in sweet pepper in Spain (Van der Blom et al. 2008) to over 40,000 ha. The use of microbials in greenhouses is briefly described by Lipa and Smits (1999), Paulitz and Bélanger (2001), Ravensberg and Elad (2001), Zheng et al. (2005), Van Lenteren (2006b), Shipp et al. (2007) and Van der Blom (2009). Some information is provided by Van Lenteren (2006b) on the use of biological control, including microbial insecticides, in various regions and countries.

Little information is available specifically on the use of biopesticides in greenhouses. Quantitative data in terms of hectares or product sales are lacking in the literature. The main products used are bioinsecticides as part of an IPM system (Table 7.6). The use of Bt for control of caterpillars represents by far the greatest use of microbial products in protected crops. Baculoviruses and nematodes are used to some extent against caterpillars. Bti and Btt are used on a very limited scale against sciarids, respectively Colorado potato beetle. Entomopathogenic fungi are chiefly used for control of whitefly, and to a lesser extent against thrips and aphids. Some use is against capsid bugs. Nematodes are used for black vine weevil larvae, sciarid larvae, and some other soil-dwelling insects or insects living in cryptic habitats such as borers. There are no microbials available for important pests such as spider mites and leafminers. Bt is the standard solution in the case of caterpillar control. For other pests, microbials are mainly used as corrective or additive measures.

Table 7.6 Entomopathogens used as bioinsecticides in protected vegetable and ornamental crops in Europe

Microorganism	Target pests	Used since	Extent
Bacillus thuringiensis	Caterpillars	1972	Wide scale use
Bacillus israelensis	Sciarids	1985	Very limited ornamentals
Bacillus tenebrionis	Colorado potato beetle	1990	Very limited
Lecanicillium muscarium	Whitefly, thrips	1981	Limited
Lecanicillium longisporum	Aphids	1982	Very limited
Beauveria bassiana	Whitefly	2002	Limited
Paecilomyces fumosoroseus	Whitefly	1999	Very limited
Metarhizium anisopliae	Black vine weevil	2008	Very limited
Spodoptera exigua NPV	Beet army worm	1994	Limited, increasing
Steinernema feltiae	Sciarids, thrips	1984	Increasing
Heterorhabditis megidis	Black vine weevil	1984	Very limited
Heterorhabditis bacteriophora	Black vine weevil	1995	Limited
Steinernema carpocapsae	Caterpillars, tipulids	2003	Limited

Microbial control of diseases is still in its infancy. A number of products is available, primarily for soil diseases, but the use of these products is limited (Paulitz and Bélanger 2001; Ravensberg and Elad 2001; Stewart 2001; Fravel 2005). The use of microbial insecticides in greenhouse crops in the Netherlands consists largely of products based on Bt, entomopathogenic fungi and nematodes (see Table 7.5). The size of the market is estimated to be around €3 million at end-user level. Spain and The Netherlands comprise the majority of the protected crops in Europe where microbial insecticides are being used. In Spain, Bt is used in greenhouse vegetables against *Spodoptera* spp. and *Helicoverpa armigera*, and since 2009, the market has more than doubled due to the sudden demand for control of *Tuta absoluta*, a new invasive pest in tomato. The total market is approximately €11–12 million at end-user level in 2009. Baculoviruses are used against *S. exigua* in sweet pepper and the sales are about €1.3 million in 2009 (J. van der Blom, Coexphal, personal communication). Estimates for entomopathogenic fungi and nematodes are difficult to provide, but sales are most likely only a few hundred thousand Euros for each group.

3.5 The Main Crops for Microbial Pesticides

Bt products account for about 70% of the bioinsecticide sales. The Bt markets obviously represent the main markets for biopesticides: forestry, field vegetables, protected crops, and the use for vector control. The use of viruses in orchards is an increasing market. Nematodes are used in a variety of crops such as protected crops, mushrooms, lawns, hardy ornamentals and forestry (Georgis 2004). Use of slug nematodes is an example of use in agricultural crops such as wheat, Brussels sprouts, and potatoes. These nematodes are also sold for use in home gardens. Entomopathogenic fungi are mainly used in protected vegetables. Biopesticides have predominantly been used in niche markets, like high value crops such as protected crops, berries, and fruit crops. Thakore (2006) stated that orchard crops represent the largest share of biopesticide use at 55%. MPCPs used in orchard systems are reviewed by Lacey and Shapiro-Ilan (2008). Biopesticides have had great successes in forestry where chemical insecticides are not practical or acceptable. This is likely to be an expanding market for biopesticides although application techniques and budgets are limiting factors. Currently, however, use in agricultural crops is increasing and this is a promising signal. Disease control products like Contans and Cedomon are used in agricultural crops such as oil-seed rape and sunflower, respectively cereals. The use in forestry is also developing with disease control products and nematodes. The organic market represents just a small proportion of the biopesticide market; I estimate less than 10%. Braverman (2007) reported that only 3% of all biopesticides used in the USA are applied in the organic market.

3.6 Successful Entomopathogenic Products

The most successful bioinsecticides are undoubtedly the Bt products. They are "the simplest to produce and use" (Burges 1981a), and applied in many crops and environments on a worldwide scale. Sales are estimated to be between €60–80 million worldwide (Federici 2007) and $100 million (Phillips McDougall 2008). The latter refers to use in agriculture and forestry. Figures reported by CPL indicated that the total world market is greater than $150 million. The market for Bt products is approximately $160 million, which represents around 60% of the total microbial insecticides (Avé 2008). The use of Bt in China and export from China is estimated to exceed $100 million (Huang et al. 2007). Little information is available on the use of Bti for control of dipteran pests. The overall worldwide market is most likely over $200 million considering all uses of Bt products. Nevertheless, sales have been declining as a result of transgenic Bt-crops and new chemical and natural insecticides. On the other hand, Avé (2008) believes that the market potential for Bt products is likely to increase in the next few years due to the demands of the regulatory agencies and the general public for safer food and for crop protection products with no detrimental effect on the environment.

The second most successful group of entomopathogens is the nematodes although there is a huge gap between sales of Bt and nematodes. Sales in Europe are estimated to be around $6 million (CPL, 2006b). Here too, it is very difficult to provide figures on the various markets where they are used such as greenhouses, outdoor ornamentals, soft fruit, mushrooms, turf, and upcoming markets like top fruit, forestry, and home gardens (Blum 2001). The most important target insects are the larvae of Sciaridae and the black vine weevil.

Baculoviruses have increasingly become popular and the use of CpGV in fruit orchards is the largest market. Their usage in greenhouses for control of *S. exigua* has also increased, mainly in Spain (Van der Blom 2009). The number of products remains small.

The use of entomopathogenic fungi has slowly grown, yet the total market in Europe is limited when compared to the sales of the other three groups of pathogens (CPL 2006b). Figures are provided in Tables 7.4 and 7.5. There are about ten products in Europe, and most of them have a small market share.

From the above it is clear that the order of success of entomopathogens is bacteria, EPNs, baculoviruses and fungi, although the order may differ per area (Fig. 7.1). I think that this ranking can be explained by comparing efficacy, costs, and user-friendliness of biopesticides to traditional chemicals. The more these parameters reflect the characteristics of a chemical pesticide, the easier the adoption of a biopesticide in the market.

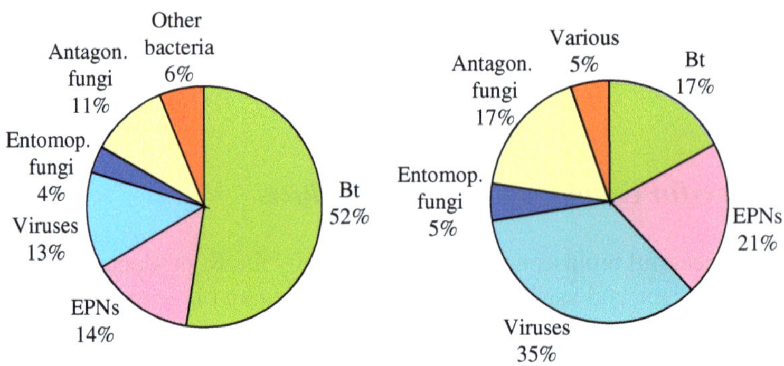

Fig. 7.1 Division of microbial pesticides in Europe (left) and in the Netherlands (right) in 2008

3.7 Sales and Profitability of Producers

Estimates of sales of microbial pesticides have been presented by a few authors. Information on profitability is not available due to confidentiality. Still, some conclusions can be drawn from the sales figures and the number of products on the market. When I compare the figures from Frost and Sullivan (2001) – $25 million for all microbials, including B.t., in Europe – with Jarvis (2001) – $16 million for all microbials world-wide excluding B.t. – and estimating the non-Bt microbials market in Europe (10–20%), this might have reached $2.5–$5 million in 2000. It is a small amount, and considering that it represents the total of 20–30 microbials (viral insecticides, fungal insecticides and fungal and bacterial fungicides) one can see that the turnover per microbial is very low. On top of this, these sales are spread over a number of countries for most products, giving small turnovers per country and relatively high marketing expenses. When I take into account development and registration costs, it is apparent that the biopesticide industry is small and has difficulty being a profitable business.

When a similar approach is taken with figures from 2005 from CPL (2006b) a comparable picture appears. The total of microbial- and nematode-based products was $43 million in 2005 at end-user level. Bt's accounted for $30 million, nematodes for $6 million and baculoviruses for $5.45 million. Fungi and other bacteria generated sales of about $2.2, respectively $2.6 million. If it is assumed that 70–80% of these figures is returned to the manufacturer, and that these overall figures represent the sales of a number of products over a number of manufacturers, it is obvious that the turnover per product is very low. Particularly in the case of fungi-based products where about ten products account for this turnover of $2.2 million, it is apparent that profitability is minimal. For the Bt products, the profitability looks much better since there are fewer products and fewer manufacturers. The same is true for the baculoviruses. Gelernter (2005) estimated annual sales to be between $100,000 and $5 million per biocontrol product.

Manufacturers of biopesticides in the developed world can be characterized as companies with an industrialized production that needs a high initial capital input. There is a limited number of companies which typically are small-sized enterprises; there are a few larger players. Companies that have existed for more than a decade seem to be successful and therefore should be profitable. Expectations for a sustainable business and financial success need, however, to be reflected by the current industry position and profile.

4 Critical Success and Failure Factors for a Biopesticide Company

4.1 Successful and Unsuccessful Companies

Numerous companies have entered the field of production and marketing of microbial pesticides over the last 40 years. This field attracted investors and entrepreneurs every where in the world. Obviously, its perspective from the outside must have been of a promising area where profits could be made in a relatively short period of time. Many companies though, once active in the field, could not meet product objectives and market projections, and registrations took too long and delayed income. As a result, many, who started optimistically, failed (Gelernter 2005; CPL 2006c). On the other hand, some companies have been in business in this area for a decade or even much longer and have been expanding. Examples from Europe are: Andermatt Biocontrol, Enema, Intrachem, Koppert, NPP, Prophyta, Verdera (formerly owned by Kemira, now part of Lallemand), and from the USA: Agraquest, Becker Underwood and BioWorks. Some have multiple areas of activity. Whether these companies are profitable is hard to say. Their survival over a long period suggests that their business is profitable. Examples of companies, large and small, which failed and stopped or sold the biopesticide activity are: Bayer, Ciba Geigy, MicroBio, MRL, Novo Nordisk, Novartis, Sandoz, Solvay, Tate and Lyle, all from Europe; and from the USA: Abbott, Biosys, CropGenetics, Ecogen, Ecoscience, Eastman Kodak, Mycogen, Mycotech, Taensa, Thermo Trilogy, Troy Biosciences, WR Grace, and many more. Surely, the reasons are various, but details have usually not been provided. Lack of revenues versus high investment has certainly been an important reason in most cases. This is exemplified by the case of Biosys, a venture capital company active in the field of nematodes in the 1980s and 1990s. Details of the company's foundation, its achievements, and its failure have been described by Georgis (2002).

The above raises two questions:

(1) why have companies decided to enter the microbial pesticide market?
Incentives are manifold. Lisansky (CPL 2006c) provided putative reasons for pesticide companies, fermentation companies, new companies funded by external capital, spin out companies, biocontrol (natural enemies) companies and commodity companies (Bt producers). He also discussed advantages and disadvantages for each type of company. Excellent perspectives have been predicted for biocontrol products by numerous authors. Obviously, each company has its own valid reasons and objectives to start in this business field. As long as the reason(s) match the company objectives, success can be achieved. I will provide two examples to illustrate this. When a company has excess fermentation capacity, it could start producing a microbial agent and sell it through a partner or distributor. The main objective is then to use the fermentation to its full capacity and spreading operating costs as efficiently as possible over various products. Success in the biopesticide market is then less essential. Another objective for the development of a biopesticide could be to safeguard an IPM system in which the company has a vested interest. This was exemplified by Koppert who in the early days of Mycotal initiated this development to ensure whitefly control with natural enemies. A product as Mycotal was used as a corrective means when control with parasitoids was less effective. In this case, profit is not just measured over the activity with Mycotal alone, but over the IPM system for whitefly control and sales of natural enemies. Of course, in most cases biopesticides are being developed in order to create a stable and profitable business.

(2) why have some companies been successful while others failed?
A company can only be successful when the product is successful, and a product is only successful when the company succeeds in marketing it for a considerable period of time and is profitable in doing so. These two conditions lead to a sustainable and growing business and that is what I consider the definition of a successful enterprise. After all, the basic objective of a company is to be profitable in order to remain and grow in business. Many companies have failed in the biopesticide area as described above; however, a considerable number is active today. Even when the product is successful, a company could still fail when it cannot meet its financial objectives. This creates two new questions: Which factors determine success or failure? Can we learn from the past and develop a pathway that leads to success?

A list of factors determining the success or failure of a biopesticide comprises the attributes of the product, and can be easily drawn up. In fact, many authors have reported the advantages and disadvantages of a biopesticide. A list of factors determining whether a company will be successful is more difficult to provide, and few authors have attempted or succeeded to do so in a systematic way, in my opinion. In the book of Gurr and Wratten (2000) where ten chapters described cases of successful biological control, the criteria for success are various, predominantly biological and even conflicting. Little attention is paid to commercial success. First of all and obviously, a product must be biologically and technically successful before it can become commercially successful. Commercial success for the manufacturer relies on the ability to sell. There is, however, a matrix of interlinked factors which are ultimately responsible for a profitable and sustainable business activity. Lisansky (CPL 2006c) provided nineteen success factors for a company to consider in order to optimize its chances of success. Companies that failed committed three main mistakes: they assumed that a biopesticide would be easy to find and develop; they expected that they would outperform others; and they underestimated budgets and time (CPL 2006c). Below I will present and discuss the critical commercial success factors which I consider the most important (Table 7.7).

Table 7.7 Success and failure factors in a novel biopesticide enterprise

Success factors	Failure factors
Do your homework well: make an accurate business plan, focus on data integrity; explore best-case and worst-case scenarios	Assume that biopesticides are easy and cheap to develop and market
Be fully committed to the project	Overestimate own capabilities and knowledge
Start small, with modest investments	Start with high VC investment and many people, e.g. with a large R & D group
Focus strongly on one product	Do it "on the side"
Work according to a proper project management process; enforce timelines, milestones and decisions	Lack of efficient project planning
Develop deep market and customer knowledge	Assume that the market will just adopt a new "wonderful" biopesticide
Allocate sufficient budget for the product development *and* for the market development	Lack of resources until sales grow; overestimation of the market size and market penetration
Estimate registration costs, time to market, and time to volume as realistically as possible	Underestimate costs of registration and time to registration in each country
Use existing distributors that are well introduced in the market; train distributors well, develop a long-term interest between partners; involve distributors early in project	Assume that distribution will be easy and simple. Shortage of demonstration trials to convince users
Develop knowledge on compatibility and integration in IPM systems	Underestimate efforts to incorporate the product into IPM systems
Develop the company step by step; expand when the market demand is well understood	Overestimation of the market potential may lead to over-expansion of the company
Balance risks, progress, and debts; estimate profit margin, break-even point, and profit	Count on quick and large sales volumes and profits

4.2 Critical Commercial Success Factors

4.2.1 Commitment from the Top

The development of a new biopesticide is a lengthy and costly challenge. Ultimately, sales need to generate a reasonable profit so that a sustained business is possible. The whole project costs many millions of Euros, and takes on average 5–8 years to come to a break-even point, and even more years to reach a sales volume that provides the desired profit. This business field requires understanding, motivation, patience, and perseverance in order to become successful. Management needs to be aware of this from the outset. "Quick profit making" is not a standard feature in this business. Many investors have experienced this and have withdrawn from this business (Gelernter 2005). It is imperative that the long and winding road to success be understood and accepted, and that the vision and commitment be present in the company top in order to achieve success.

4.2.2 The Business Plan and Allocation of Financial Resources

An accurate business plan with realistic profit projections is a prerequisite for the project. The calculation of the potential revenues from 6 to 10 years in the future is a difficult task and must be done conservatively. Minimum to maximum sales need to be projected accurately and carefully, and extensive knowledge of the market is critical. Furthermore, it is imperative to establish as accurately as possible the resources that the development of the business or the product will require; not only the costs of initiating, but also the costs of staying in business. It is important to realize that it will take a considerable amount of time before there is any return on investment. This means that sufficient resources need to be available to cover all costs until sales can eventually pay back these costs and generate profit. A common mistake of many failed businesses has been to underestimate the necessary time and resources. Product development and registration costs are obvious expenses and most companies realize this. The requirements and costs involved for market adoption and reaching sales volume are often underestimated or even neglected, and many companies have failed in this phase in the past (CPL 2006c). Business plans should allow for a low penetration percentage of the market, particularly in the first years. The financial support must be guaranteed during the product developmental period *and* the initial sales period.

4.2.3 Qualified Management and Personnel

A good scientist is not necessarily an able entrepreneur. If the company is a spin-off from a research organization, the emphasis should be given to management skills. If a large company is new to the field of biopesticides, the management does not always understand this type of business. The management should fit with the type and size of the company or invest sufficient time to understand the field of biopesticides. In new businesses, poor management is often cited as the principal reason for failure. The owners of a new small company frequently lack relevant business and management expertise in areas such as finance, purchasing, selling, production, and the management of employees. All these aspects are crucial to the survival and success of the business. The development of a biopesticide requires a multi-disciplinary team with sufficient knowledge and expertise in the areas of research, production, registration and marketing. Ideally, a company will have all these experts in-house for a well-planned and time-efficient developmental project. The lack of careful, methodical planning could be expensive and jeopardize the project.

4.2.4 Knowledge of the Market

In-depth knowledge of the pest and the potential market is elementary and seems so obvious. But many companies have overestimated the potential share of the market and as a consequence failed dramatically. Pest damage figures cannot be transferred directly to market potential, a grower is willing to spend about one-third of the damage for control measures (CPL 2006a). Poor understanding and insufficient knowledge about the pest problems, customers, cropping systems, current practises, prices, competition, and the requirements to promote a new product will lead to errors and slow penetration in the market. Generally, biopesticides need to be implemented as part of an IPM system and this can be complex. Time and money for the study of the correct positioning of the product and the education of the users is often neglected in business plans. Many companies have had an unrealistic expectation of the true market potential and their ability to sell the product. Biopesticide markets are generally niche markets. Annual sales may range from a few hundred thousand Euros to a few million Euros (Ravensberg and Elad 2001; Gelernter 2005). Unrealistic sales expectations caused many companies to abandon biopesticides. Experience with biopesticides tells us that adoption in the market is a slow process; biopesticides are received with some scepticism and have to prove themselves. The increase in sales does not come overnight, but takes considerable efforts and time.

4.2.5 Insight into the Competition

Biopesticides have to conquer their position in the market and justify their place. Chemical pesticides are still the main competition due to price and efficacy, and also because of grower's routine, and ease of use. Users must be convinced that a new product achieves similar results as their usual practises. It is essential to understand the competition well and to develop ways to position the biopesticide as a replacement or as an alternative product. This needs demonstrations, education, and comparative trials. Because of competition, the market share may be small initially and only increase with efforts over time. Be conservative in market share assumptions, I would say, around a few percent. This may increase, but the market will have to be divided between many products. Initially, the focus should be on the development of suitable and accessible markets with few competitive products.

4.2.6 Other Business Factors

When a company does everything right and the product is a good product, does this guarantee a sustainable business? Obviously, Bt's are successful products that are well accepted by users and are the most sold biopesticides. Still, large companies have stopped their activity with Bt and sold that activity. Examples from the past are Sandoz, Novartis, Novo Nordisk, and Abbott. From the outside, these companies appeared to be successfully producing and marketing these products. Corporate decisions, however, led to the abandonment of the biopesticide field completely. VC companies are under pressure to generate revenues within a short timeline which may lead to leaving the business even if the product has potential. A company's objectives, whatever they are, may have a decisive impact on their perception of success. Many external factors continuously influence the company and its activities in the market with biopesticides. This is an ever-changing environment to which the company must adapt and be flexible in order to maintain success with biopesticides. This requires advanced entrepreneurial skills and expertise, and a close view of the market.

5 Factors that Influence the Decision to Develop and Commercialize Microbial Pesticides

5.1 Product and Market Related Factors

Factors that determine the decision whether to initiate the commercialization of a new biopesticide need to be understood extremely well. The difficulty regarding this decision is that it must be made when most information is only indicative. The decision should in most cases be market-driven. Is there a pest problem that asks for a solution, for a new product? And if so, how big is this market and what is a grower willing to spend to solve the pest problem? A decision could, on the other hand, also be product-driven when an innovative research idea can be transformed into a valuable product that has a good chance to penetrate an existing market because of its novel character with improved control, or other desired features. The key factors concern market need, market size, cost price and market price, time to market, and time to volume. These factors determine when return on investment can be obtained and whether a company can become profitable. Market size and competitiveness were regarded by Burges (1981b) as vital. According to Lisansky and Hall (1983) profit-rate and market size determined the willingness of a company to develop a product. Lisansky (1985) identified the need, the market size and the competing products as subjects for a feasibility study on which the decision to proceed would be based. Carlton (1990) considered four aspects to be relevant for the decision to commercialize a biological product: market need, market size, price, and competition. Later, Lisansky (CPL 2006c) stated that key to the biopesticide business is a cost-competitive product *and* understanding the market. I believe that the efficacy of the product is the most relevant factor. Without reliable and consistent efficacy, no product will become successful. Further, the market factors are determinative and an accurate analysis of the market situation is crucial. The following five factors should be carefully investigated. They are similar to the ones mentioned by Carlton (1990), but I have not factored competition separately in, it is an inherent dimension of all mentioned market aspects.

5.1.1 Market Demand

The need for a product must be estimated at the start of the project and over a period of 5–10 years in the future when the product will reach the market. Elemental is the answer to the question: "Is there a need or a place in the market for a new product?" Product development is interesting when there is a demand for a solution for a certain pest or disease. The actual need for a new or an alternative product must be as concrete as possible. A confirmed market demand is the ideal situation, but this is rarely the case. When the need is expected in the future, it will make the decision more uncertain. On the other hand, when a solution is needed directly, the situation may have changed once the foreseen product is ready to be launched. This takes many years and in the meantime other solutions may reach the market. This can never be fully foreseen and the decision to commercialize is always risky in this sense. This aspect should certainly be taken into consideration. When a new product can be developed that offers an opportunity, a novelty that has advantages over existing products, this may be another reason to develop and commercialize a product. Both cases could be the starting point for a new activity.

Market demand could be generated by various factors: occurrence of a new pest, lack of pesticides, resistance to available pesticides, and withdrawal of pesticides. These factors are directly related to the pest. Indirect factors from society can also play a role. Examples are human health and environmental concerns, residue concerns by supermarkets and consumers, and political motives. Socio-economic considerations are relevant for a society, but for a manufacturer of biopesticides these "green credentials" should not be driving the decision-making process. In some cases, political, and societal, and environmental factors have been determinative as in the development of Green Muscle for locust control (Lomer et al. 1999; Douthwaite et al. 2001). This is however, not a standard situation in the commercial biopesticide business. Some of the market incentives may be more concrete than others; a realistic estimation of these incentives is difficult, but still useful. The end-point must be the grower's demand and incentive to choose the new biopesticide.

5.1.2 Market Size

Biopesticides are still niche products. Bt products could be considered an exception, but their market share is still small within the entire pesticide business. Thus, when a company considers whether or not to initiate a biopesticide development, a market study is extremely important. The size of the market is determined by the importance of the pest(s), the number of crops in which this pest is a serious problem, and the extent of these crops. The size of the market in hectares, and the number of applications per season (together often called "super hectares"), and the competing products, with their application costs, need to be estimated at best. The spectrum of pests that can be controlled with the product has a great influence on the market potential. Does the new product fill a hole in the market or does it have to penetrate a market with existing products? The size of the market and the potential growth of the market profoundly impact the decision on commercialization. Potential sales should be large enough to allow for return of investment. Market shares for biopesticides will only be partial and a careful estimate of the market share should be made, including the time to reach a projected market share. Great market shares have frequently been expected, but proven completely out of scope resulting in a failure of the company (Cross and Polonenko 1996; Gelernter 2005). Market shares should be estimated in a conservative way, and best-case and worst-case scenarios should be examined. Biopesticides have rarely been developed for the major agricultural crops, rather for small crops or sectors.

5.1.3 Profit Margin

The cost price and the potential selling price determine whether the product can be sold in a profitable way. The calculation of the cost price has been provided in detail in Chapter 3. The selling price depends on the demand of the product in the market and the price of other competing control measures, in other words, the market value. Pricing should also allow for margins for distributors. A selling price can be established in two ways. The first is to see what margins are required to make a profit on the product. A second is to see what the market value is and what price will be accepted in the market. The product price is often set high for the first years after its launch in order to obtain the return on investment within a short period. Later on, the price may become lower because of competition and in order to expand the area of use. Most biopesticides are used in high-value crops, which generally contain a limited acreage. These crops, however, allow growers to spend more money per unit area on, amongst others, crop protection, and therefore offer an interesting niche market for biopesticides. Nevertheless, biopesticides also face competition from both synthetic chemical pesticides as well other IPM control measures, including other biocontrol agents: both arthropods and microbial agents. Favourable properties of biocontrol products such as less or no residue, short or no re-entry periods, safety to humans and the environment, etc, are difficult to exploit in a selling price. Competition is predominantly based on traditional factors; efficacy and costs. "Feel-good factors" do not automatically allow charging higher prices. Selling prices should therefore be estimated based on a realistic approach to the current market coupled with traditional product performance characteristics.

5.1.4 Time to Market

The developmental time of a product and the time to registration need to be estimated as accurately as possible. The first is in the hands of the manufacturer, at least to a large extent, and an estimate can be made. The second depends on the regulatory authorities and predictions have proven very difficult to make. In general, registration of the active ingredient has taken more than 5 years in Europe, and many years more to obtain national authorizations that open the market. More details on the registration and the time periods are given in Chapter 5. For entomopathogenic nematodes, registration is different and time to market much faster. Average periods for microbials from start to market are provided below. It should be realized that it will take many years before the product is on the market and that the situation has changed considerably in many ways. The pest problem may have been partially solved by other products, new cultural practices, resistant cultivars, etc. It is only possible to foresee these developments in a limited way. An efficient and affordable product, however, will find its place in a changing environment.

5.1.5 Time to Volume

Once a product has been registered, sales can begin. The adoption of a new product in the market is usually slow and needs a substantial effort of the producer and his distributors to initiate and increase sales. The details of this process are provided in Chapter 6. Time to volume is pivotal with regard to return on investment; the faster the better. Within a 3-year period of sales, profits should pay back the investment for the development and registration expenses of the product. Subsequent sales should generate margins that allow for new product developments or other investments. Product development should only be initiated when the projected sales and profits are sufficient to guarantee return on investment within an acceptable time period. This determines the success of a company. Investors will require a sound business plan where these factors are realistically provided. Time to volume is often neglected in business plans or assumed to happen automatically. This, however, is a misconception and market penetration can take many seasons. Users need to obtain positive results from their first trialling experiences with the product before they use it on their whole acreage and continue to buy the product. The lifetime of a MPCP should be also considered. The projected length of the lifetime depends on the efficacy, user satisfaction, and price of the product. Resistance is unlikely although not impossible. Competition will change overtime, but new uses may be developed and new opportunities may come along. The lifetime is difficult to predict, and it cannot be used as a solid criterion in investment decisions.

5.2 The Decision Process

If the above-mentioned market aspects have been investigated and analysed in a feasibility study, a decision should be taken. However, for any given product and for any company the results of the market analysis may lead to a different outcome depending on the company's strategy and objectives. Therefore, it is not possible to give exact decision criteria. It is possible though to start product development and take "go/no go"-decisions along the way. At any phase in the project, company management should be courageous enough to make such a decision and perhaps decide to stop the project. Considerations may differ per company. For an agro-chemical company, the situation is quite different than for a small start-up company. Froyd (1997) discussed the reasons for a large agrochemical company, American Cyanamid, to develop biopesticides, in the mid 1990s. He considered the less complicated registration requirements, rapid market entry, possibility to become active in a new market segment, residue-free products, public opinion, and public funding for a part of the development important incentives. Other requirements considered relevant were product performance and profitability for the company. The company believed in the potential of genetically improved baculoviruses, but thought that field performance of biofungicides needed much more research. Today this company is not active in biocontrol products; apparently the potential was insufficient to reach the expected economic results. This picture has been seen with many of the large agrochemical companies. The incentives for small companies can be characterized by passion and commitment. Many survived despite the modest sales and profits. They learned to take the right decisions and found a balance between their costs and revenues, and they continued in speciality markets (Gelernter 2005).

A positive decision to develop a new product will please the persons involved because that was the intention from the beginning, and it fits with the (usual) goals of the company: develop new products and markets and grow. It is often seen that people focus on the "best-case" scenario, but a "worst-case" scenario should also be analysed. A negative decision is, in my experience, much harder to take since all those involved have the mindset of starting this new project. Making such a "no go"-decision is difficult and requires solid and objective arguments to make everybody understand and accept, despite certain disappointment. It needs courage to make such a decision and doubting too long will bring unnecessary costs, particularly when such a decision has to be taken in an advanced stage in the project. But management should realize that "*a correct kill is a success*- it just saved the company a bag of money and a heap of trouble" (Cooper and Edgett 2006). Many companies struggle on and in the end are not successful. Efforts put in one project exclude working on another one, and this is a crucial element to keep in mind in the decision process. The positive attitude to develop a new product is a potential pitfall for a company and one tends to continue with something that was started. A regular critical evaluation of "where do we stand" during a project is an absolute need in order to keep on track with budgets, timelines and goals, followed by a "go/no go"-decision. Most companies who failed showed overconfidence on their own performance, and blamed the products or the market for not being successful rather than being critical of themselves (CPL 2006c).

Taking the right decisions is vital for a company, and it is often a process full of struggles. This is not just the case for biopesticide companies, but for all companies that develop new processes and new products. The development of a MPCP can be considered a technology development (TD) project. These projects are usually characterized by many uncertainties and "difficult to predict"-outcomes. To facilitate decision-making, a systematic process has been developed for moving a new product project through the various stages and steps from idea-to-sales, and making decisions between these stages, at the gate of a new stage. This Stage-Gate process is a management tool that facilitates outlining the process in steps, and making the right decisions at the right time (Cooper et al. 2002; Cooper 2007; www.stage-gate.com). To make the decision process more objective and quantifiable, scorecards can be developed that help rating and prioritizing TD projects. It needs to be custom-designed for each individual company and product. This systematic approach can be used to decide whether or not to start a project, or to decide between two or more projects. Tables 7.8, 7.9 and 7.10 provide examples that could be used in the biopesticide industry.

This exercise forces management to define success criteria and to commit to the project when these criteria are met and evaluated at moments of "go/no go". The Lubilosa project seems to have worked with a similar system where funding had to be approved for each successive phase (www.lubilosa.org). Adopting a Stage-Gate process ensures a better and more transparent decision-making process within an organization and could lead to a more successful business.

Table 7.8 An example of a scorecard rating various elements of the development of a biopesticide, as a tool to estimate the probability of a technical success © 2002 Robert G. Cooper, Scott J. Edgett, Elko J. Kleinschmidt. Reprinted by permission of Basic Books, a member of the Perseus Books Group

	Rating scale				
Key factors	1	4	7	10	Rating
Technical "Gap"	Large gap between current expertise and objective; must invent new technology	"Order of magnitude" change proposed	Step-change short of "order of magnitude"	Incremental improvement; most technology present	
Product complexity	Difficult to define; many hurdles	Easy to define; many hurdles	A challenge, but "do-able"	Straight forward	
Technology skill base	Technology new to the company; (almost) no skills	Some R&D experience, but probably insufficient	Selectively practiced in company	Widely practiced in company	
Availability of people and facilities	No appropriate people/ facilities; must hire/build	Acknowledged shortage in key areas	Resources are available, but in demand, must plan in advance	People/ facilities immediately available	

Modified after Cooper et al. (2002).

Table 7.9 An example of a scorecard rating the various elements of the market of a biopesticide, as a tool to estimate the probability of commercial success © 2002 Robert G. Cooper, Scott J. Edgett, Elko J. Kleinschmidt. Reprinted by permission of Basic Books, a member of the Perseus Books Group

	Rating scale				
Key factors	1	4	7	10	Rating
Market demand	Extensive market development required; no apparent demand	Demand must be highlighted for customers; product tailoring and market penetration required	Clear relationship between product and need; market penetration needed	Product immediately responsive to customer need; no other control means available	
Market size	Partial niche market	Small market	Modest market	Large market potential	
Competitive intensity	High	Moderate/high	Moderate/low	Low	
Development and commercial skills	Must develop; new to company	Must develop beyond current limited use	Need to tailor to proposed program	Already in place	
Commercial assumptions	Low probability/low impact	Low predictability/low impact	High probability/high impact	High predictability/high impact	
Regulatory, social and political impact	Negative	Neutral	Somewhat favourable (e.g. less chemicals available)	Positive impact on high-profile issues (e.g. food safety, no residues)	

Modified after Cooper et al. (2002).

6 Is There a Successful Business Model for a Biopesticide Company?

6.1 Business Models in the Biopesticide Industry

Friedman (1990) described three business models prevalent in the field of biocontrol. The first is the owner equity model in which the owner invests to a limited level and development is gradual. The second type is the social investment model where large investments are conducted before any sales occur and development can be quick. Funding from governmental and non-profit organizations supports this model. An example is the Lubilosa project. The third model is the venture capital

Table 7.10 An example of a scorecard rating various financial elements of a biopesticide and the income expectations © 2002 Robert G. Cooper, Scott J. Edgett, Elko J. Kleinschmidt. Reprinted by permission of Basic Books, a member of the Perseus Books Group

	Rating scale				
Key factors	1	4	7	10	Rating
Sales (5-year cumulative income from first sales)	< €1 M	< €5 M	< €10 M	< €20 M	
Technology payback period	> 10 years	8 years	5 years	< 3 years	
Time to market	> 7 years	5 years	3 years	< 1 year	
Certainty of return/profit estimates	Low; pure guess, < 20% probability	40% probability	70% probability	Highly certain, > 90% probability	

Modified after Cooper et al. (2002).

(VC) investment model with levels of investment that greatly exceed any sales, and the developments are large and fast. In this model patent-based technology is a prerequisite to safeguard the investment from competition. In his opinion, the best model was a combination of a VC company in collaboration with public research and governmental funding. Friedman illustrated this model with Biosys where he was working at the time. In the end, Biosys, however, was not successful due to unrealistic market projections and had to be sold (Georgis 2002). Many more VC companies and large agrochemical companies left the field of biocontrol mainly because market forecasts were unrealistic (Gelernter 2005).

The three models mentioned by Friedman can still be recognized today, although his second model is rarely seen. I distinguish a fourth model as the large multinational company model that projected biopesticides as an attractive new opportunity for profit and established a business unit to develop this potential. Two categories can be separated: (1) the large agrochemical companies that envisaged new products next to their chemical crop protection products, and (2) large pharmaceutical or food companies, new to the crop protection area, that hoped to exploit their expertise and technology in production for developing biopesticides. Usually, resources were allocated internally to investigate this new activity. Examples of the first group are Bayer, Cyanamid, Monsanto, and BASF; examples of the second group are Tate & Lyle, Novo Nordisk, and Abbott. Almost none of these companies are any longer active in this field. Lisansky (CPL 2006c) described six basic types of companies involved in biopesticides; his categories have many similarities with the ones mentioned above. His sixth type is a commodity production company from the Far East.

Currently, most biopesticide companies are small to medium-sized organizations and originate from the owner equity model. The best performers seem to follow an incremental, stepwise, and manageable growth of the organization. They are cautious with large investments and only invest when the market demand is foreseeable. Relatively small growth can be controlled, the changes are still manageable, and new activities can be well integrated into the existing organization. They strive for profitability and try to avoid large debts and risks. Expansion is, in general, financed in a conservative way, usually by banks or small private investors, but not by venture capital. Shares are usually held within the company itself and/or by some external shareholders. If possible, the majority of shares are kept by the owners for reasons of influence and decision-making. The motto seems to be "stick to what you are good in" and "expand your core business when opportunities are discovered and considered low risk". Core business relates to products and markets in which understanding and experience are built up over a considerable period of time. New markets are entered step by step and without taking large risks. This strategy seems to give the best chance for survival and growth over time in the current world of crop protection.

6.2 Characteristics of Successful Biopesticide Companies

Successful companies can be characterized by their commitment and belief in the products they make. Their enthusiasm and entrepreneurial skills help them succeed. Within the field of biopesticides it is pivotal to find your niche in the market where competition is limited. This seems to be the strategic decision that allows a company to establish its position and to gradually become successful. Examples are Becker Underwood and Enema which specialized in the production and commercialization of entomopathogenic nematodes. Others focussed on production methods for certain organisms. This is illustrated by Valent BioSciences which specialized in Bt production, and Prophyta which specialized in fungal spore-based products. Andermatt seems to have chosen to become a specialist in baculoviruses. A strong focus on a particular market can also be a key factor that makes a business grow as demonstrated by Bioworks (Evans 2004a).

Others have chosen for a broad range of products and seem to be successful in that way. This is illustrated by Koppert which produces natural enemies, microbial products and bumblebees; by Intrachem which produces and markets microbial products, natural enemies and plant nutrients; and by Certis which commercializes chemical as well as biological pesticides. Some companies are only active in biopesticides, some have broader activities in agriculture; others have several business areas, even not closely related areas, such as Becker Underwood and Novozymes. Many players in this field have not survived in the past, particularly when they were founded with venture capital, or when biopesticides were a side-activity. Few VC companies are active nowadays, Agraquest is one of them and they seem to do well. The large agrochemical and pharmaceutical multinationals failed and have left this field. The most common mistakes made were unchecked assumptions of how easy and cheap it is to develop and sell a biopesticide, overestimation of their own capabilities, followed by under-budgeting of the new activity (CPL 2006a).

6.3 Key Components of a Successful Business Model

Several business models could be successful; there is not one straightforward model that guarantees a profitable business. Key factors are to have a responsible, dedicated and committed management with core competence in the products and the target markets. Successful companies are more carefully investigating the market potential than before and this is key-information, besides the ability to supply cost-competitive products (CPL 2006a). Biopesticides are unique products in many ways, and understanding these products and the way to commercialize them is indispensible in order to succeed. Cost-effective and reliable products with a demand in the market have a high probability of becoming successful. Crucial is the ability of the company to survive years when income is limited and to be able to accept a stepwise growth of the organization. Gelernter (2005) called the smaller companies value-driven entrepreneurs and considered them as the core of the future successes in biological control. The current portrait of the biocontrol companies proves this to be right.

Biopesticides can be developed by many types of companies, from small spin-offs from universities to large multinationals in agrochemicals, and even other businesses. The owner equity model, however, performs the best at present since it implicitly comprises all the required characteristics: drive, motivation, passion, determination, perseverance, and knowledge and skills. Together with this model, there is one approach, one strategy that is obviously the best option. This has been proven many times in the last 50 years. This approach is the incremental growing company which has an excellent understanding of the product and a realistic view on the market and that is financially able to cover the developmental period until revenues come in effectively. And it keeps a focus on slow and steady growth as the optimum, after careful research and an accurate market analysis, avoiding large risks. Growth is often perceived in business as success, but too rapid expansion has caused many bankruptcies. Still, growth is required in order to establish a firm position in the market and to be come stronger with respect to competitors.

7 Critical Success and Failure Factors for a Microbial Pesticide

Successful commercialization of a biopesticide poses a great number of challenges for a company of which many are described in this book. Success cannot be described as just one factor, but as an overall outcome of a number of attributes of a product, and the role it plays in the crop protection market. The decisive criteria are first, customer satisfaction based on the balance between reliable and consistent efficacy of the product in relation to the costs and ease of use, and second, profitability for the manufacturer. When these two conditions give rise to a sustainable business, I consider that a success from a manufacturer's point of view. Above I have presented success and failure factors for a company. I will present these factors separately for a product. Which factors determine success and failure of a product? Can we learn from the past and can we identify factors that are determinative in the path to a successful product?

7.1 Requirements for a Successful Microbial Pesticide

Microbial pesticides are often compared with synthetic chemical pesticides and in that context many authors have reported the advantages and disadvantages of microbial pesticides. The disadvantages are frequently highlighted as the reason why biopesticides only cover such a small percentage of the total crop protection market. Requirements for and constraints of microbial products have been presented by many authors. For biopesticides in general see Cross and Polonenko (1996) and Marrone (2007). For bioinsecticides, I refer to Jaronski (1986), Lisansky and Coombs (1992), Froyd (1997), Straus and Knight (1997), Butt and Copping (2000), Hokkanen and Menzler-Hokkanen (2000), Navon (2000), Van der Pas et al. (2000) and Bateman (2004). Comparable constraints have been presented for nematodes by Kaya (1986), Georgis (1992, 2004), Grewal et al. (2005) and Georgis et al. (2006). Similar limitations have been presented for biofungicides (Powell and Faull 1988; De Vrije et al. 2001; Paulitz and Bélanger 2001; Stewart 2001; Krause et al. 2006). For bioherbicides requirements are even more stringent (Cross and Polonenko 1996; Hallett 2005; Ash 2009). Powell and Faull (1988) have presented a list of scientific and commercial requirements for aspects such as control, registration, and commercialization. Many of these topics have been discussed before in the previous chapters. The list illustrates the development of such a product as a daunting exercise, and Powell and Faull (1988) even said "that it will take nothing short of a miracle to obtain a successful BCA". More than two decades later, there is still some truth to this comment. Surely, it is an enormous challenge, but a successful BCA can be developed.

The factors of success and failure of microbial insecticides have been analysed by a number of authors during a SIP symposium in 1998 in Japan (Gelernter and Evans 1999). The key issues addressed were: (1) factors (biological, economical, environmental, political, educational, and others) that have the greatest influence; (2) a framework to predict success; and (3) pathogen/pest/cropping system that hold the most promise for success. Several cases were analysed such as soil-dwelling pests (Jackson 1999), pests in vegetables (Gelernter and Trumble 1999), migratory pests (Lomer 1999), pests in forestry (Evans 1999), and pests of turf grass (Grewal 1999). The main influencing factors were considered biological and economical, wherein scientists viewed the biological as decisive, while industry and farmers saw the economical aspect as the most relevant. The answer to the second issue was that the best chance for success is when a multi-disciplinary team with enough resources develops a MPCP analogue to the method of the agrochemical industry. Third, a number of sectors were identified as promising for a successful exploitation of a microbial pesticide: forestry, urban pest control, high value horticulture and vector control. Gelernter and Lomer (2000) analysed a number of other cases of microbial insecticides. They rated success by means of five parameters: technical efficacy, practical efficacy, commercial viability, sustainability, and public benefit. They came to the conclusion that commercial viability was the most essential factor. Commercial viability was defined as the generation of sufficient profit for a manufacturer to sustain itself in biopesticides; this was also perceived as the most difficult.

In such analyses, aspects of the product, the company, and the market are often mixed up. I have separated those, and above I have made an analysis of factors influencing company performance. Regarding products, I will present the factors that influence success by means of distinguishing true product attributes, and external factors from the market and society in general. Analysing these factors forces the developers to critically consider a product's pro's and con's, and to try to exploit the positive features and improve the limiting features. Further, a forecast on success of a product depends on its attributes that are most appreciated in the market, particularly by customers. Key factors that determine the predictability of a successful product are identified below. Some examples are provided of technological breakthroughs that led to the creation of new and successful biopesticides as well as ideas for new technologies and foreseen products.

7.2 Strengths, Weaknesses, Opportunities and Threats

A list of factors that determine the success or failure comprises the attributes of a biopesticide. The chance for success is not only dependent on the product, but also on the environment in a broad sense in which it is to be used. These internal and external factors are often referred to as the SWOT (strengths, weaknesses, opportunities and threats) factors, and a SWOT analysis can be made for each product. Such an analysis identifies the internal and external factors that have a favourable and unfavourable influence on the chance to achieve success. In Table 7.11 these factors are provided for biopesticides in general. When a specific new product is being developed, it is more interesting to list these SWOT factors specifically for this particular product. It provides more insight into the product in a broader context. A disadvantage of these factors is that they are only qualitative which could lead to subjective interpretations and valuing of these factors. An attempt should be made to rank them in order of relevance. A SWOT analysis can be a useful tool, and together with the five market considerations mentioned above, which are quantitative parameters, the potential of the product can be determined in a more reliable way.

The list of weaknesses of biopesticides is long, this cannot be ignored. And chiefly for this reason the use of biopesticides is restricted to niche markets at present. On the other hand, biopesticides have unique strengths, and the opportunities to exploit these will increase. Some weaknesses can turn into strengths, for instance, a small host spectrum makes them safe to use with natural enemies. Although many factors need to be considered by a company developing these products, a small number of factors can be decisive for its success. Honesty about the product is critical; over-promising will backfire at some point. Users may decide to use a biopesticide for just a small number of reasons. Examples are a short re-entry period or pre-harvest interval. Demands from the retailer to deliver residue-free produce may be another incentive for a grower to choose for a biopesticide, regardless of some disadvantages of the product. Still, costs and efficacy remain the most important features for a grower.

Table 7.11 Strengths, weaknesses, opportunities and threats for biopesticides

STRENGTHS • Specific host range • Unique mode of action (EPNs) • Excellent tool in IPM systems • No residue (exempt of MRL) • No or short pre-harvest interval • No or short worker re-entry interval • Compatible with natural enemies, pollinators • Compatible with other microbial pest control agents • Probability of (cross-) resistance low • Excellent tool in resistance management programmes • Safe for humans and the environment • Safe for plants • Approved for organic production	WEAKNESSES • Efficacy moderate and variable • Relatively high end-user price • Narrow target spectrum • Speed of kill slow and insect stage-dependent • Short residual activity • Sensitive to abiotic factors • Limited storage stability (at low temp.) • Refrigeration requirement • More complicated application technology, spraying necessary, good coverage essential • Not very user-friendly • Incompatibility with chemical pesticides • Knowledge-intensive products that demand know-how transfer • Often only work as part of IPM programme
OPPORTUNITIES • Growing demand for residue-free or minimum residue food • Withdrawal and reduction of use of chemicals due to new regulations • Developmental costs low to reasonable • Combination with chemicals extend the product life of risk-resistant pesticides • BCAs are a natural resource and offer sustainable crop protection • Environmental benign, no environmental pollution due to production and use • Increase or maintain biodiversity since they are more target-specific than chemicals • Consumer awareness • Harmonization of registration • Organic food production • Growing of sustainable agriculture and IPM • Increasing availability of high quality biopesticides, thereby increasing users' confidence in biopesticide use in general • Increasing markets where use of chemicals is forbidden, such as forestry, amenity areas, home and garden, natural environments, etc.	THREATS • Novel safer chemical pesticides • Growers' scepticism based on expectation level of chemicals • Small market sizes • Transgenic crops • Increasing regulatory burden • Biodiversity and access benefit sharing regulations • Bioterrorism regulations • Fear of microbes by the public • Limited public research funding • Lack of user education and extension services • Weak economic position of farmers • Unregistered ineffective "snake oils" • Import of low quality products

7.3 How to Predict Success of a Microbial Pesticide?

A pathway can be developed that substantially increases the chance to achieve a successful microbial pest control product. But there is no guarantee. Each product and each company will face all the challenges from idea to launch to repeated sales. The steps along that pathway have been treated in this book, and the focus should be on developing an efficacious product that can compete on price with other crop protection means. An ideal biopesticide does not exist and the limitations must be known and accepted, and communicated in the market. How the product is marketed is pivotal to its success. But the product must demonstrate a number of key characteristics that convinces not only users, but also marketing and salespeople, including distributors and pest control advisers. Key components of a successful product are reliable and consistent field performance, a balanced cost price-effectiveness that can compete with chemicals, high and stable quality, user-friendliness, availability at all times, and sufficient knowledge on the product's activity that can be transferred to the user where necessary. When these conditions are met, customer satisfaction can be obtained and repeated sales foreseen, which will make the product successful.

7.4 Breakthroughs in Technology Lead to Successful Products

Scientists are constantly trying to elucidate more fundamental aspects of entomopathogens with regard to their mode of action, virulence factors, mass production, etc., in order to improve the efficacy and other aspects of biopesticides. Some prominent examples illustrate that a new technical discovery can lead to the development of successful biopesticides that were previously not in sight.

- The production of entomopathogenic nematodes in an artificial medium instead of in insects was discovered by Bedding (1984). This gave rise to a more cost-effective and industrial mass-production system, followed by a commercialization of nematodes for broad use. This step was later followed by the ability to produce nematodes in submerged fermentors. Scaling-up advantages reduced the cost price again and nematodes became affordable pest control agents (Georgis 2002; Ehlers 2007);
- Formulation of fungal spores of *M. anisopliae* in mineral oil demonstrated long shelf-life at ambient temperatures, an improved adherence, and an extended activity at extreme low humidity conditions (Bateman et al. 1993). This discovery opened the possibility to utilize this fungus against desert locusts under arid conditions, and in the Lubilosa project the product Green Muscle was developed for large scale use (Langewald and Kooyman 2007). Following this invention, two other oil-based mycoinsecticides were developed, Naturalis-L and Botanigard ES. However, these oil-based formulations have not resulted in a wide use of these products.

New molecular techniques enable researchers to investigate regulation of genes, stability factors, immune response factors, etc. Some new discoveries have been reported that may improve a particular aspect of a pathogen so that its potential to become a successful product increases considerably. I will present a few examples.

- The discovery of a synergist, a new protein that increased the effectiveness of Bt's (Abdullah et al. 2009). This product could widen the spectrum and the commercial use of Bt. It is said to enhance the speed of controlling the larvae, increase efficacy against older larvae, control insect species that are not susceptible to current Bt products, and remain viable and effective on the treated plants for extended periods. The product will be marketed by Insectigen, USA, as BtBooster, and has great potential according to the company;
- Prospects of genetic improvement of pathogenicity of fungal entomopathogens have been discussed by St. Leger and Screen (2001). Examples of improved strains have been documented (Lu et al. 2008; Pava-Ripoll et al. 2008) and these strains have been offered to biopesticide companies in Europe for development into products. These improvements in terms of virulence lead to lower LD50 values and possibly to more effective and cheaper products. Genetic improvement is a promising road for all microbial control agents. For now, this is not an option in the EU, but it could be developed for other regions, or for later in the EU;
- The use of Bt in transgenic crops has become a multi-billion Euros business which indirectly emerged from the insect pathologists' research and knowledge (Federici 2007). Although still controversial, this development reduces the use of conventional pesticides and offers enormous potential for a more sustainable agriculture. Other entomopathogens may be used in a similar way.

In some areas, the industry is awaiting a breakthrough due to some serious constraints, for example, with the labour-intensive mass production of baculoviruses.

- In vitro production of baculoviruses has been investigated for a long time. At present this technology is not yet commercially feasible for these viruses (see Chapter 3) and in vivo production is still the only method. A breakthrough in this area would increase the potential for viruses enormously;
- Genetic modification of baculoviruses could improve the speed of kill and lower the required dose. This development needs to go hand in hand with in vitro production so that production remains economically feasible. In vivo production of modified viruses would give low yields (Inceoglu et al. 2001; Szewczyk et al. 2006). On the other hand, acceptance and regulation of a genetically modified organism is at present a huge hurdle (Vlak 2009).

Broader use of existing products could provide companies with a larger market and improved return on investment. Examples are entomopathogenic fungi to control plant diseases and nematodes as well (Goettel et al. 2008), the use of entomopathogenic fungi as endophytes against insects (Vega 2008) and plant diseases (Ownley et al. 2008), and the use of microbial disease control agents that also demonstrate insecticidal activity (Keel and Maurhofer 2009).

New discoveries could widen the potential of entomopathogens considerably. Improved products that are active in a more cost-effective way will satisfy customers and will be more competitive with chemicals. Those products will improve the economic situation for biocontrol companies. More basic and applied research is needed to achieve this.

8 Developmental Costs and Time of a Microbial Pest Control Product

8.1 Total Developmental Costs

The most wanted information for decision-makers in companies who contemplate the development of a new microbial pest control products is what are the total costs? There is no simple straightforward answer to this question, but an indication must be provided for a sound business decision. Each case is unique, but it is possible to provide a general figure for a certain type of entomopathogen and its foreseen purpose. For each company, the investment will be different since expertise, facilities and equipment, R & D, and presence in the market will be unique too. Many authors have attempted to come up with an indication of the total costs for various kinds of microbial pest control products. The difficulty with many of these estimates is the failure to clearly describe which expenses are included in the amount reported. Are registration expenses counted, and if so, to what extent are in-house expenses, contract research expenses, and registration costs in various countries included? I have provided more detailed estimates for registration costs in Chapter 5. The estimates in the literature range from developmental costs, usually R & D and registration costs, up to total investment costs for setting up a biopesticide company. A general estimate has been made by Lisansky (CPL 2006c): "A company that wants to go into biopesticides needs to be able to allocate $30 million". He analysed that on average, R & D costs will be around $6 million, not including registration and marketing costs. An economic analysis for the commercial production of Btk has been presented by Rowe and Margaritis (2004). This is one of the few studies that investigated in depth the economic feasibility of manufacturing Bt based insecticides by a state-of-the-art bioprocess design on a large scale. The investment needed to produce 8–15% of the world demand for Bt's was approximately CA$18 million for a stand-alone plant, including equipment costs of about CA$12 million. They calculated the rate of return on the total capital investment versus co-varying the product selling price and the annual production scale for two different fermentation processes. The outcome strongly relies on the correct market and selling price assumptions. This approach could be used for obtaining capital and operating cost estimates for a preliminary budget approval as in a business plan. Figures were reported concerning the Lubilosa project: "Green Muscle was commercially available after 12 years of research involving at least 40 scientists and costing £15 million" (Moore 2008). Calculations for a small biopesticide plant in developing countries such as Nicaragua (Grimm 2001) and Madagascar (Swanson 1997) demonstrated that an investment of less than one million Euros could be sufficient.

The above-mentioned amounts include the establishment of a new company, and to make an estimate of those costs is very difficult. Amounts for product development are considerably lower. Woodhead et al. (1990) estimated the costs for a microbial biopesticide to be under $5 million, but it is not clear whether the registration costs of $2.5–3.5 million are included in this figure. Törmälä (1995) assumed that "overall development cost may be anything between $1–10 million" based on R & D budgets of some biocontrol companies. According to Cross and Polonenko (1996) total development costs of a bioherbicide and registration in Canada are in the range of $5–8 million. Marrone (1999) mentioned costs to be between $2–4 million (in the USA) in 2–4 years of development. The developmental costs of a Bt insecticide are estimated to be $3–5 million (Navon 2000). Jarvis (2001) estimated development costs for a microbial around $3 million, and when registration costs of at least $0.5 million are added, total costs are about $4 million. More recent figures have been provided by a few authors involved in commercial activities. Krause et al. (2006) estimated the registration and commercialization of a new biopesticide to cost €6–8 million over the course of 6–8 years. The development and registration of a baculovirus product will need an investment of several million Euros according to Kessler et al. (2008) and it will take 4–8 years before any revenues will return to the company. Furthermore, the restricted host range allows for small markets and this makes it "extremely difficult" and "a financial risk and a tightrope walk" for small and medium size companies to become successful (Kessler et al. 2008). Lucarotti et al. (2007) estimated that the development from isolation to registration has cost around CA $6 million for the balsam fir sawfly NPV product Abietiv. The costs of developing a new microbial control agent typically exceed $25 million according to Gelernter (2007). A current estimate from Agraquest claims that developing and registering a biopesticide globally costs about $15–20 million (CPM 2009b). In a company survey in the REBECA project among 52 biocontrol companies, registration expenses amounted to an average of almost €2 million. R & D expenditure has been €15 million in one case (Hokkanen 2007).

Figures provided in the literature and consultant reports range from $2 million to $30 million. Clearly, this is useless as an estimate for those considering the initiation of the development of a new microbial control agent. If an established biocontrol company with experience and available production facilities, researchers, and marketing and sales people, starts with a new product, costs will be substantially lower than when a new company needs to be founded, built, equipped and staffed. In the first case I estimate that full developmental costs range between €2 and €5 million, in the second case an investment of €10–15 million will be needed until sales reach the break-even point. In both situations total research and registration costs could be considerably lower, in the order of a few million Euros, when much information has been generated by public research. I will discuss the role of public research below. In case a new company starts with the labour-intensive production system versus the capital-intensive model (see Chapter 3), for example with in-vitro production of baculoviruses or production of fungi in bags, starting up costs can be reduced by several millions. By comparison, developmental costs for a chemical pesticide range between $180 and $200 million for an established R & D agrochemical multinational and it takes about 9–10 years before it can be launched (see e.g. www.agro.basf.com; www.ecpa.be).

Obviously, each situation is different depending on the company, the pathogen, the target pest(s) and the sector(s) and countries in which the product will be

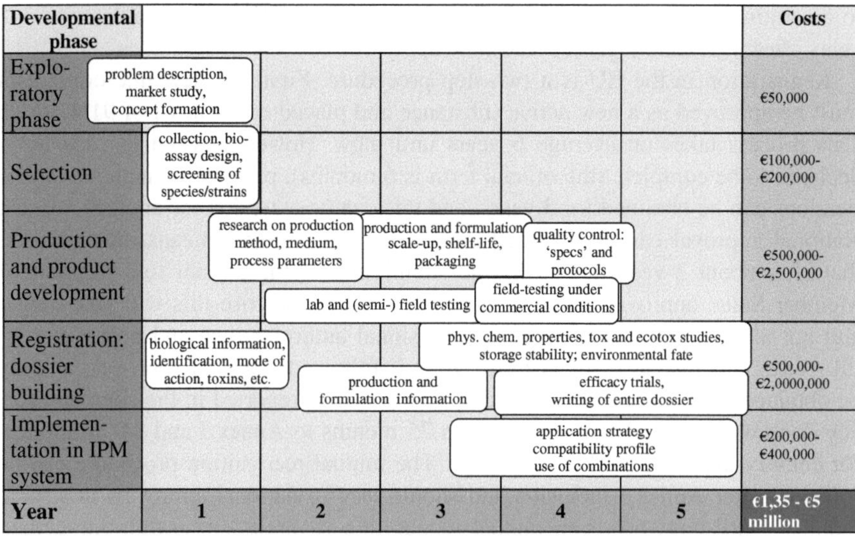

Fig. 7.2 Developmental scheme of a microbial pest control product

registered and marketed. Above-provided estimates are only general. A better way to estimate the costs could be by dividing the developmental process into phases and calculate the costs per phase. An example for developmental time and costs without registration fees, is provided in Fig. 7.2.

8.2 Time to Market

The development time and the time it takes to obtain registration of the plant protection product in one or more countries determines when selling can begin and a company starts to receive income from the product. This period must be overcome by income from other company activities or by external resources. The length of this period influences the investment through interest, and the longer the period, the riskier the project. Changes in the market over this period such as a changed pest problem, new competitive products, resistant cultivars, etc., also influence the potential of the product. Therefore, it is pivotal to make the most accurate estimate of this period. The development period can be estimated at best by the researchers and depends on the available information on the new strain, the experience of the developers, and the size of the team. Few authors report on the development time for a product. The development of Green Muscle took about 12 years which is very long (Moore 2008). The development of Invade (*Serratia entomophila*) took 4 years (Jackson et al. 1992). Marrone (1999) mentioned 2–4 years. I believe that, on average, a product can be developed in 2–4 years. The generation of the registration dossier should start as early as possible, and will overlap the developmental period to a certain extent. This will take about 2 years. In total, this accumulates to 3–5 years after which the registration can be applied for.

Registration in the EU is a two-step procedure. First, the new microorganism must be approved as a new active substance and placed on Annex I of 91/414/EC. This process takes on average 6 years until now. However, when the dossier is declared to be complete (the official term is 6 months), provisional national authorizations can be obtained for 3 years, and this can be extended for another 3 years. National approval can be obtained after about 1 year. This means, theoretically, that after about 2 years following the submission of the dossier to a Rapporteur Member State, approval to sell could be granted. In practise this will take longer and not all Member States approve provisional authorizations and prefer to wait till Annex I inclusion. If the last procedure is followed, national approval can only be obtained after more than 7–8 years. This was also reported in the company survey done within REBECA: greater than 75 months to Annex I and 24–36 months for country approval (Hokkanen 2007). The mutual recognition procedure can be followed after Annex I inclusion and should lead to national approvals in several months. Until now, however, mutual recognition is rarely successful. Altogether, the time to market can be extremely long and this period is usually underestimated. Surely, the management of a company likes to see products being developed as quickly as possible and optimistic estimates have a better chance for a positive decision than long-lasting, riskier scenarios: a case of wishful thinking. In the USA, the registration period is much shorter (Hauschild and Speiser 2007; Kiewnick 2007). Therefore, time to market is quicker, it is estimated to be 2–4 years (Marrone 1999). A realistic picture of the time from idea to market in the EU is between 5 and 7 years when provisional registrations can be obtained. Registrations in a large part of the EU after Annex inclusion will take some years more.

The long period to market in Europe is very difficult for companies to bridge and it deters many companies from developing biopesticides or from registration. It is surely a reason why so few products are available in the European market compared to the American market. The period from idea to market introduction is a key factor in business success. The shorter the period, the greater is the chance to reach the market ahead of competitors and the earlier a break-even point can be reached and profit can be generated.

8.3 The Role of Public and Industrial Research in the Development of Microbial Pesticides

Generally, the initial research on a microbial control agent is performed by public research. This could comprise the discovery and screening of strains, the identification, the unravelling of the mode of action, the host range spectrum, even basic research on production and formulation, initial application and efficacy testing. In the process of the commercial development of a new biocontrol agent this research is extremely important. Very few companies are able to perform this early work themselves because of a lack of knowledge and resources. On the other hand, the focus of a public research institute is not the same as that of a commercial enterprise. Often, particularly in the past, this has resulted in uncoordinated research efforts that did not favour the development and use of biopesticides (Dent 1997). An example is the development of a baculovirus for control of *S. exigua* in the Netherlands while no company could be found to commercialize it. Only many years later when there was a market demand the product was registered and marketed (Smits and Vlak 1994).

Some researchers have expressed self-criticism with regard to their attitude towards research on biocontrol agents and are calling upon colleagues to maximize efforts to attain more commercial successes (Fravel 1999; Stewart 2001). Collaboration in an early phase when commercialization is foreseen is highly desired. In this way, research goals can be compared and adjusted in order to prevent omissions and overlap in research. Several authors have advocated partnership between public and private organizations (Carlton 1990; Riba et al. 1996; Waage 1997; Butt 2000; Ravensberg and Elad 2001; Whipps and Lumsden 2001; Gelernter 2005, 2007; Fravel 2005; Krause et al. 2006; Cherry and Gwynn 2007; Ash 2009; Peters 2009). Fortunately, we see this kind of collaboration more and more and governmental funding increasingly requires involvement of industry and relevance for society. The value of public research results can be millions of Euros and will help biocontrol companies to develop products successfully. In some cases, part of the revenues has to be paid back, and most companies agree to this once revenues allow for it.

In my experience, it is not always easy to establish collaborative projects since the goals of scientists (often demanded by funding agencies) and industry researchers can diverge. The first are mainly oriented towards the organism and new knowledge that can be published. The second focus on market- and product-driven research with a strong emphasis on cost-effectiveness, registration, market size, and the investment needed. A balance must be found between the scientific research topics and the industry requirements. Disclosure and publication of data can be a delicate issue and clear agreements need to be made in advance. The same is true for exploitation of project results, patents and licenses. Nevertheless, I believe that collaboration between scientists and industry is crucial in order to develop these products, and contacts between these two groups are necessary from the outset of the development of a microbial. Such collaboration is very valuable for a biocontrol company, and fruitful relationships with public institutes are indispensable for the industry. On the other hand, a company must take it own decisions and stick to its own agenda, even when this is not always appreciated by scientists.

The discovery of new strains is a sensitive issue that has come up frequently. Researchers often prefer to work with their own, locally found strains, and will present them as better adapted, and therefore providing superior control under the local circumstances. Is a new strain significantly better so that it justifies developing a new product? This may or may not be true. It is, however, impossible to develop multiple products based on local strains for different regions, given the time and costs for such a development. Therefore, company management has to make choices and work with one strain that is best suited for many circumstances. This is an unavoidable compromise. Products can only be developed for relatively large markets.

Interactions between public and private scientists are numerous as can be seen within the IOBC, the SIP and at many conferences. Many biocontrol companies have relationships and cooperative projects with research institutes and universities. I assume that for the large majority of biocontrol products, public research has been the basis of the development and that this will continue in the near future. As a result of university research, new biocontrol companies have been established as spin-off companies, and close ties between the two organizations remain in place. Recent examples are Bio-ferm, Austria, and Bionext, Belgium.

On the other hand, research organizations tend to patent new inventions with the goal of licensing technology to companies. However, this often had the opposite effect of deterring small biocontrol companies from developing the products because of royalty payments. This trend is mainly seen in the USA and needs readjusting (Gelernter 2007).

In the current market, biopesticide companies are still relatively small, and the assumption that they can afford to develop new products on their own is unrealistic. The collaboration with public research is essential if our society wishes to have continued progress in this field (Gelernter 2007).

9 Distribution and Sale Strategies

9.1 Various Models for Distribution

There are several options for selling biopesticides. The choices that can be made depend on the size and nature of the manufacturer. A large agrochemical company may use its own outlets for distribution. A small biopesticide company generally does not have its own distribution organization, and needs to assign other companies as their distributors. The same is true for manufacturers outside agriculture, such as pharmaceutical companies. Few biopesticide companies have developed their own distribution organization. Examples are Koppert and Intrachem, but both have more activities next to microbial pesticides that justify such an organization. The advantage is that subsidiaries are committed to their own products while others may be less so. Microbial pest control products require more attention and user education than chemical products do, and this demands a versatile and flexible distributor. Big agrochemical distributors may have a better understanding of the markets in general, but they lack the ability to deal with niche products. The use of large agrochemical companies as distributors has not resulted in successful sales of biopesticides according to Lysanski (CPL, 2006c). Collaboration with small local companies has been more successful, although sales have rarely become large. Companies with their own infrastructure have done the best (CPL, 2006a). Pricing and margins play an important role too in this choice. The extent of the target market and crops is also a determinative factor. Different distribution strategies are required for greenhouse crops versus top fruit or versus agricultural crops, due to the size of the markets and the nature of crop protection in those sectors. Greenhouse crops face an almost year-round pest and disease problem that needs frequent interventions and visits of IPM consultants. Direct contact with the growers has been the cornerstone of the success of biocontrol with natural enemies in protected crops (Bolckmans 1999), and the same can be said for microbials.

9.2 The Role of a Distributor

The distribution of crop protection products may be set up very differently per country and/or per sector. The role of the distributor can be limited to the ordering, invoicing and shipping on the one hand; to the storage, shipping, and provision of intensive support on an IPM system and cultivation techniques to the grower on the other hand. In the case of biopesticides, proper storage and transport are essential elements due to the vulnerability of the products, particularly at high temperatures. For biopesticides in greenhouses, a distribution system that has been established for natural enemies and pollinators (Ravensberg 1992; Bolckmans 1999) can be used. The advantage is that these distributors are accustomed to handling sensitive products and they call often directly on growers to provide technical advice.

The proper use of a biopesticide needs a careful implementation in an IPM programme and the knowledge needed for this needs to be disseminated to distributors too, this is described in detail in Chapter 6. Distributors must be aware of this crucial function, and they must be trained and updated on a regular basis. Distributors play a vital role in increasing the grower's awareness of the product and how to use it. If a distributor is not willing to play this role, then a producer should considered not to choose him for the sale of a biopesticide. Further to this role, storage facilities and keeping track of expiry dates of products are also a task for distributors. Finding distributors that are able and willing to handle microbial products can be a challenge, but it is an imperative factor in the chain from producer to end-user (Benuzzi 2004). In the Netherlands, distributors are organized in the branch association Agrodis and the primary goals are maintaining the intermediary function between producer and end-user, improving the flow and transfer of knowledge, strengthening the role of the adviser, stimulating the use of IPM, and increasing the commitment of advisers (www.agrodis.nl). Distributors with this kind of commitment are needed in the field of biopesticides, and since distributors that call on growers are often considered key-influencers of growers' decisions, they must be closely involved.

Few authors have discussed the role of distribution in the success of biopesticides. But this aspect of the use of biopesticides should not be underestimated. The perception of biopesticides can be a barrier to the adoption of such products, and the role of distributors and pest control advisors is pivotal for a successful biopesticide (Marrone 2007). A new product is typically tried on a small scale on the grower's premises under supervision of the distributor's advisor, and the relationship between both determines whether a grower is willing to try this or not. Marrone (2007) recommended that biocontrol companies invest in the education of customers and distribution channel partners to improve adoption of biopesticides. Another important factor is the feedback from the distributor to the manufacturer on the customer's satisfaction or complaints. Complaints need to be handled with care and with the correct follow-up. New uses may also be developed by distributors. In some cases, manufacturers ask local distributors to register their product in countries not well known to the product owner. Such relationships between producer and distributor are seen in international collaborations, and a much closer business partnership is required in such situations.

9.3 The Distribution Channel

Distributors generally sell both biopesticides and chemical pesticides. The latter are easier in terms of storage, shelf-life, ease of use, and advice. Moreover, market sizes are larger. All this makes selling pesticides a profitable business for distributors. To interest distributors to sell more complicated products such as biopesticides is not easy. Novel products are interesting for distributors to demonstrate that they are well-informed and active in finding new and better solutions for their customers. Nevertheless, profit margins should be high enough to persuade distributors to start selling biopesticides. Biopesticide manufacturers have to calculate this in when they make market projections based on end-user prices. The distribution channel should be as short as possible.

Ideally, a producer markets his product directly to the end-user for two reasons: it keeps the costs down, and knowledge transfer is direct and efficient. Every extra link in the distribution channel increases the costs due to margins and reduces the quality of knowledge transfer. To have just one partner between producer and user is optimal because of these two factors, and because of practical reasons. In the market analysis of Cross and Polonenko (1996) for a microbial herbicide they assumed that biopesticides are usually sold as agrochemical products and follow a four-step distribution path; manufacturer→wholesaler→distributor→retailer→end-user. Further, they assumed that every distribution step required a 100% marking up because of a profit margin. This renders the end-user price much too high. Such a distribution channel should be avoided for biopesticides. In general, biopesticides are sold in a short channel from manufacturer→distributor→end-user, or with an extra partner between manufacturer and distributor for international sales. This keeps the channel short, efficient and affordable. When an extra link is used in the distribution channel, their task is different from the distributor who calls on customers. This last role requires more knowledge and work and therefore a higher margin than a wholesaler. The situation, however, may be different in various countries due to established ways of working, and adaptation to the local distribution system should be considered.

Direct sales by a manufacturer are generally limited since only a limited number of growers can be reached. Often, a biopesticide company is small and new, and a sales team is not available. Developing a large sales team is extremely costly. A distribution network using established local distributors that have a strong relationship with their customers is therefore the most preferred and used option for selling biopesticides. And it is also the most likely and practical way to start sales quickly after registration has been obtained. Actually, the distribution channel should be involved long before the product can be sold; proper and timely introduction of the product to the distributors increases the chances of a successful product launch in the market. Identification of effective distributors is the key to success; they should not be considered as just a link in the process from manufacturer and end-user. They are essential in the promotion of the product and should be regarded as real partners in a collaborative project.

10 Conclusions and Recommendations

10.1 Successful Biopesticide Companies and Products

In more than 30 years of commercialization of biopesticides, many companies have been active in this field and many products have been developed and sold in the crop protection market. The historical picture presents outstanding failures as well as ongoing success stories. Particularly in the 1990s in the USA, large venture capital biopesticide companies failed to develop a sustainable business. Most of the world's largest agrochemical companies had a brief encounter with biocontrol, but they all abandoned the field. Small and committed entrepreneurs have been able to continue, albeit with difficulties. Investment has been high and risky, time to registration too long, and adoption in the market too slow. Nevertheless, a few dozen companies seem to be successful, and new ones are regularly founded. Several hundred biopesticides have been developed, some very successfully, but the majority are small products, however, and many never saw the light of day. Many ideas are still on the shelves of scientists and are not being developed further by companies. The development and commercialization of biopesticides is ongoing, maybe now more then ever, and apparently it is still an attractive business field. On the other hand, the development of a biopesticide is still perceived as a risky adventure illustrated by titles of papers such as: Developing new baculovirus products or "How to walk a tightrope" (Kessler et al. 2008), and "The long and winding road- discovery to commercial product: are we there yet?" (Brownbridge 2008). In this chapter I have identified the commercial factors that are critical in the route to success for a company and a product. There is no guaranteed route to success, but lessons from the past should be taken seriously, and recommendations for do's and don'ts are provided below that will stimulate and help new projects, products and companies to be successful.

10.2 Five Determinants for Success or Failure

To develop a successful product and to become a profitable company in the area of biopesticides is an enormous challenge. The many critical aspects described in this chapter provide a matrix of product attributes and market features that need to be profoundly understood and weaved together. The complicated commercialization route demands biological and technical expertise, understanding of the market, and entrepreneurial intelligence, next to commitment and perseverance. On the other hand, if the prospects for a product are not convincing, management must be so courageous to abandon the project in time, to prevent serious loss of money and time. This may be the most difficult decision since everyone involved is focussed on developing a new product. Therefore, it is elementary to set up a stage-gate system, and to make decisions transparent based on quantifiable factors. Resources can only be used once, and it is essential for a company to spend them on the right project. The lack of resources or ill-judged use of them has caused many companies to fail (Lisansky 1997; Gelernter 2005). Decisive factors that determine success or failure are:

- the expenses and time for development and registration;
- the quality of the product in terms of efficacy and price, measured in customer's satisfaction;
- the size of the market(s), and a realistic estimate of the sales volume;
- the margin between full cost price and the end-user price that provides a sustainable and profitable business and resources to grow and develop new products;
- a company's strategy regarding expansion and taking financial risks.

These five factors are briefly discussed below and recommendations are deduced which, in my opinion, when followed, will increase successful commercialization of a biopesticide.

10.2.1 Developmental Expenses and Time

The estimation of the developmental and registration costs and time to market must be as accurate as possible for a realistic business plan. Due to the fact that each biopesticide is unique, and its potential markets different, only a rough estimate can be provided. The accuracy also depends very much on the company's experience in biopesticide development and registration procedures. Estimates are provided above. Initially, the uncertainties will be many, but as time progresses and investments increase, the uncertainty level should drop. If this is not the case, the decision to continue should be re-evaluated seriously since the financial risks may become dangerous for a company. The total expenses are high for a small company and often a barrier. Collaboration with public research is essential and should be set up from the outset of a project. The idea for a new product often originates from a research organization and co-development with the help of public resources will reduce costs and time, and increases the chance for success considerably.

10.2.2 Product Quality

Biopesticides face a number of limiting factors which need to be considered during the commercialization. These are biological (mode of action, dependence on environmental conditions), technical (production, formulation, shelf-life), and economical and market-related (competition with chemicals, efficacy, costs, sales potential) (Van der Pas et al. 2000). Some of these are hard to overcome and are inherent to entomopathogenic organisms. The ideal biopesticide does not exist, and we have to accept that biopesticides do have advantages and disadvantages. Both need attention in the developmental phase. The advantages should be exploited where possible; the disadvantages should be recognized early on and improved where possible. Neglecting those can cause disappointments when it is too late.

A product should fulfil the following conditions: reliable and consistent field performance, a balanced cost price-effectiveness that can compete with chemicals, high and stable quality, user-friendliness, and availability at all times. Products are successful when customers are satisfied with their performance and use them repeatedly. Biopesticides are price-sensitive, but more and more, other attributes are appreciated in the market. Good products should offer economic benefit and value for the grower. So-called weaknesses of biopesticides (Table 7.11) can also be advantages, particularly in IPM systems and in food crops. In positioning the product, these "weak" attributes can possibly be exploited in the market. A profound understanding of the strong and weak aspects of the product will allow an optimal market positioning. But a good product is not necessarily a successful product. It is the demanding task of a company to promote and sell the product in substantial amounts.

10.2.3 Market Development

The success of a product is decided in the market, and only there. Many companies underestimate this part of the business and have not succeeded because of that. Lisansky (1985) stated that "selling the products is the most difficult task of all", and this is probably still true. Most biopesticide companies are technology-based and product-driven and have sufficient skills to develop good quality products. Marketing skills are less often available and are sought outside the company, with distributors. But this is often only seriously undertaken once the product is developed. This is generally too late, and as it takes years to reach sales volumes, this could bring the company in trouble. As rightly said by Moore regarding the slow uptake of Green Muscle, "The lesson that inventing the better mouse-trap is not enough has been understood" (Moore 2008). Similarly, Cross and Polonenko (1996) concluded, after having developed a highly efficacious bioherbicide that "the possession of great technology doesn't necessarily guarantee commercial success". The adoption of a new product in the market must be taken seriously in an early phase of the project. Market research should be carried out prior to the product development stage, and the outcome could lead to a "no go"-decision of a project. Market research should be continued since the market is always in motion. When necessary, the original plan must be adapted and decisions re-evaluated.

Early involvement of distributors is indispensible. The distribution channel is a key aspect in successful commercialization. Marketing and distribution services are indispensible to reach the end-user. Understanding the product, and distributor and grower training is decisive in the promotion and selling of the product. A comprehensive and excellent analysis of why the commercialization of a bioherbicide failed in Canada is provided by Cross and Polonenko (1996). They concluded that in particular their unrealistic estimate of the market size, and of the market penetration percentage, was the reason for failure. This analysis exemplifies how projects are often run, and I have frequently seen this approach taken, unfortunately. A solid market analysis is the foundation of a successful market introduction. Effective marketing requires that a considerable part of the resources be allocated for the product's field development and this aspect should not be underestimated. Market research should continue after product introduction in order to identify new opportunities that could maximize revenues from the product.

10.2.4 Profit Margin Is a Must

In most cases biopesticides need to compete with chemicals and generally they are more expensive. Market penetration and sales volume is imperative for return on investment and for a profitable business. The market price needs to allow a profitable margin for the manufacturer and distributor. Lowering the market price may increase sales, but jeopardizes the profit margin. A balance needs to be found, and other attributes of the product need to be emphasized in the market to promote the use of the product. This requires an inventive marketing strategy and the involvement of all stakeholders without increasing the marketing and sales expenses too much. Profit is an absolute necessity for a company and determinative of the success or failure ultimately. Therefore, profit calculation is vital early in the project and must be strongly weighed in "go/no go"-decisions.

10.2.5 Successful Business Model

All kinds of companies from very small to large multinational companies have been active in biopesticides. Those that have developed a sustainable business are mostly small enterprises, specialized, and committed to a certain niche market, particularly when they first started. Examples are companies who only produce nematodes, or baculoviruses; or just for a certain niche market like greenhouse vegetables, or outdoor ornamentals. Apparently, *it is not just what and where a company does it, but mostly how a company does it.* It is essential to understand the route from the first step in the project up to the marketing and sales of the product. The prerequisite is a full commitment to the whole route and the ascertainment of resources. Companies active with microbial control agents, where registration is required, as well as companies active with entomopathogenic nematodes have demonstrated that this approach performs well. A small company who knows its products and markets in depth, who steadily develops the company in terms of expansion into new markets and new products, and who carefully judges the risks it takes, is the successful business model today.

On the other hand, it is remarkable that three large Japanese companies are successfully active in the biopesticide field through their affiliates. These are Arysta LifeScience, Mitsui & Co, and Sumitomo Chemical, through respectively Natural Plant Protection (NPP) in France, Certis in USA and Europe, and Valent BioSciences (VBS) in the USA. These parent companies produce and market chemical pesticides and they position pesticides in IPM programmes next to their biological products. There seems to be a difference in strategy; where Certis is active in both biopesticides and chemicals, NPP and VBS are solely active in biopesticides. The success of these Japanese companies in biopesticides may be explained by their long-term vision and commitment, that is a general characteristic of Japanese business. This illustrates again that success in biopesticides depends on "how a company does it" and that long-term commitment is determinative. Businesses focussed on short-term money-making did not and will not succeed in biopesticides to date.

10.3 Recommendations for Successful Commercialization of a Microbial Pesticide

Each biopesticide development and commercialization route is different and each company has its own strategy and objectives. A general pathway to a successful product can be provided, but it needs to be tailored in each case. New product development is a difficult process in any industry and one of the weakest facets is effective project selection and resource allocation (Cooper and Edgett 2006). Learning from the mistakes made in the past by others as well as inside one's own company is essential. I have deduced ten operational recommendations from the biopesticide industry and from other industries which face similar challenges (Table 7.12). Following these general recommendations will, in my opinion, increase the chances of developing and selling a biopesticide in a profitable and sustainable way for a biopesticide company.

10.4 Requirements for a Company to be Successful

For a company to be successful, it is necessary to have a vision, to have the required knowledge, skills and resources, and to develop an appropriate business plan. If any of these requirements is insufficient or lacking, the project may end in confusion, anxiety or frustration and eventually failure. A company must develop a strategy, and follow it. Making choices at the right time is a crucial management attribute. When a company wants to develop new products and grow, it must have a critical

Table 7.12 Ten recommendations for the commercialization process of a biopesticide

- Ensure that the project fits the company's strategy and capabilities
- Draw up a proper business plan with accurate information and market data
- Be conservative in the estimation of the market potential
- Set clear objectives, and allocate resources
- Obtain genuine commitment from all decision-makers in the company
- Install a systematic idea-to-launch-to-sale-volume process, and make the gates work
- Develop a rating system for quantifiable and objective decisions
- Define success criteria and "go/no go"-decision moments up front
- Involve all functional fields in the decision team: R & D, M & S, finance, and corporate management
- Continuously evaluate progress, expenses and expected revenues

look inside, and consider whether the current organization is fit for its future outlook. It is vital to have the internal organization structures and procedures in place when growth is the objective. If only growth is pursued, this can easily frustrate an organization and give rise to large difficulties. Growth and internationalization must be encouraged in predictable markets that can be developed without taking too high risks. Continuous improvements on all terrains are necessary to stay competitive. Costs must be vigilantly observed, and progress strictly monitored, new markets developed as speedily as possible to increase income. Profitability is required, so new projects can be started and new products developed and registered. Only profitability can ensure return on investment, new investments and new projects for the future, and that a company can remain competitive, innovative and sustainable. All this is a great challenge for all those involved, and at the same time also highly motivating and fascinating. This should drive entrepreneurs to continue in the field of biopesticides and to make it a successful and sustainable business.

References

Abdullah, M.A.F., S. Moussa, M.D. Taylor, and M.J. Adang, 2009. *Manduca sexta* (Lepidoptera: Sphingidae) cadherin fragments function as synergists for Cry1A and Cry1C *Bacillus thuringiensis* toxins against noctuid moths *Helicoverpa zea, Agrotis ipsilon* and *Spodoptera exigua. Pest Manag. Sci.* 65(10): 1097–1103.

Alves, S.B., R.M. Pereira, R.B. Lopes, and M.A. Tamai, 2003. Use of entomopathogenic fungi in Latin America. In: R.K. Upadhyay (ed), *Advances in microbial control of insect pests.* Kluwer/Plenum, New York, NY. pp. 193–211.

Arif, B.M., 2005. A brief journey with insect viruses with emphasis on baculoviruses. *J. Invertebr. Pathol.* 89(1): 39–45.

Ash, G.J., 2009. The science, art and business of successful bioherbicides. *Biol. Control* 52(3): 230–240.

Avé, D., 2008. *Bacillus thuringiensis* – based products: forever young. *Paper presented at the 41th annual meeting of the society for invertebrate pathology*, August 3–7, 2008. University of Warwick, Coventry.

Bailey, K.L., S.M. Boyetchko, and T. Längle, 2009. Social and economic drivers shaping the future of biological control: a Canadian perspective on the factors affecting the development and use of microbial biopesticides. *Biol. Control* 52(3): 221–229.

Bateman, R., 2004. Constraints and enabling technologies for mycopesticide development. *Outlooks Pest Manag.* 15(2): 64–69.

Bateman, R.P., M. Carey, D. Moore, and C. Prior, 1993. The enhanced infectivity of *Metarhizium flavoviride* in oil formulations to desert locusts at low humidity's. *Ann. Appl. Biol.* 122: 145–152.

BCC Research, 2006. *The new biopesticide market.* Report CHM029B. Wellesley, MA.

Bedding, R.A., 1984. Large scale production, storage and transport of the insect-parasitic nematodes *Neoaplectana* spp. and *Heterorhabditis* spp. *Ann. Appl. Biol.* 104: 117–120.

Beegle, C.C. and T. Yamamoto, 1992. History of *Bacillus thuringiensis* berliner research and development. *Can. Entomol.* 124(4): 587–616.

Benuzzi, M., 2004. What will be the future for BCAs? The industry's point of view on problems in developing BCAs. *IOBC/WPRS Bull.* 27(8): 429–431.

Besse, S., 2008. Preventive and curative efficacy of Ostrinil against the palm borer *Paysandisia archon* (Burmeister, 1880). *Paper presented at the 3rd annual biocontrol industry meeting (ABIM) Lucerne*, Switzerland, October 20–21, 2008. http://www.abim-lucerne.ch/archive/conference2008. Cited May 23, 2009.

Blum, B., 2001. Entomopathogenic nematodes (EPNs), a market survey. In: J.M. Grunder (ed), *Quality control of entomopathogenic nematodes. COST Action 819.* Agroscope FAW/Berti Druck AG, Wädenswil, Rapperswil. pp. 78–84.

Bolckmans, K., 1999. Commercial aspects of biological pest control in greenhouses. In: R. Albajes, M.L. Gullino, J.C. van Lenteren, Y. Elad (eds), *Integrated pest and disease management in greenhouse crops.* Kluwer, Dordrecht. pp. 310–318.

Braverman, M., 2007. Biopesticide market and opportunities in North America. *Paper presented at the 2nd annual biocontrol industry meeting (ABIM) Lucerne*, Switzerland, October 22–23, 2007. http://www.abim-lucerne.ch/archive/conference2007. Cited May 21, 2009.

Brownbridge, M., 2008. The long and winding road – discovery to commercial product: are we there yet? *Paper presented at the 41th annual meeting of the society for invertebrate pathology*, August 3–7, 2008, University of Warwick, Coventry

Burges, H.D., 1981a. *Microbial control of pests and plant diseases 1970–1980.* Academic Press, London.

Burges, H.D., 1981b. Strategy for the microbial control of pests in 1980 and beyond. In: H.D. Burges (ed), *Microbial control of pests and plant diseases 1970–1980.* Academic Press, London. pp. 797–836.

Burges, H.D., June, 2001. *Bacillus thuringiensis* in pest control. *Pesticide Outlook* 12: 90–97.

Burges, H.D. and N.W. Hussey (eds), 1971. *Microbial control of insects and mites.* Academic Press, London.

Butt, T.M., 2000. Fungal biological control agents. *Pestic Outlook*-October 2000: 186–191.

Butt, T.M., 2002. Use of entomogenous fungi for the control of insect pests. In: F. Kempken (ed), *The Mycota XI, Agricultural applications.* Springer, Berlin. pp. 111–134.

Butt, T.M. and L.G. Copping, October, 2000. Fungal biological control agents. *Pesticide Outlook* 11: 186–191.

Butt, T.M., J.G. Harris, and K.A. Powell, 1999. Microbial insecticides. The European scene. In: F.R. Hall, J.J. Menn (eds), *Biopesticides: use and delivery.* Humana Press, Totowa, NJ. pp. 23–44.

Butt, T.M., C. Jackson, and N. Magan, 2001. *Fungi as biocontrol agents: progress, problems and potential.* CAB International, Wallingford.

Carlton, B.C., 1990. Economic considerations in marketing and application of biocontrol agents. In: R.R. Baker, P.E. Dunn (eds), *New directions in biological control: alternatives for suppressing agricultural pests and diseases.* A.R. Liss, New York, NY. pp. 419–434.

Carlton, B.C., 1996. Development and commercialization of new and improved biopesticides. *Ann. N. Y. Acad. Sci.* 792: 154–163.

Cherry, A.J. and R.L. Gwynn, 2007. Perspectives on the development of biological control agents in Africa. *Biocontl. Sci. Technol.* 17(7): 665–676.

Cooper, R.G., 2007. Managing technology development projects. *IEEE Eng. Manag. Rev.* 35(1): 67–76.
Cooper, R.G. and S. Edgett, 2006. Ten ways to make a better portfolio and project selection decisions. *PDMA Visions Magazine*, June 2006. www.stage-gate.com. Cited on August 3, 2009.
Cooper, R.G., S.J. Edgett, and E.J. Kleinschmidt, 2002. *Portfolio management for new products*, 2nd edition. Perseus Books, Reading, MA.
Copping, L.G., 1998. *The biopesticide manual*. BCPC Publications, Farnham.
Copping, L.G., 2001. *The biopesticide manual*, 2nd edition. BCPC Publications, Farnham.
Copping, L.G., 2004. *The manual of biocontrol agents*, 3rd edition. BCPC Publications, Alton, IL.
Copping, L.G., 2009. *The manual of biocontrol agents*, 4th edition. BCPC Publications, Alton.
Copping, L.G. and J.L. Menn, 2000. Review: biopesticides: a review of their action, applications and efficacy. *Pest Manag. Sci.* 56: 651–676.
Coté, J.-C., 2007. How early discoveries about Bacillus thuringiensis prejudiced subsequent research and use. In: C. Vincent, M.S. Goettel, G. Lazarovits (eds), *Biological control – a global perspective*. CAB International, Wallingford. pp. 169–178.
CPL, 2006a. *Biopesticides 2007*. CPL Business Consultants, Wallingford.
CPL, 2006b. *Biopesticides 2007, Vol. 4. Europe: the biopesticide market*. CPL Business Consultants, Wallingford.
CPL, 2006c. *Biopesticides 2007, Vol. 5. A how to do it guide to biopesticides*. CPL Business Consultants, Wallingford.
CPM, 2009a. Other news and markets. *Crop Protection Monthly*, April 30, 2009: 16.
CPM, 2009b. Biopesticides. *Crop Protection Monthly*, June 30, 2009: 9–10.
Cross, J.V. and D.R. Polonenko, 1996. An industry perspective on registration and commercialization of biocontrol agents in Canada. *Can. J. Plant Pathol.* 18: 446–454.
De Faria, M.R. and S.P. Wraight, 2007. Mycoinsecticides and mycoararicides: a comprehensive list with worldwide coverage and international classification of formulation types. *Biol. Control* 43: 237–256.
Dent, D.R., 1997. Integrated pest management and microbial insecticides. In: H.F. Evans (ed), *Microbial insecticides: novelty or necessity? BCPC Symposium Proceedings* No. 68, Coventry, April 16–18, 1997. BCPC, Farnham. pp. 127–138.
De Vrije, T., N. Antoine, R.M. Buitelaar, S. Bruckner, M. Dissevelt, A. Durand, M. Gerlagh, E.E. Jones, P. Lüth, J. Oostra, W.J. Ravensberg, R. Renaud, A. Rinzema, F.J. Weber, and J.M. Whipps, 2001. The fungal biocontrol agent *Coniothyrium minitans*: production by solid-state fermentation, application and marketing. *Appl. Microbiol. Biotechnol.* 56: 58–68.
Dillon, A.B., M.J. Downes, D. Ward, and C.T. Griffin, 2007. Optimizing application of entomopathogenic nematodes to manage large pine weevil, *Hylobius abietis* L. (Coleoptera: Curculionidae) populations developing in pine stumps, *Pinus sylvestris*. *Biol. Control* 40(2): 253–263.
Douthwaite, B., J. Langewald, and J. Harris, 2001. Development and commercialization of green muscle biopesticide. *Impact*. International institute of tropical agriculture, Ibadan, Nigeria. www.lubilosa.org. Cited August 10, 2009.
Droby, S., M. Wisniewski, D. Macarisin, and C. Wilson, 2009. Twenty years of postharvest biocontrol research: is it time for a new paradigm? *Postharvest Biol. Technol.* 52: 137–145.
Ehlers, R.-U., 2003. Entomopathogenic nematodes in the European biocontrol market. *Comm. Appl. Biol. Sci. Ghent Univ.* 68(4a): 3–16.
Ehlers, R.-U., 2007. Entomopathogenic nematodes: from science to commercial use. In: C. Vincent, M.S. Goettel, G. Lazarovits (eds), *Biological control – a global perspective*. CAB International, Wallingford. pp. 136–151.
Evans, H.F., 1997. The role of microbial insecticides in forest pest management. In: H.F. Evans (ed), *Microbial insecticides: novelty or necessity? BCPC Symposium Proceedings* No. 68, Coventry, April 16–18, 1997. BCPC, Farnham. pp. 29–40.
Evans, H.F., 1999. Factors in the success and failure of microbial insecticides in forestry. *Integr. Pest Manag. Rev.* 4: 295–299.

Evans, J., 2004a. *Biopesticide, biocontrol and semiochemical markets*. Agrow Report DS246. PJB Publications, London.
Evans, J., 2004b. Shifting perceptions on biopesticides. *Agrow* No 455, September 3rd, 2004: 18–21.
Evans, J., 2008. Biopesticides: from cult to mainstream. *Agrow Magazine* October 27, 2008: 11–14.
Federici, B.A., 2005. Insecticidal bacteria: an overwhelming success for invertebrate pathology. *J. Invertebr. Pathol.* 89: 30–38.
Federici, B.A., 2007. Bacteria as biological control agents for insects: economics, engineering, and environmental safety. In: M. Vurro, J. Gressel (eds), *Novel biotechnologies for biocontrol agent enhancement and management*. Springer, Dordrecht. pp. 25–51.
Fravel, D., 1999. Hurdles and bottlenecks on the road to biocontrol of plant pathogens. *Australas. Pl. Pathol.* 28(1): 153–157.
Fravel, D.R., 2005. Commercialization and implementation of biocontrol. *Annu. Rev. Phytopathol.* 43: 337–359.
Friedman, M.J., 1990. Commercial production and development. In: R. Gaugler, H.K. Kaya (eds), *Entomopathogenic nematodes in biological control*. CRC Press, Boca Raton, FL. pp. 153–172.
Frost and Sullivan, 2001. European biopesticides market. *Report 3905-14*, London.
Froyd, J.D., 1997. Can synthetic pesticides be replaced with biologically-based alternatives? – an industry perspective. *J. Ind. Microbiol. Biotechnol.* 19: 192–195.
Gelernter, W.D., 2005. Biological control products in a changing landscape. *Proceedings of the BCPC Int. Congress – Crop Science & Technology 2005,* October 31–November 2, 2005, Glasgow. pp. 293–300.
Gelernter, W.D., 2007. Microbial control in Asia: a bellwether for the future? *J. Invertebr. Pathol.* 95: 161–167.
Gelernter, W.D. and H.F. Evans, 1999. Overview: factors in the success and failure of microbial insecticides. *Integr. Pest Manag. Rev.* 4(4): 279.
Gelernter, W.D. and C.J. Lomer, 2000. Success in biological control of above-ground insects by pathogens. In: G. Gurr, S. Wratten (eds), *Biological control: measures of success*. Kluwer, Dordrecht. pp. 297–322.
Gelernter, W.D. and J.T. Trumble, 1999. Factors in the success and failure of microbial insecticides in vegetable crops. *Integr. Pest Manag. Rev.* 4(4): 301–306.
Georgis, R., 1992. Present and future prospects for entomopathogenic nematode products. *Biocontrol Sci. Technol.* 2: 83–99.
Georgis, R., 2002. The Biosys experiment: an insider's perspective. In: R. Gaugler (ed), *Entomopathogenic nematology*. CAB International, Wallingford. pp. 357–372.
Georgis, R., 2004. Current and prospective markets for entomopathogenic nematodes. *Int. J. Nematol.* 14(1): 1–8.
Georgis, R., A.M. Koppenhöfer, L.A. Lacey, G. Bélair, L.W. Duncan, P.S. Grewal, M. Samish, L. Tan, P. Torr, and R.W.H.M. van Tol, 2006. Successes and failures in the use of parasitic nematodes for pest control. *Biol. Control* 38: 103–123.
Glare, T.R. and M. O'Callaghan, 2000. *Bacillus thuringiensis: biology, ecology and safety*. Wiley & Sons, Chichester.
Goettel, M.S., M. Koike, J.J. Kim, D. Aiuchi, R. Shinya, and J. Brodeur, 2008. Potential of *Lecanicillium* spp. for management of insects, nematodes and plant diseases. *J. Invertebr. Pathol.* 98(3): 256–261.
Grewal, P.S., 1999. Factors in the success and failure of microbial control in turfgrass. *Integr. Pest Manag. Rev.* 4: 287–294.
Grewal, P.S., R.-U. Ehlers, and D.I. Shapiro-Ilan, 2005. Critical issues and research needs for expanding the use of nematodes in biocontrol. In: P.S. Grewal, R.-U. Ehlers, D.I. Shapiro-Ilan (eds), *Nematodes as biocontrol agents*. CAB International, Wallingford. pp. 479–489.
Grimm, C., 2001. Economic feasibility of a small-scale production plant for entomopathogenic fungi in Nicaragua. *Crop Prot.* 20: 623–630.

Guillon, M., 2004. Current world situation on acceptance and marketing of biological control agents (BCAs). www.ibma.ch. Cited May 7, 2009.

Gurr, G., and Wratten, S. (eds), 2000. *Biological control: measures of success*. Kluwer, Dordrecht.

Gwynn, R.L., 2009 Biopesticide product gap analysis and evaluation to support development policy for biopesticides for use in integrated soft fruit production. *HDC Report SF 104*. HDC, Kenilworth, UK.

Hall, F.R., and Menn, J.J. (eds), 1999. *Biopesticides: use and delivery*. Humana Press, Totowa, NJ.

Hallett, S.G., 2005. Where are the bioherbicides? *Weed Sci.* 53: 404–415.

Harwood, R.W.J., M.S.K. Lee, S.G. Lisansky, and R. Quinlan, 2007. Current worldwide markets for biopesticides and success factors for the business. *Proceedings of the XVIth International Plant Protection Congress/BCPC International Congress – Crop Science & Technology 2007*, October 15–18, 2007, Glasgow. pp. 598–599.

Hauschild, R. and B. Speiser, 2007. A comparison of legislation and current practise regarding evaluation and registration of plant protection products based on micro-organisms, plant extracts and pheromones in the EU, USA, Canada and Australia. *Report prepared in the course of WP 2 of the REBECA project, December 2007*. www.rebeca-net.de. Cited May 3, 2009.

Hokkanen, H., 2007. Survey among industry concerning biological plant protection products. *Paper presented at the REBECA workshop on Balancing the benefits and costs of regulating biological plant protection products,* May 6–8, 2007 Porvoo, Finland. www.rebeca-net.de. Cited May 1, 2009.

Hokkanen, H.M.T. and I. Menzler-Hokkanen, 2000. Field performance of biopesticides. *BCPC Symposium Proceedings No. 74: Predicting Field Performance in Crop Protection*, BCPC Publications, Farnham. pp. 77–86.

Huang, D.-F., J. Zhang, F.-P. Song, and Z.-H. Lang, 2007. Microbial control and biotechnology research on *Bacillus thuringiensis* in China. *J. Invertebr. Pathol.* 95: 175–180.

Hunter-Fujita, F.R., P.F. Entwhistle, H.F. Evans, and N.E. Crook, 1998. *Insect viruses and pest management*. Wiley & Sons, Chichester.

Hynes, R.K. and S.M. Boyetchko, 2006. Research initiatives in the art and science of biopesticide formulations. *Soil Biol. Biochem.* 38: 845–849.

Inceoglu, A.B., S.G. Kamita, A.C. Hinton, Q. Huang, T.F. Severson, K. Kang, and B.D. Hammock, 2001. Recombinant baculoviruses for insect control. *Pest Manag. Sci.* 57: 981–987.

Jackson, T.A., 1999. Factors in the success and failure of microbial control agents for soil dwelling pests. *Integr. Pest Manag. Rev.* 4: 281–285.

Jackson, T.A., J.F. Pearson, M. O'Callaghan, H.K. Mahanty, and M.J. Willocks, 1992. Pathogen to product – development of Serratia entomophila (Enterobacteriaceae) as a commercial biological control agent for the New Zealand grass grub (Costelytra zealandica). In: T.A. Jackson, T.R. Glare (eds), *Use of pathogens in scarab pest management*. Intercept, Andover. pp. 191–198.

Jaronski, S.T., 1986. Commercial development of Deuteromycetous fungi of arthropods: a critical appraisal. In: R.A. Samson, J.M. Vlak, R. Peters (eds), *Fundamental and applied aspects of invertebrate pathology*. Foundation of the Fourth International Colloquium of Invertebrate Pathology, Wageningen. pp. 653–656.

Jarvis, P., 2001. Biopesticide industry strives to meet full potential. *Agrow* No. 387, October 26th, 2001. PJB Publications, London. pp. 23–25.

Kaya, H.K., 1986. Constraints associated with commercialization of entomogenous nematodes. In: R.A. Samson, J.M. Vlak, R. Peters (eds), *Fundamental and applied aspects of invertebrate pathology*. Foundation of the Fourth International Colloquium of Invertebrate Pathology, Wageningen. pp. 661–664.

Kaya, H.K., M.M. Aguillera, A. Alumai, H.Y. Choo, M. de la Torre, A. Fodor, S. Ganguly, S. Hazir, T. Lakatos, A. Pye, M. Wilson, S. Yamanaka, H. Yang, and R.-U. Ehlers, 2006. Status of entomopathogenic nematodes and their symbiotic bacteria from selected countries or regions of the world. *Biol. Control* 38: 134–155.

Keel, C. and M. Maurhofer, 2009. Insecticidal activity in root-associated, plant-beneficial pseudomonads. *Bull. IOBC/WPRS* 45: 181.

Kessler, P., M. Benuzzi, and F. Mayoral, 2008. Developing new baculovirus products or "how to walk a tightrope". *IOBC/WPRS Bull.* 31: 50–53.

Kiewnick, S., 2007. Practicalities of developing and registering microbial biological control agents. *CAB Rev.: Perspectives Agric. Vet. Sci. Nutr. Nat. Res.* 2(No. 013): 1–11.

Krause, M.S., A.C.R.C. Vanachter, and T.J.J. de Ceuster, 2006. Commercial research and development of disease-suppressive microorganisms. *IOBC/WPRS Bull.* 29(2): 63–66.

Kunimi, Y., 2007. Current status and prospects on microbial control in Japan. *J. Invertebr. Pathol.* 95: 181–186.

Lacey, L.A. and D.I. Shapiro-Ilan, 2008. Microbial control of insect pests in temperate orchard systems: potential for incorporation into IPM. *Annu. Rev. Entomol.* 53: 121–144.

Ladurner, E., M. Benuzzi, and S. Franceschini, 2008. Btk strain EG 2348: a valuable tool for the control of a wide range of Lepidopteran pests. *Paper presented at the 3rd annual biocontrol industry meeting (ABIM) Lucerne,* Switzerland, October 20th–21st, 2008. http://www.abim-lucerne.ch/archive/conference2008. Cited May 23, 2009.

Lambert, B. and M. Peferoen, 1992. Insecticidal promise of *Bacillus thuringiensis*. *BioScience* 42(2): 112–121.

Langewald, J. and C. Kooyman, 2007. Green Muscle, a fungal biopesticide for control of grasshoppers and locusts in Africa. In: C. Vincent, M.S. Goettel, G. Lazarovits (eds), *Biological control – a global perspective*. CAB International, Wallingford. pp. 311–318.

Li, Z., S.B. Alves, D.W. Roberts, M. Fan, I. Delalibera, J. Tang, R.B. Lopes, M. Faria, and D.E.N. Rangel, 2010. Biological control of insects in Brazil and China: history, current programs and reasons for their successes using entomopathogenic fungi. *Biocontl. Sci. Technol.* 20(2): 117–136.

Lipa, J.J. and P.H. Smits, 1999. Microbial control of pests in greenhouses. In: R. Albajes, M.L. Gullino, J.C. van Lenteren, Y. Elad (eds), *Integrated pest and disease management in greenhouse crops*. Kluwer, Dordrecht. pp. 295–309.

Lisansky, S.G., 1985. Production and commercialization of pathogens. In: N.W. Hussey, N. Scopes (eds), *Biological pest control. The glasshouse experience*. Blanford Press, Poole. pp. 210–218.

Lisansky, S.G., 1997. Microbial biopesticides. In: H.F. Evans (ed), *Microbial insecticides: novelty or necessity? BCPC Symposium Proceedings* No. 68, Coventry, April 16–18, 1997. BCPC, Farnham. pp. 3–19.

Lisansky, S.G. and J. Coombs, 1992. Technical improvements to biopesticides. *Proceedings BCPC – Pests and Diseases 1992*, 1(4C): 345–350.

Lisansky, S.G. and J. Coombs, 1994. Developments in the market for biopesticides. *Proceedings BCPC – Pests and Diseases 1994*, 3(4C): 1049–1054.

Lisansky, S.G. and R.A. Hall, 1983. Fungal control of insects. In: J.E. Smith, D.R. Berry, B. Kristiansen (eds), *The filamentous fungi, vol. 4: fungal technology*. Edward Arnold, London. pp. 327–345.

Lomer, C.J., 1999. Factors in the success and failure of microbial agents for control of migratory pests. *Integr. Pest Manag. Rev.* 4: 307–312.

Lomer, C.J., R.P. Bateman, D. Dent, H. De Groote, O.-K. Douro-Kpindou, C. Kooyman, J. Langewald, Z. Ouambama, R. Peveling, and M. Thomas, 1999. Development of strategies for the incorporation of biological pesticides into the integrated management of locusts and grasshoppers. *Agric. Forest Entomol.* 1: 71–88.

Lord, J.C., 2005. From Metchnikoff to Monsanto and beyond: the path of microbial control. *J. Invertebr. Pathol.* 89: 19–29.

Lu, D., M. Pava-Ripoll, Z. Li, and C. Wang, 2008. Insecticidal evaluation of *Beauveria bassiana* engineered to express a scorpion neurotoxin and a cuticle degrading protease. *Appl. Microbiol. Biotechnol.* 81(3): 515–522.

Lucarotti, C.J., G. Moreau, and E.G. Kettela, 2007. Abietiv^{TM} a viral biopesticide for control of the balsam fir sawfly. In: C. Vincent, M.S. Goettel, G. Lazarovits (eds), *Biological control – a global perspective*. CAB International, Wallingford. pp. 353–361.

Marrone, P.G., 1999. Microbial pesticides and natural products as alternatives. *Outlook Agric.* 28(3): 149–154.

Marrone, P.G., 2007. Barriers to adoption of biological control agents and biological pesticides. *CAB Rev.: Perspectives Agric. Vet. Sci. Nutr. Nat. Res.* 2(No. 051): 1–12.

McCoy, C.W., R.A. Samson, D.G. Boucias, L.S. Osborne, J.E. Pena, and L.J. Buss, 2009. *Pathogens infecting insects and mites of citrus*. LLC Friends of Microbes, Winter Park, FL.

Montesinos, E., 2003. Development, registration and commercialization of microbial pesticides for plant protection. *Int. Microbiol.* 6: 245–252.

Moore, D., 2008. A plague on locusts – the Lubilosa story. *Outlooks Pest Manag.* 19(1): 14–17.

Moscardi, F., 1999. Assessment of the application of baculoviruses for control of lepidoptera. *Annu. Rev. Entomol.* 44: 257–289.

Navon, A., 2000. *Bacillus thuringiensis* insecticides in crop protection – reality and prospects. *Crop Protect.* 19: 669–676.

Nefyto, 2008. Samenvatting van de afzet (per actieve stof) van gewasbeschermingsmiddelen 2007. www.nefyto.nl. Cited June 14, 2009.

Ownley, B.H., M.R. Griffin, W.E. Klingeman, K.D. Gwinn, J.K. Moulton, and R.M. Pereira, 2008. *Beauveria bassiana*: endophytic colonization and plant disease control. *J. Invertebr. Pathol.* 98: 267–270.

Pallett, K., December, 2005. R & D functions of the crop protection industry. *Outlooks Pest Manag.* 16(6): 242–243.

Paulitz, T.C. and R.R. Bélanger, 2001. Biological control in greenhouse systems. *Annu. Rev. Phytopathol.* 39: 103–133.

Pava-Ripoll, M., F.J. Posada, B. Momen, C. Wang, and R.J. St Leger, 2008. Increased pathogenicity against coffee berry borer, *Hypothenemus hampei* (Coleoptera: Curculionidae) by *Metarhizium anisopliae* expressing the scorpion toxin (AaIT) gene. *J. Invertebr. Pathol.* 99(2): 220–226.

Payne, C.C., 1987. Current uses and future prospects for microbial pest control agents. *Med. Fac. Landbouww. Rijksuniv. Gent* 52(2a): 113–123.

Peters, A., 2009. Research needs for entomopathogenic nematodes – an industry's perspective. *Bull. IOBC/WPRS* 45: 35–39.

Phillips McDougall, 2008. *Natural products insecticides*. Edinburgh, Scotland.

Powell, K.A. and J.L. Faull, 1988. Commercial approaches to the use of biological control agents. In: J.M. Whipps, R.D. Lumsden (eds), *Biotechnology of fungi for improving plant growth*. Cambridge University Press, Cambridge. pp. 259–275.

Ravensberg, W.J., 1992. Production and utilization of natural enemies in Western European glasshouse crops. In: T.E. Anderson, N.C. Leppla (eds), *Advances in insect rearing for research & pest management*. Westview Press, Boulder, CO. pp. 465–487.

Ravensberg, W.J. and Y. Elad, 2001. Current status of biological control of diseases in greenhouse crops – a commercial perspective. *IOBC/WPRS Bull.* 25(1): 225–231.

Reinecke, P., W. Andersch, K. Stenzel, and J. Hartwig, 1990. BIO1020, a new microbial insecticide for use in horticultural crops. *Proc. BCPC – Pest Dis. 1990* 2(5): 49–54.

Riba, G., S. Poitout, and C. Silvy, 1996. Public policy and partnership in research on biological control in France. *Entomophaga* 41(3/4): 425–434.

Rodgers, P., 1993. Potential of biopesticides in agriculture. *Pestic. Sci.* 39: 117–129.

Rodríguez, L.C. and H.M. Niemeyer, 2005. Integrated pest management, semiochemicals and microbial pest-control agents in Latin American agriculture. *Crop Prot.* 24: 615–623.

Rosell, G., C. Quero, J. Coll, and A. Guerrero, 2008. Biorational insecticides in pest management. *J. Pestic. Sci.* 33(2): 103–121.

Rowe, G.E. and A. Margaritis, 2004. Bioprocess design and economic analysis for the commercial production of environmentally friendly bioinsecticides from *Bacilllus thuringiensis* HD-1 *kurstaki*. *Biotechnol. Bioeng.* 86(4): 377–388.

Shah, P.A. and M.S. Goettel, 1999. *Directory of microbial control products and services. Society of insect pathology*. www.sipweb.org/publications.cfm. Cited June 16, 2009.

Shipp, L., D. Elliot, D. Gillespie, and J. Brodeur, 2007. From chemical to biological control in Canadian greenhouse crops. In: C. Vincent, M.S. Goettel, G. Lazarovits (eds), *Biological control – a global perspective*. CAB International, Wallingford. pp. 118–127.

Skovmand, O., 2007. Microbial control in Southeast Asia. *J. Invertebr. Patholog.* 95: 168–174.

Smits, P.H. and J.M. Vlak, 1994. Registration of the first viral insecticide in the Netherlands: the development of Spod-X, based on *Spodoptera exigua* nuclear polyhedrosis virus. *Med. Fac. Landbouw. Rijksuniv. Gent.* 59(2a): 385–392.

Starnes, R.L., C.L. Liu, and P.G. Marrone, 1993. History, use and future of microbial insecticides. *Am. Entomol.* 39: 83–91.

Sterk, G., K. Bolckmans, and J. Eyal, 1996. A new microbial insecticide, *Paecilomyces fumosoroseus* strain Apopka 97, for the control of the greenhouse whitefly. *Proceedings BCPC – Pests and Diseases 1996,* November 18–21, 1996, Brighton. pp. 461–466.

Stewart, A., 2001. Commercial biocontrol – reality or fantasy? *Australas. Pl. Pathol.* 30: 127–131.

St. Leger, R. and S. Screen, 2001. Prospects for strain improvement of fungal pathogens of insects and weeds. In: T.M. Butt, C. Jackson, N. Magan (eds), *Fungi as biocontrol agents: progress, problems and potential*. CAB International, Wallingford. pp. 219–237.

Straus, E.E. and J.D. Knight, 1997. Microbial insecticides: can farmers be persuaded to use them. In: H.F. Evans (ed), *Microbial insecticides: novelty or necessity? BCPC Symposium Proceedings* No. 68, Coventry, April 16–18, 1997. BCPC, Farnham. pp. 141–150.

Swanson, D., 1997. Economic feasibility of two technologies for production of a mycopesticide in Madagascar. *Memoirs Entomol. Soc. Canada* 171: 101–113.

Szewczyk, B., L. Hoyos-Carvajal, M. Paluszek, I. Skrzecz, and M. Lobo de Souza, 2006. Baculoviruses – re-emerging biopesticides. *Biotech. Adv.* 24: 143–160.

Tabaluk, T. and K. Gaznik, 2007. *Directory of microbial pesticides for agricultural crops in OECD countries*. Agriculture and agri-food Canada. Catalogue no. A42–107/2007E-PDF. www.organicagcentre.ca. Cited May 23, 2009.

Thakore, Y., 2006. The biopesticide market for global agricultural use. *Industr. Biotechnol.* 2(3): 194–208.

Törmälä, T., 1995. Economics of biocontrol agents: an industrial view. In: H.M.T. Hokkanen, J.M. Lynch (eds), *Biological control: benefits and risks*. Cambridge University Press, Cambridge. pp. 277–282.

Van der Blom, J., 2009. Microbiological insecticides against lepidopteran pests in greenhouse horticulture in Almeria, Spain. *Bull. IOBC/WPRS* 45: 59–62.

Van der Blom, J., A. Robledo, S. Torres, J.A. Sanchez, and M. Contreras, 2008. Control biológico de plagas en Almería: revolución verde después de dos decadas. *Phytoma España* 198: 42–48.

Van der Pas, R.K., C.J. Waddington, and W.J. Ravensberg, 2000. Commercialisation of a microbial pesticide: "challenges and constraints". *Bull. IOBC/WPRS* 23(2): 15–18.

van Lenteren, J.C., 2006a. The area under biological control and IPM in greenhouses is much larger than we thought. *Sting* 29: 7. http://web.agrsci.dk/plb/iobc/sting/sting29.pdf. Cited May 20, 2009.

van Lenteren, J.C., 2006b. *The IOBC internet book of biological control*. http://www.unipa.it/iobc/download/IOBC%20InternetBookBiCoVersion4October2006.pdf. Cited May 20, 2009.

Vasantharaj David, B., 2008. Biotechnological approaches in IPM and their impact on environment. *J. Biopesticides* 1: 1–5.

Vega, F.E., 2008. Insect pathology and fungal endophytes. *J. Invertebr. Pathol.* 98: 277–279.

C. Vincent, M.S. Goettel, G. Lazarovits (eds), 2007. *Biological control – a global perspective*. CAB International, Wallingford, CT.

Vlak, J., 2009. Genetic engineering of baculoviruses for biocontrol: dead end or bright future? *Bull. IOBC/WPRS* 45: 517–518.

Waage, J.K., 1997. Biopesticides at the crossroads: IPM products or chemical clones? In: H.F. Evans (ed), *Microbial insecticides: novelty or necessity? BCPC Symposium Proceedings* No. 68, Coventry, April 16–18, 1997. BCPC, Farnham. pp. 11–19.

Whipps, J.M. and K.G. Davies, 2000. Success in biological control of plant pathogens and nematodes by microorganisms. In: G. Gurr, S. Wratten (eds), *Biological control: measures of success*. Kluwer, Dordrecht. pp. 231–270.

Whipps, J.M. and R.D. Lumsden, 2001. Commercial use of fungi as plant disease biological control agents: status and prospects. In: T.M. Butt, C. Jackson, N. Magan (eds), *Fungi as biocontrol agents: progress, problems and potential.* CAB International, Wallingford. pp. 9–22.

Woodhead, S.H., A.L. O'Leary, D.J. O'Leary, and S.C. Rabatin, 1990. Discovery, development, and registration of a biocontrol agent from an industrial perspective. *Can. J. Plant Pathol.* 12: 328–331.

Wraight, S.P., M.A. Jackson, and S.L. de Kock, 2001. Production, stabilization and formulation of fungal biocontrol agents. In: T.M. Butt, C. Jackson, N. Magan (eds), *Fungi as biocontrol agents: progress, problems and potential.* CAB International, Wallingford. pp. 253–287.

Zheng, L., Y. Zhou, and K. Song, 2005. Augmentative biological control in greenhouses: experiences from China. *Proceedings of the 2nd International Symposium on Biological Control of Arthropods*, September 12–16, 2005, Davos, Switzerland. USDA Forest Service, FHTET 2005–08. pp. 538–545.

Chapter 8
Roadmap to Success and Future Perspective

Abstract The development of a microbial pest control product requires a structured project plan. The building blocks of the entire process are identified and essential aspects highlighted. The roadmap to successful development and commercialization of a microbial pest control product is amply illustrated in newly designed flow diagrams. Diagrams are presented for the screening phase, the product development phase, and the implementation phase. Relevant input criteria and requirements are provided. Output information leads to consecutive steps and "go/no go"-decisions. A future perspective on the biopesticide market is presented with limiting and promoting factors and trends. The most important drivers are concerns about food safety and the environment, new research and technology, changes in the regulatory climate, and the arrival of exotic and invasive pests. The biopesticide industry has reached a level of maturity and critical mass that forms a solid base for further expansion. The future of the market size and growth is estimated, and an outlook of the way forward is presented.

1 Introduction

The development and commercialization of a microbial pest control product is an extremely complex process. The total project from idea to market entry can take many years, and costs may amount to a few million Euros. Therefore, it is necessary to have from the outset an extensive overview on how such a project will be developed, and which critical issues will be encountered. A complete and structured project plan is required which oversees all phases, all decisions, and the consequences thereof. Such a complete project plan has not yet been described in the literature. This study aimed to produce a complete roadmap for the development and commercialization of an entomopathogenic microbial pest control product, with an emphasis on commercial and economic issues.

I will briefly summarize the most important results of my study and highlight the pivotal issues. Diagrams are presented illustrating the relevant building block and

W.J. Ravensberg, *A Roadmap to the Successful Development and Commercialization of Microbial Pest Control Products for Control of Arthropods*, Progress in Biological Control 10,
DOI 10.1007/978-94-007-0437-4_8,

routes in the entire developmental process. Further, I will report the limiting and promotional factors and trends, and present a future perspective on the biopesticide industry and market. Suggestions are provided to overcome obstacles to a successful uptake of biopesticides.

1.1 Identification of Selection Criteria

In the first phase of the project the objectives must be established (Chapter 2). It must become clear what type of product will be developed, against which pest(s), and in which cropping system(s). An elaborate description of the pest problem provides direction to the search for a microbial pest control agent. The collection of entomopathogenic pathogens is the first concrete step in the process, followed by a screening process to select the best candidate strain(s). The first level of selection is the type of pathogen: bacteria, fungi, viruses, protozoa and entomopathogenic nematodes. The second level of selection is at the species and strain level. Biological features of a potential agent are crucial, and these need to be evaluated in this selection phase along with economic factors. I consider relevant selection criteria for a commercial MPCA to be: mortality, production efficiency, and safety. Favourable features related to mortality and economics are: a low LC50 dose, a broad host range, good persistence, low sensitivity to environmental influences, and a quick speed of kill. Strains with positive productivity indications as a high yield and low production costs are promising. A selective mode of action suggests safety to humans and non-target organisms and this may keep registration costs low. The absence or presence of metabolites will greatly influence registration expenses too.

Efficacy tests used in the selection process need to reflect the ultimate commercial conditions. A tiered approach is recommended (see Table 2.3). The gathered information allows the developing team to make a "go/no go"-decision: i.e. have strain(s) been found that offer sufficient promise for the following phases in the developmental process? The best strain(s) will be a compromise of biological parameters and economic factors in the decision whether to continue, and economic considerations should be leading.

1.2 Critical Parameters in Product Development

The following step is the examination of the feasibility of an economic mass production of the selected strain(s) and the development of a stable product (Chapter 3). Biological and technical options together with economics (Table 3.2) determine the choice of the production system. Preferably, it will be an in vitro process because that offers more control of the production process than an in vivo process. In both, production efficiency and cost-effectiveness are the key-factors.

The challenge in process development is the extrapolation and realization from laboratory scale techniques to commercial and large-scale technology. I propose therefore that production research is initiated as early as possible in large-scale bioreactors. Economy of scale is essential to decrease costs in three areas: capital, labour, and materials. Improvements in production technology as well as the reduction of production costs should be a continuous research objective. Equipment should be versatile allowing the production of various insect pathogens or other biocontrol agents.

Formulation research is required for the development of a stable product that delivers effective pest control. Formulation is linked with the production system, the medium composition, and the down-stream process. Formulation serves four functions: stability of the propagules, effective delivery, on site protection of the propagules, and safety for the applicator. These four functions need to be investigated and can be optimized by manipulation of process parameters in the production phase as well as in the down-stream phase. Medium composition influences the quality of the propagules, and co-formulants can protect the propagules and improve other product properties. Formulation can improve efficacy, shelf-life, and user-friendliness of the product. Solutions for the four functions are not always compatible and the final formulation is a compromise between all these demands. Formulation is specific to a certain type of pathogen (Table 3.5) and its specific use, so it demands a case by case approach.

Field testing links all the phases in the developmental process (Table 3.7). Its results provide information on whether the selected strain is sufficiently effective, on the quality of the produced propagules, on the formulation, on the application strategy, on efficacy required for registration, on the implementation of the product in an IPM system, on compatibility, and on the marketability of the final product. Results from field tests provide a continuous circle of feedback that allows improving aspects of each of the steps of the entire process. Field tests must allow the final selection of the best strain. The method of testing is crucial and it must reflect "real world" conditions. Often, the relevance of field testing is underestimated and it is insufficiently executed. Field testing, however, is an imperative step in the product development and sufficient resources must be allocated to conduct high quality field tests.

A cost price model for biopesticides is presented with costs factors involved from production to product (Table 3.9), and from product to market (Table 3.10). I report key considerations for an economic analysis of a biopesticide production (Table 3.12). I recommend making an analysis of production economics and the calculation of final product costs as early as possible in the developmental process for a reliable decision on whether the product could be produced in a profitable manner.

Key factors that determine the feasibility of an economical mass production and formulated product are investments, economy of scale, capacity usage, stability of production process, production yield and application dose, product shelf-life, level of efficacy, user-friendliness, full product cost price and end-user price.

1.3 Quality Control Procedures

Quality control (QC) provides feedback on the production and formulation processes and the final product. The continuous process of improvements will ultimately decrease costs and improve performance of the production system and the product (Chapter 4). Production control and process control refer to internal quality control of the production, and ensure a stable production process with a minimum of failures. Product control refers to the quality of the final product that will leave the factory. Parameters checked per batch are the number of effective propagules, microbial purity, presence of toxins, technical properties, and efficacy. Product specifications must be met until the end of the claimed shelf-life. Post-shipment product quality control is particularly warranted for nematode-based products.

A biocontrol company should implement QC procedures, at the same time they will be required for registration. Both industry and regulators face the lack of officially recognized standard QC criteria and methods. Guidance documents need to be developed as soon as possible. I argue, however, that a standardization of efficacy for all microbial pesticides should not be established since this has not succeeded for Bt's after 30 years of research, comparative testing, and discussions. Further, I recommend that the tolerance range for the number of active propagules be set at $\pm 25\%$. Biocontrol companies should ensure that product quality is maintained through the entire distribution chain and that end-users receive high quality products. In that way, both the biocontrol industry and its customers benefit from proper QC.

1.4 Implementation in an IPM System

The implementation strategy of the product in an IPM programme is a basic element of the use of any MPCP (Chapter 6). The positioning of a product implies a good understanding of all components of an IPM system and their interactions with the new product, in any given cropping system. This phase requires a considerable amount of research which should be conducted before market launch. I have experienced that many companies underestimated or even neglected this part of product development. The application strategy, compatibility, and knowledge transfer to the user are key factors for a cost-effective implementation of a microbial pest control product, and are determinative for rapid product adoption.

The first element is an optimal application strategy of the product itself. This has to be studied for each type of entomopathogen, target pest(s) and cropping system. Label recommendations need to be established for proper use and advice by sales people, and for registration. This usually receives enough attention from product developers since it has a direct impact on the overall field results of the product.

The second element is the incorporation of the microbial pest control product in an IPM system. For proper positioning in an IPM system many interactions have to be identified and investigated, but often, only limited research is conducted due to lack of availability of facilities and restricted budgets. In my opinion, this phase is paramount for good market introduction, and needs appropriate resources.

The third element of implementation is a carefully designed adoption strategy. The benefits of the new product need to be demonstrated to the grower. Efficient knowledge transfer and training are essential. I recommend that all stakeholders participate in this process, particularly those in the distribution channel, but I emphasize on direct involvement of growers.

I suggest the study of alternate and simultaneous applications of two types of microbial pest control products. I consider using multiple biopesticides a useful strategy when larval stages have a different susceptibility towards two pathogens, when speed of kill may be enhanced, when insect stages are in different locations on the plant, when two pathogens increase insect susceptibility, and when multiple pests are targeted. Combinations of a microbial with a chemical or another pest control means offer increased control, but this is not often applied in practice either. The use of pollinators as vectors for entomopathogens is another potential combination. These possibilities warrant further research since they allow a reduction of use of chemical pesticides and an enhanced biocontrol efficacy.

Field resistance has been reported for Bt and for one baculovirus. Thus, resistance management must become part of the application strategy, and biopesticide producers need to take this seriously.

The methodology used in compatibility tests for chemical as well as biological control agents may influence the outcome, and the relevance for field conditions is often debatable. I propose a tiered approach that results in reliable data for commercial conditions (Table 6.3).

Successful implementation of a MPCP depends on how well relevant interactions are studied and translated into practical recommendations for the grower. This phase continues after market introduction. It requires a continuous effort from producer, distributor and customer to ensure that product adoption will increase, and satisfied customers will remain using the new product in their IPM system.

1.5 Key Success and Failure Factors in the Commercialization Process

Commercialization is the final and most difficult step in the development and the introduction in the market of a microbial pest control product (Chapter 7). This is not just the case for bio-insecticides, but also for microbials for control of diseases, weeds, and postharvest pest and diseases. I estimate that the global biopesticide market is approximately \$500–\$700 million in 2009 while the total pesticide market was about \$40 billion (CropLife 2009). Biopesticides are predominantly used in protected crops, orchards, forestry, and for vector control; use in agricultural crops is just beginning. *Bacillus thuringiensis* products are the most successful. My estimate is that commercialization of baculoviruses and EPNs is a profitable business for the manufacturers. Sales of both groups are expanding. Business in entomopathogenic fungi seems to be marginal or unprofitable. I have identified the common success and failure factors for a biopesticide company (Table 7.7).

The biggest mistake companies still make today is a misjudgement of the potential market size and the expected market adoption rate. They underestimate the efforts, time, and costs to introduce a novel product successfully in the market. I recommend that product developers use realistic insights in the target markets, and that decisions should also take conservative market projections into account. I propose the use of a stage-gate process with objective, quantifiable, and transparent tools in decision-making which can assist companies in understanding the key issues and moments. Score-cards as presented in Tables 7.8, 7.9 and 7.10 will help in this process. Many companies discover too late that the developed product does not live up to its expectations. A "no-go"-decision is also a valuable decision, and time and money can then be spent on a better project. A careful approach in biopesticide activities explains that the model that currently performs best seems to be a small company which pursues an incremental and manageable growth of the organization.

Microbial pesticides have limitations as well as strengths (Table 7.11). A SWOT analysis of these products together with an analysis of external factors assists in recognizing the product's value in the market. The challenge is to position the product in the market with its pro's and con's and not to "overpromise".

Total developmental costs and time to market are significant factors of a company's success. Costs amount to €10–15 million for a company that still needs to be built; while in an existing company, costs may reach €5–10 million for a biopesticide project. Time to market including registration is 5–7 years (Fig. 7.2). Collaboration between private and public organizations is imperative, and reduces time and expenses for a company. I recommend both parties to work together from the outset of the project so that poor transfer of ideas from the lab to the market can be significantly improved.

Economics is the key to success. I have identified five determinants for successful commercialization. These are: (1) acceptable expenses and time to market; (2) a high quality product that provides function and value to the customer; (3) a sufficiently large market; (4) a profit margin that allows development of new markets and products; and, (5) the appropriate business approach. Only when these conditions are met can a sustainable and profitable business be run.

1.6 Salient Regulatory Topics

Registration (Chapter 5) is not an integral part of the chain of steps in the developmental process of a MPCP, but it can be seen as an umbrella process that provides a framework, with conditions and limits, in which a new product can be developed. This is particularly the case with microorganisms, but much less with entomopathogenic nematodes. The registration dossier is built up with information obtained from each step in the developmental process. The entire developmental process and the final product must be considered in light of registration as many aspects may have consequences and increase registration costs. Product developers must be well aware of the impact of registration requirements. These requirements differ between countries. Guidance documents are lacking as well as clear criteria for many studies. This renders the development of a product for an international market difficult and complex. I recommend the familiarization of the registration requirements from the outset. When these are not clear, a pre-submission consultation meeting with the evaluation authority should be requested. This will minimize trouble, time, and costs in a later phase, for both the applicant as well as the regulator.

Registration is perceived as the main hurdle to the development of a biopesticide. And surely it is a great challenge, for newcomers, and even for experienced companies. In my experience the generation of a dossier for a microorganism is not the greatest challenge, but the inconsistency and unpredictability of the product approval timeframe and costs are the most difficult to manage. This can significantly impact company growth, and even survival. Therefore, there is an urgent need to streamline the registration process and make it more transparent and predictable in terms of costs and time. Regulators and industry need to continue the dialogue that was started with the REBECA project, and assist each other improving this awkward situation.

Import and export regulations affect the use of microorganisms. Here as well, many different rules apply depending on the country, particularly for nematode-based products. The same is true for other invertebrate BCAs. At the risk of repeating myself, regulators and industry need to begin a dialogue to keep trade in biopesticides manageable. A similar case is developing around Access and Benefit Sharing which may create a further impediment for biocontrol (Cock et al. 2009).

Investment and IP with regard to MPCPs can be protected by a patent. It may be worthwhile to investigate this option in order to safeguard the position of a company. Nevertheless, patents are expensive and whether to apply for a patent needs careful consideration. Registration also offers protection.

Registration and regulations are a great challenge for biopesticide companies. On the one hand, they should safeguard them in the market from unlawful products. On the other hand, many products are illegally sold in the market and often are products with poor quality. Authorities should inspect this aspect more rigorously since it has a negative effect on biocontrol in general.

2 Roadmap to a Successful Microbial Pest Control Product

A new product development project is extensive and it is difficult to oversee. The process consists of hundreds of various aspects and steps. An analysis of the various phases facilitates overseeing the entire process. I distinguish an exploratory phase in which an enterprise contemplates and investigates the development of a MPCP. It is predominantly a theoretical phase in which essential business information is gathered from the market. Together with considerations reflecting the internal capabilities, this forms the basis for the decision whether to proceed to the executive phase. This first phase requires the knowledge of what needs to be conducted in the following phases. During these, there must be a constant feedback to the objectives of the project.

In Fig. 8.1 a schematic view is presented which illustrates the main influences and processes in the development and commercialization of a MPCP. The external factors from governments, society, and the market influence the strategy, mission and vision of the biopesticide company. The company's strategy, expertise, and facilities govern the direction and design of the development and positioning of the product. The development process is divided into a number of phases that are addressed in sequence. However, parts of processes can be dealt with simultaneously. It is not a simple stepwise linear process. Parallel working is important since this saves time. Planning is paramount for a time- and cost-efficient route to the market. In the optimization of processes it is necessary to go back to the preceding phase to improve the outcome until a satisfactory answer is provided. This feedback loop frequently occurs as illustrated in Figs. 8.2, 8.3 and 8.4. Each answer or result impacts the following processes. Between phases, decisions should be made whether to continue with the next phase. Commercialization as defined in this study is an overall process that sets conditions for the product and the investments, the costs of the product, and margins in the market.

Registration provides an external set of requirements and conditions. It is also influenced by the internal constraints of registration, i.e. how much is a company able to spend on registration studies and how much of this can be implemented as part of the cost price still allowing an acceptable final cost price. The number of

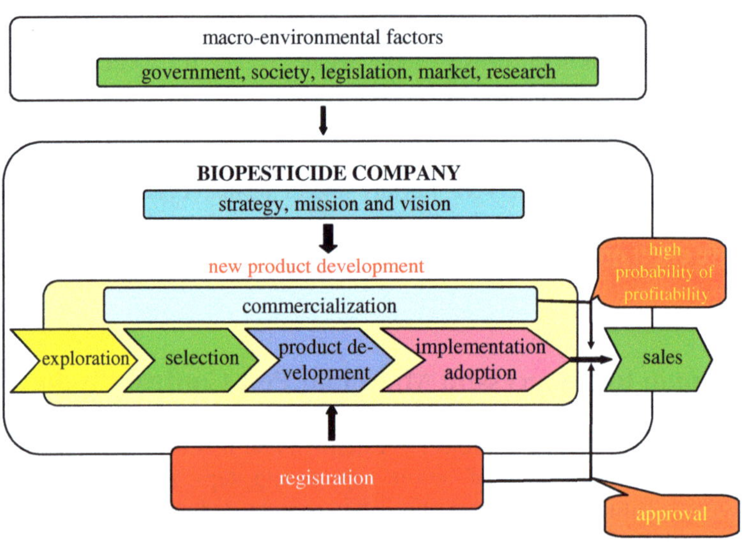

Fig. 8.1 General model which illustrates the main external influences and internal processes and phases in the development and commercialization of a MPCP

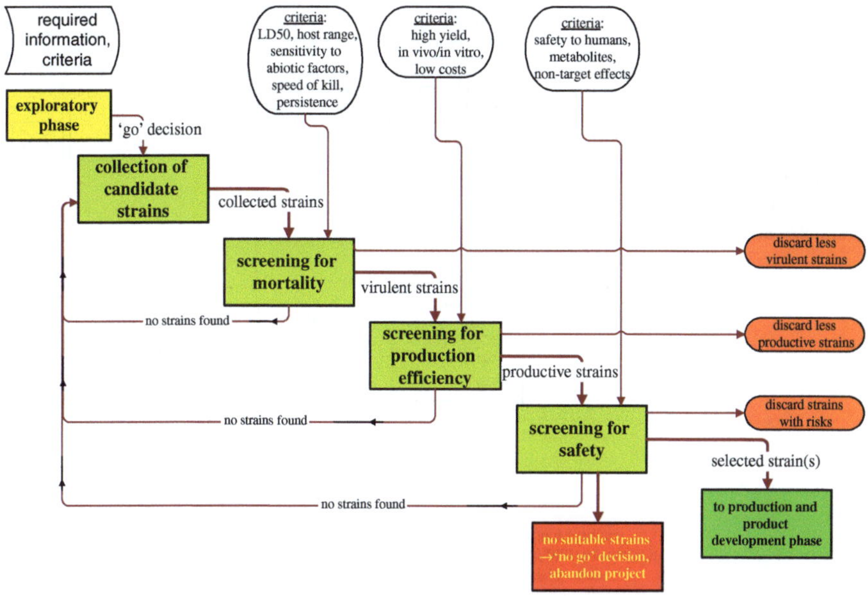

Fig. 8.2 The phases and steps of the screening process and the selection criteria for the consecutive steps

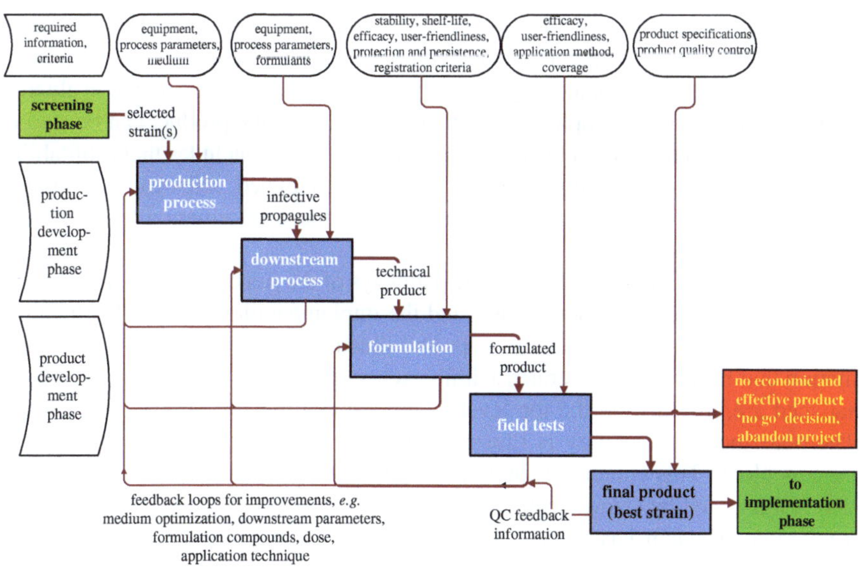

Fig. 8.3 The phases and steps in the production and product development process and the criteria for the consecutive steps

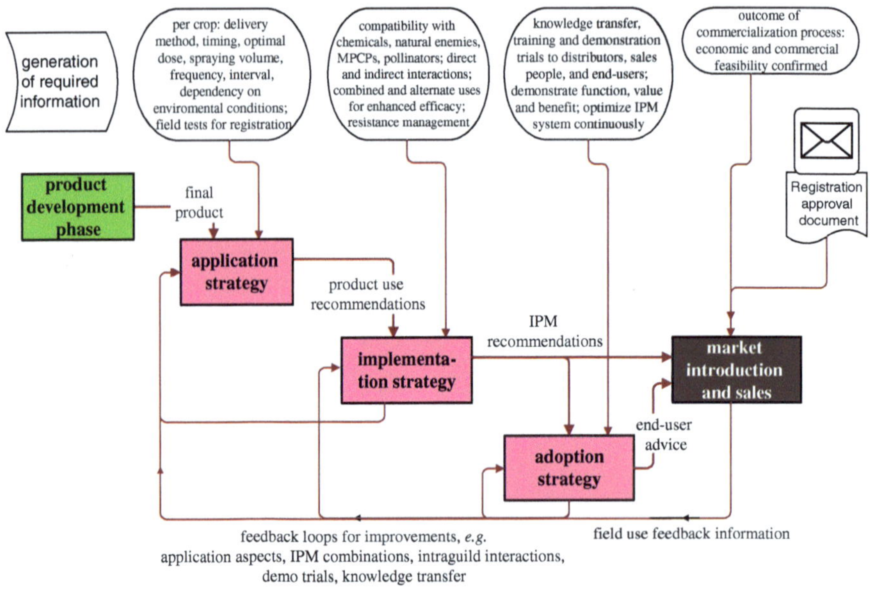

Fig. 8.4 The phases and steps of the implementation phase and the information that needs to be generated

target crops as well as the number of countries in which sales are anticipated influence registration costs. The generation of the registration dossier should be initiated early in the process. This is not a distinct phase at some point in the process, but a continuous activity that provides input and demands output for the registration dossier. When field testing has been completed with satisfactory results, external toxicology and ecotoxicology studies can be conducted. During the evaluation period by the authority, implementation and adoption of the product can be studied and prepared. Finally, when the company's management concludes that profitability can be achieved, and when regulatory approval is obtained, sales can be initiated. After-sales service is imperative, and provides feedback information too that could improve the product and the strategy of use.

Figures 8.2, 8.3 and 8.4 illustrate the pathways of the selection phase, the production and product development phase, and the implementation phase. The diagrams show the interactions between the steps and illustrate the entire process. The information that needs to be generated within each step, and the criteria that need to be satisfied are provided on the top level as input conditions into the activity blocks. The outcome of an activity is illustrated by an output arrow on the left of each process block. This outcome is input for the following activity. Returning information is illustrated by arrows from the underside of blocks. At the end of each phase a no/no-go decision has to be made before entering the next phase. Continuous feedback information is imperative for improvements and optimization in any process. The diagrams present the essential building blocks of the development and commercialization of a MPCP. Its stepwise structure outlines the entire process architecture and provides a roadmap for product developers.

3 Future Perspective

The driving forces in pest management are varied, and they continue to develop and change. The main forces that influence biological control and IPM are legislation, environmental and food safety concerns, science and technology, and economics (Uri 1998; Dent 2000; Marrone 2007; Bale et al. 2008; Bailey et al. 2009). Governmental policy has direct and indirect influences on crop protection through legislation, funding of research and extension, environmental programmes, trading policies, biodiversity topics, etc. (Uri 1998; Chandler et al. 2008a). Some are favourable, others unfavourable for the use of biopesticides.

An analysis of macro-environmental factors and trends assists a company in understanding its business position, the market, and potential directions for new developments. At the present, many external factors occur simultaneously which cause pest management policy to change in such a way that this offers an improved climate for the biopesticide industry. I will present the major factors and trends that limit or stimulate biological control. Suggestions are provided to improve wider adoption of microbial pest control products. I will sketch an outlook on the market, and on the way forward with biopesticides and other biologically-based alternatives for crop protection.

3.1 Limiting Factors and Threats

3.1.1 Political Factors

As long as conventional pesticides are cheaper than microbial disease management products, they will continue to dominate the market for pesticides. Pimentel (2005) reported on the major hidden environmental and economic costs of the application of synthetic pesticides. Biopesticides do not cause these costs and this should be part of the equation when a new policy for crop protection is designed. Menzler-Hokkanen (2006) argued that a mechanism for price support in favour of biological control should be installed because of the socio-economic benefits associated with the replacement of chemicals by biocontrol to the society. This good idea should be considered by policy-makers.

The establishment of an agreement on Access Benefit Sharing regarding genetic resources could impede collection of new BCAs and form a threat to biocontrol (Cock et al. 2009). Policy-makers should enable research and industry to continue with the development of sustainable pest management methods by providing a practical and workable framework.

3.1.2 Societal and Market-Related Trends

The broad use of biocontrol with natural enemies and microorganisms may include a risk since any insect or "germ" on food is perceived as highly undesirable by the public. Good communication will be necessary to avoid such a "food-scare" or a negative association of "man playing with nature" (Ritson and Kuznesof 2006).

Biopesticides need to break away from the single-technology approach of chemical pesticides and be implemented as key components of an IPM system. They should not be used in the same conceptual model as chemicals, and they should not be expected to out compete chemicals. Growers also need to change their use patterns in crop protection and think more about truly integrated approaches. The grower's difficulty in adoption of biological alternatives is the increased knowledge to make it function as well as the costs. This is part of a larger problem of declining returns in agriculture. Growers need to receive better returns for their product if they are going to spend more capital on their inputs such as more biological and natural solutions for their pest and disease problems. If the consumer wants sustainable agriculture and safe food, he will have to be willing to pay more for his food.

3.1.3 Research Needs

Continued research efforts are required for a deeper understanding of the mode of action, ecological constraints and requirements, and other fundamental aspects of entomopathogens. More applied research is also needed on application methods, formulation and shelf-life, on interactions in the field, and combined uses of microbials. Although Bt's are the most successful bioinsecticides, the market still asks for "faster killing", "higher efficacy against mid-late instars", "longer field life", and "rain-fastness tolerant" Bt products (Georgis and Warrior 2007). Better and more competitive biopesticides could be developed with more in-depth knowledge of entomopathogens. More funds are required for this kind of research.

Institutional scientists need to better understand registration requirements, and help industry with registration since in many areas they are the only experts. They have a role to play to provide regulators with sufficient relevant biological data to allow them to install a more relaxed regulation. An example is the paper of Scheepmaker and Butt (2010) which demonstrated that entomopathogenic fungi all decline to natural background levels after application and therefore do not pose a threat to the environment. Funding agencies should support this kind of work more.

The assumption that the small biocontrol companies can support the bulk of the research is unrealistic as most have limited budgets. Collaboration between public and private sector needs to be stimulated. Institutions, however, increasingly require support from the private sector. Instead, more governmental support is needed for both public and private organizations to develop these products. The political responsibility is to further the development of biopesticides if society desires a reduction of the use of chemical pesticides.

3.1.4 Regulatory Issue

Although regulatory innovation is taking place, its pace is much too slow. Governments need to give higher priority to a faster and wider adoption of biopesticides and other alternatives to accomplish economic and environmental sustainability of modern agriculture. Targeted active stimulation is required to achieve successful uptake of biological pesticides in the market. The needs for the realization of the envisioned crop protection has been assessed for the EU, and the outcome indicated the necessity for more research, more extension, and for harmonization in regulations to mitigate the current barriers for adoption of alternatives. An interesting recommendation is the "co-funding of the purchase of consultancy advice" (Chandler et al. 2008b). These proposals need to be transferred into concrete measures.

3.2 Promotional Factors and Trends

3.2.1 Societal and Market-Related Factors and Trends

Environmental groups and consumer organizations are increasing the pressure on retailers and growers to reduce the level of residues on fresh food. As a result, several supermarkets have even set chemical pesticide residue levels that surpass officially allowed levels, thereby limiting the number of active ingredients. GlobalGap, the global organization of food retailers and suppliers, has set standards for growers as "major musts" for crop protection measures including the use of IPM (www.globalgap.org). The role of retailers becomes *the* driving force from the market to reduce the use of synthetic pesticides. Biological alternatives could fill the gap.

Biopesticides are becoming a preferred option for pest control due to resistance to chemicals, restricted entry intervals for workers, and residue concerns. They are also safe for beneficial organisms, for the crop, and for the environment. Growers' awareness of these benefits is increasing and this stimulates the adoption of biopesticides.

Organic production of food has experienced worldwide growth surpassing 32 million hectares in 2007, and the area continues to expand (www.organic-world.net). It is Benuzzi's (2009) and my own experience that only about 10% of the sales of biopesticides occur in organic agriculture. The use of microbials in organic production is allowed, but not recommended by the major regulations and standards (Speiser et al. 2006). I do not expect this to change dramatically, and biopesticides will continue to be mainly used in conventional agriculture.

Genetically modified crops were cultivated on more than 125 million hectares in 2008 (Stein and Rodríguez-Cerezo 2009). Genetic engineering, however, is not a solution for all pests and diseases. A synergy between transgenic crops and biocontrol is promising since fewer chemicals are used, and secondary pests and diseases require sustainable solutions (Kos et al. 2009; Lundgren et al. 2009). This option deserves more research and field testing with existing biocontrol products.

3.2.2 Scientific and Technological Trends

Research continually delivers new ideas and options for biocontrol. Recently new opportunities were proposed for well-known fungal insect pathogens (Vega et al.

2009). They may be used as antagonists of plant pathogens (Ownley et al. 2008; Kim et al. 2009) and for control of nematodes (Goettel et al. 2008; Shinya et al. 2008), as endophytes against insects (Vega 2008), and as plant growth promoting agents (Vega et al. 2009). Microbial disease control agents can also affect herbivorous insects through triggering the plant's defence system (Van Wees et al. 2008; Keel and Maurhofer 2009). Broader use of existing products offers companies larger markets and a better return on investment.

Strain improvement by hybridization of fungal insect pathogens may deliver strains with a broader host range and better persistence on the plant (Aiuchia et al. 2007; Yamada et al. 2009). New discoveries in production technology with *Metarhizium anisopliae* producing microsclerotia offer potential for improved control of soil pests (Jackson and Jaronski 2009). The discovery of a synergist, a new protein that increased the efficacy of Bt's (Abdullah et al. 2009), could widen the spectrum and the commercial use of Bt.

Genetic improvement is a promising avenue for all microbial control agents and it has led to the development of more effective, faster acting baculoviruses, bacteria, fungi, protozoa, and nematodes at lab scale (Narayanan 2002). Genetic improvement of pathogenicity has been demonstrated for bacteria (Federici 2007), for fungal entomopathogens (Lu et al. 2008; Pava-Ripoll et al. 2008) as well as for baculoviruses (Inceoglu et al. 2006; Szewczyk et al. 2006). Strain improvement by transgenic technology in nematodes has been shown for enhancements in environmental tolerances (Burnell 2002). Genetically improved MPCAs may become an option in the longer term in the EU. They could be developed today for other countries, like the United States, where they are less controversial.

3.2.3 Changes in the Regulatory Climate

Registration is still one of the greatest challenges for biopesticide companies, particularly in the EU. This is caused by the application of an inappropriate synthetic pesticides model (Chandler et al. 2008a), and by the double procedure of evaluation of the active ingredient by the EU and product registrations on national level. The new Regulation EC/1107/2009 (EC 2009a) allows for zonal authorizations and mutual recognition. A category of low risk active substances has been established, but requirements and a procedure still have to be specified. These changes are expected to facilitate registration for biopesticides.

Initiatives as the Biopesticide Scheme (UK) and the project Genoeg (NL) accomplished a substantial number of new approvals of biopesticides. The example of the regulatory innovation in the UK (Greaves 2009) should be followed up by other EU member states to facilitate procedures. Similar projects in Canada, by the Pest Management Centre (Bailey et al. 2009), and in the USA, by the IR-4 project (www.ir4.rutgers.edu), assist applicants in registering their product and boosted approvals of biopesticides.

The work currently undertaken by OECD to develop appropriate guidance documents and test protocols for biopesticides are crucial for harmonization of the regulatory process on a global scale. Ideas from the REBECA project should be implemented in a timely manner in official procedures and criteria. IBMA continues to lobby for a new registration model specifically developed for biopesticides (see Position Paper at www.ibma.ch). A step forward is the registration of baculoviruses at species level instead of at isolate level.

The new Directive 2009/128/EC (EC 2009b) established a framework to achieve "sustainable use of pesticides". Member States must "encourage the development of Integrated Pest Management and of alternative approaches or techniques in order to reduce dependency on the use of pesticides". National Action Plans have to be developed in which the new crop protection aims will be laid down. Implementation is mandatory as from 2014. Both the new Regulation and Directive will impact the availability and use of synthetic pesticides. IPM will become the prevailing paradigm in pest management and this will create a momentum for increased use of biopesticides. Still, there is a need to increase awareness of biocontrol for policy-makers, and the biopesticide industry can play an active role here by participating in the design of the National Actions Plans.

3.2.4 Biodiversity and Environmental Concerns

The extensive use of chemical insecticides and fungicides in agriculture has a negative effect on biodiversity. Insecticides also reduce the biological control potential (Geiger et al. 2010). Through the implementation of the Convention of Biodiversity governments are required to improve conservation and sustainable use of biological diversity and to mitigate negative impacts and to support sustainable agriculture. The Directive 2009/128/EC (EC 2009b) also states that due to the environment and biodiversity the use of pesticides should be minimized and that "biological control means shall be considered in the first place" (article 12). These regulations will further reduce the use of chemicals and promote biologically-based alternatives.

3.2.5 Trends in the Biopesticide Industry

Biopesticide manufacturers have made considerable progress in production efficiency, formulation, quality control, field efficacy, application strategies and marketing in the last decade. Products have become more reliable and cost-effective, and markets have expanded. Manufacturers realized that products must offer true economic benefit and value to the grower. Companies educated distributors and users, focussing on function and value, by which a wider adoption has been gained.

A substantial biopesticide business has been created which forms the foundation for further progress. Furthermore, every year new companies are established. Industry has reached a sufficient level of maturity and critical mass that will help to increase and accelerate future developments. For companies that have gained experience it becomes easier to develop and register new products. Successful companies will expand and cover larger markets. When more products are available and used by growers, the users will become more confident with microbials, and this will help adoption of new products.

Noteworthy is the recent renewed interest of the large agrochemical industry in biopesticides. They seem to complement their portfolio with microbial products, perhaps to be ready when the changes in society today will force them into other directions. Some striking examples are the activities of Bayer, BASF and several Japanese companies. In 2009, Bayer CropScience acquired Agrogreen's microbial product Bionem (*B. firmus*) for control of nematodes (CPM 2009a), and their post-harvest biofungicide Shemer (*Metschnikowia fructicola*) (CPM 2010). Bayer sells already XenTari (*Bt* subsp. *aizawai*) and Bio1020 (*Metarhizium anisopliae*) in some countries. In April 2009, Agraquest assigned world wide distribution rights to BASF for their biofungicide Serenade (*Bacillus subtilis*) (CPM 2009b), and to Bayer for the US home and garden market (www.agraquest.com). Three large Japanese companies, ArystaLifeSciences, Mitsui, and Sumitomo Chemical have played an active role in biopesticides for many years through their subsidiaries, respectively NPP, Valent BioSciences, and Certis.

I expect a greater involvement of large chemical companies in IPM and biocontrol products in the near future.

3.2.6 New Invasive Pests

New invasive pests offer an opportunity for the use of microbial pesticides when no chemicals are available or registered, or because of resistance to chemicals. Recent examples in Europe are the western corn root worm *Diabrotica virgifera virgifera* (Pilz et al. 2009), and the South American tomato pinworm *Tuta absoluta*. Bt is applied in large quantities against *T. absoluta* (J. van der Blom, Coexphal, personal communication) and nematodes may be used (Morton et al. 2009). New non-agricultural pests are the red palm weevil *Rhynchophorus ferrugineus* against which Bt (Manachini et al. 2009) and nematodes (Llácer et al. 2009) are used, and the South American palm borer *Paysandisia archon* (Nardi et al. 2009) against which nematodes are used as well as *Beauveria bassiana* (Besse 2008).

4 Concluding Remarks

In conclusion, many factors influence the adoption of biopesticides in the marketplace. Changes in regulations, and in political, cultural and social perceptions all determine the demand for sustainable crop protection agents. Research and technological discoveries create new possibilities for development of better products. Globalization opens new markets and offers potential usages, but also creates more problems such as invasive pests and diseases. The current macro-environmental trends support the assumption that demand for alternative crop protection products will grow rapidly to replace conventional chemical pesticides. We have heard similar stories many times before, but they never materialized in terms of large sales of biopesticides. But this time, they all seem to fall in place, and the incentive to reduce chemical pest control is really coming from the market. I am confident that this will create a true demand.

4.1 Market Size and Growth

Biopesticides have been promised a bright future by many scientists and industrial investors since the beginning of their development. Reality is, however, disappointing. Sales of microbial pesticides were $268 million in 2005, and the market projection was $750 million by 2015 (CPL 2006). Thakore (2006) expected biopesticides to reach 4.2% of the total pesticide market in 2010. My estimate is that today microbial sales are still less than 2% of the total market. Positive is, however, the growth rate of biopesticides which is about 10–15% per annum, and this is likely to continue. Until today biopesticides have predominantly been used in niche markets of high value crops. But expansion in agricultural crops is reality, particularly for disease control. Other areas of increased usage are forestry, public amenity areas, orchards, and vector control.

I envision growth in the use of bacteria. In addition to Bt, bacteria for plant disease control (*Bacillus subtilis, Pseudomonas* spp.) show promise. Secondly, baculoviruses are gaining terrain in orchards and also in greenhouse vegetables. The use of entomopathogenic nematodes is growing steadily for control of new target pests in orchards and in greenhouses. The development of the market for entomopathogenic fungi is the most difficult task. On the other hand, the use of fungal antagonists for plant disease control, such as *Trichoderma* spp. is rapidly expanding. Because of food safety and environmental issues new markets are expected to arise in China, India, North Africa, the Middle East, and in South America.

Growth of the biocontrol industry with the current business model with small and medium-sized enterprises will continue. The large agrochemical companies will also exploit activities with biopesticides, currently as distributors, and maybe soon as producers. An alternative is that they will acquire existing biopesticide manufacturers. The distinction between chemicals and microbials is likely to blur as metabolites from microorganisms are developed as pesticides such as spinosad and spinetoram, and microbials are developed whose activity is mainly based on the associated metabolites such as Serenade. I foresee the distinction between the two industries to disappear over time.

4.2 The Way Forward

Increased legislation on the use of pesticides combined with reduced availability of synthetic pesticides, demands for sustainable agriculture and low-residues have become reality and offer great opportunities for, amongst others, biopesticides. The societal need for biological products is, however, something very different from actual customer demand. The putative desire from society for sustainable products still leaves the important task to manufacturers to decide which product they will develop. Political and societal trends can be strong promotional factors, but they should not be determinative. Specific market driven demand must remain the target for a company.

"Biopesticides: the next revolution?" was the title of an article by Lisansky (1989). This revolution never occurred, and I do not think that we will see a spectacular growth soon. Biopesticides will, however, become a substantial part of the use of all crop protection products, and come closer to meeting their potential. I predict a steady and continued growth for the next decades. I am convinced that the biological products developed today will form a substantial part of the crop protection means of the future. Biopesticide companies need to grasp this chance and remain committed to the development of high quality cost-effective products. I am confident that the systematic roadmap with a strong focus on economics and market introduction proposed in this study will assist developers of biopesticides in accomplishing their goals.

References

Abdullah, M.A.F., S. Moussa, M.D. Taylor, and M.J. Adang, 2009. Manduca sexta (Lepidoptera: Sphingidae) cadherin fragments function as synergists for Cry1A and Cry1C Bacillus thuringiensis toxins against noctuid moths Helicoverpa zea, Agrotis ipsilon and Spodoptera exigua. *Pest Manag. Sci.* 65(10): 1097–1103.

Aiuchi, D., Y. Baba, K. Inami, R. Shinya, M. Tani, K. Kuramochi, S. Horie, and M. Koike, 2007. Screening of *Verticillium lecanii* (*Lecanicillium* spp.) hybrid strains based on evaluation of pathogenicity against cotton aphid and greenhouse whitefly, and viability on the leaf surface. *Jpn. J. Appl. Entomol. Zool.* 51(3): 205–212.

Bailey, K.L., S.M. Boyetchko, and T. Längle, 2009. Social and economic drivers shaping the future of biological control: a Canadian perspective on the factors affecting the development and use of microbial biopesticides. *Biol. Control* 52(3): 221–229.

Bale, J.S., J.C. van Lenteren, and F. Bigler, 2008. Biological control and sustainable food production. *Phil. Trans. R. Soc. B* 363: 761–776.

Benuzzi, M., 2009. NO-RESIDUE: an assortment of biocontrol agents and other organic plant protection tools for valuable and residue free crop protection. *Paper presented at the 4th Annual Biocontrol Industry Meeting (ABIM),* Lucerne, Switzerland, October 19–20, 2009. www.abim-lucerne.ch/archive/conference2009. Cited December 30, 2009.

Besse, S. 2008. Preventive and curative efficacy of Ostrinil against the palm borer *Paysandisia archon* (Burmeister, 1880). *Paper presented at the 3rd Annual Biocontrol Industry Meeting (ABIM),* Lucerne, Switzerland, October 20–21, 2008. www.abim-lucerne.ch/archive/conference2008. Cited May 23, 2009.

Burnell, A., 2002. Genetics and genetic improvement. In: R. Gaugler (ed), *Entomopathogenic nematology*. CAB International, Wallingford. pp. 241–262.

Chandler, D., G. Bending, J. Clarkson, G. Davidson, S. Hall, P. Mills, D. Pink, D. Skirvin, P. Neve, R. Kennedy, J.M. Greaves, W.P. Grant, and R.H. Collie, 2008b. The consequences of the 'cut off' criteria for pesticides: alternative methods of cultivation. *Policy Department B, Structural and Cohesion Policies, DG Internal Policies of the Union. P/B/AGRI/IC/2008_168*. www.europarl.europa.eu/activities/committees/studies.doc. Cited January 2, 2010.

Chandler, D., G. Davidson, W.P. Grant, J. Greaves, and G.M. Tatchell, 2008a. Microbial biopesticides for integrated crop management: an assessment of environmental and regulatory sustainability. *Trends Food. Sci. Technol.* 19: 275–283.

Cock, M.J.W., J.C. van Lenteren, J. Brodeur, B.I.P. Barratt, F. Bigler, K. Bolckmans, F.L. Cônsoli, F. Haas, P.G. Mason, J. Roberto, and P. Parra, 2009. Do new access and benefit sharing procedures under the convention on biological diversity threaten the future of biological control? *BioControl* 55(2): 199–218.

CPL, 2006. *Biopesticides 2007*. CPL Business Consultants, Wallingford.

CPM, 2009a. Bayer acquires biological control agents. *CPM* March 31 2009: 12.

CPM, 2009b. BASF to distribute Serenade. *CPM* March 31 2009: 12.

CPM, 2010. Bayer acquires biofungicide from Agrogreen. *CPM* January 30 2010: 14.

CropLife, 2009. *Facts and figures – the status of global agriculture (2008–2009).* www.croplife.org/library. Cited January 16, 2010.

Dent, D., 2000. *Insect pest management*, 2nd edition. CAB International, Wallingford.

EC, 2009a. Regulation (EC) No 1107/2009 of the European Parliament and of the Council of 21 October 2009 concerning the placing of plant protection products on the market and repealing Council Directives 79/117/EEC and 91/414/EEC. *Official J. Euro. Union* L 309(52): 1–50.

EC, 2009b. Directive 2009/128/EC of the European Parliament and of the Council of 21 October 2009 establishing a framework for Community action to achieve the sustainable use of pesticides. *Official J. Euro. Union* L 309(52): 71–86.

Federici, B.A., 2007. Bacteria as biological control agents for insects: economics, engineering, and environmental safety. In: M. Vurro, J. Gressel (eds), *Novel biotechnologies for biocontrol agent enhancement and management*. Springer, Dordrecht. pp. 25–51.

Geiger, F. et al., 2010. Persistent negative effects of pesticides on biodiversity and biological control potential on European farmland. *Basic Appl. Ecol.* 11(2): 95–105.

Georgis, R. and P. Warrior, 2007. Why companies don't pick up new strains? *Paper presented at the 40th annual meeting of the Society for Invertebrate Pathology*, Université Laval, QC, Canada, August 12–16, 2007. www.sipweb.org/MCD/BtWorkshop.../Georgis_Bt Workshop_SIP 2007.pdf. Cited December 21, 2009.

Goettel, M.S., M. Koike, J.J. Kim, D. Aiuchi, R. Shinya, and J. Brodeur, 2008. Potential of *Lecanicillium* spp. for management of insects, nematodes and plant diseases. *J. Invertebr. Pathol.* 98(3): 256–261.

Greaves, J., 2009. Biopesticides, regulatory innovation and the regulatory state. *Public Policy Adm.* 24(3): 245–264.

Inceoglu, A.B., S.G. Kamita, and B.D. Hammock, 2006. Genetically modified baculoviruses: a historical overview and future outlook. In: B.C. Bonning (ed), *Insect viruses: biotechnological applications. Advances in virus research 68*. Academic Press, San Diego, CA pp. 323–360.

Jackson, M.A. and S.T. Jaronski, 2009. Production of microsclerotia of the fungal entomopathogen *Metarhizium anisopliae* and their potential for use as a biocontrol agent for soil-inhabiting insects. *Mycol. Res.* 113(8): 842–850.

Keel, C. and M. Maurhofer, 2009. Insecticidal activity in root-associated, plant-beneficial pseudomonads. *Bull. IOBC/WPRS* 45: 181.

Kim, J.J., M.S. Goettel, and D.R. Gillespie, 2009. Evaluation of *Lecanicillium longisporum*, Vertalec® against the cotton aphid, *Aphis gossypii*, and cucumber powdery mildew, *Sphaerotheca fuliginea* in a greenhouse environment. *Crop Prot.* 29(6): 540–544.

Kos, M., J.J.A. van Loon, M. Dicke, and L.E.M. Vet, 2009. Transgenic plants as vital components of integrated pest management. *Trends Biotechnol.* 27(11): 621–627.

Lisansky, S., 1989. Biopesticides: the next revolution? *Chem. Ind.* 15: 478–482.

Llácer, E., M.M. Martínez de Altube, and J.A. Jacas, 2009. Evaluation of the efficacy of *Steinernema carpocapsae* in a chitosan formulation against the red palm weevil, *Rhynchophorus ferrugineus*, in *Phoenix canariensis*. *BioControl* 54(4): 559–565.

Lu, D., M. Pava-Ripoll, Z. Li, and C. Wang, 2008. Insecticidal evaluation of *Beauveria bassiana* engineered to express a scorpion neurotoxin and a cuticle degrading protease. *Appl. Microbiol. Biotechnol.* 81(3): 515–522.

Lundgren, J.G., A.J. Gassmann, J. Bernal, J.J. Duan, and J. Ruberson, 2009. Ecological compatibility of GM crops and biological control. *Crop Prot.* 28: 1017–1030.

Manachini, B., P. Lo Bue, E. Peri, and S. Colazza., 2009. Potential effects of *Bacillus thuringiensis* against adults and older larvae of *Rhynchophorus ferrugineus*. *Bull. IOBC/WPRS* 45: 239–242.

Marrone, P.G., 2007. Barriers to adoption of biological control agents and biological pesticides. *CAB Reviews: perspectives in agriculture, veterinary science, nutrition and natural resources.* 2, No. 051: 1–12.

Menzler-Hokkanen, I., 2006. Socioeconomic significance of biological control. In: J. Eilenberg, H.M.T. Hokkanen (eds), *An ecological and societal approach to biological control.* Springer, Dordrecht. pp. 13–25.

Morton, A., L. Batalla-Carrera, and F. García-del-Pino, 2009. Susceptibility of the tomato leaf miner, *Tuta absoluta* to entomopathogenic nematodes. *Bull. IOBC/WPRS* 45: 417–419.

Narayanan, K., 2002. Microbial control of insect pests: role of genetic engineering and tissue culture. In: O. Koul, G.S. Dhaliwal (eds), *Microbial pesticides*. Taylor and Francis, New York. pp. 117–179.

Nardi, S., E. Ricci, R. Lozzi, F. Marozzi, E. Ladurner, F. Chiabrando, N. Isidoro, and P. Riolo, 2009. Use of entomopathogenic nematodes for the control of *Paysandisia archon* Burmeister. *Bull. IOBC/WPRS* 45: 375–378.

Ownley, B.H., M.R. Griffin, W.E. Klingeman, K.D. Gwinn, J.K. Moulton, and R.M. Pereira, 2008. *Beauveria bassiana*: endophytic colonization and plant disease control. *J. Invertebr. Pathol.* 98: 267–270.

Pava-Ripoll, M., F.J. Posada, B. Momen, C. Wang, and R.J. St Leger, 2008. Increased pathogenicity against coffee berry borer, *Hypothenemus hampei* (Coleoptera: Curculionidae) by *Metarhizium anisopliae* expressing the scorpion toxin (AaIT) gene. *J. Invertebr. Pathol.* 99(2): 220–226.

Pilz, C., S. Keller, U. Kuhlmann, and S. Toepfer, 2009. Comparative efficacy assessment of fungi, nematodes and insecticides to control western corn rootworm larvae in maize. *BioControl* 54(5): 671–684.

Pimentel, D., 2005. Environmental and economic costs of the application of pesticides primarily in the United States. *Env. Dev. Sustain.* 7: 229–252.

Ritson, C. and S. Kuznesof, 2006. Food consumption, risk perception and alternative production technologies. In: J. Eilenberg, H.M.T. Hokkanen (eds), *An ecological and societal approach to biological control*. Springer, Dordrecht. pp. 47–64.

Scheepmaker, J.W.A. and T.M. Butt, 2010. Natural and released inoculum levels of entomopathogenic fungal biocontrol agents in soil in relation to risk assessment and in accordance with EU regulations. *Biocontl. Sci. Technol.* 20(3): 503–552.

Shinya, R., D. Aiuchia, A. Kushidac, M. Tania, K. Kuramochia, and M. Koike, 2008. Effects of fungal culture filtrates of *Verticillium lecanii* (*Lecanicillium* spp.) hybrid strains on *Heterodera glycines* eggs and juveniles. *J. Invertebr. Pathol.* 97(3): 291–297.

Speiser, B., E. Wyss, and V. Maurer, 2006. Biological control in organic production: first choice or last option? In: J. Eilenberg, H.M.T. Hokkanen (eds), *An ecological and societal approach to biological control*. Springer, Dordrecht. pp. 27–46.

Stein, A.J. and E. Rodríguez-Cerezo, 2009. The global pipeline of new GM crops: implications of asynchronous approval for international trade. *Technical Report* by the Joint Research Centre of the European Commission, EUR 23486 EN. Office for Official Publications of the European Communities, Luxembourg.

Szewczyk, B., L. Hoyos-Carvajal, M. Paluszek, I. Skrzecz, and M. Lobo de Souza,, 2006. Baculoviruses – re-emerging biopesticides. *Biotech. Adv.* 24: 143–160.

Thakore, Y., 2006. The biopesticide market for global agricultural use. *Industr. Biotechnol.* 2(3): 194–208.

Uri, N.D., 1998. Government policy and the development and use of biopesticides. *Futures* 30(5): 409–423.

Van Wees, S.C.M., S. van der Ent, and C.M.J. Pieterse, 2008. Plant immune responses triggered by beneficial microbes. *Curr. Opin. Plant. Biol.* 11: 443–448.

Vega, F.E., 2008. Insect pathology and fungal endophytes. *J. Invertebr. Pathol.* 98: 277–279.

Vega, F.E., M.S. Goettel, M. Blackwell, D. Chandler, M.A. Jackson, S. Keller, M. Koike, N.K. Maniania, A. Monzón, B.H. Ownley, J.K. Pell, D.E.N. Rangel, and H.E. Roy, 2009. Fungal entomopathogens: new insights on their ecology. *Fungal Ecol.* 2: 149–159.

Yamada, K., D. Aiuchi, M. Koike, and T. Masuda, 2009. Possibility of microbial control using entomopathogenic fungi *Lecanicillium* spp. hybrid strains and *Beauveria bassiana* against the diamondback moth. *IOBC/WPRS Bull.* 45: 331–334.

Index of Entomopathogen Species

W.J. Ravensberg, *A Roadmap to the Successful Development and Commercialization of Microbial Pest Control Products for Control of Arthropods*, Progress in Biological Control 10, DOI 10.1007/978-94-007-0437-4, © Springer Science+Business Media B.V. 2011

Subject Index

MIX
Papier aus verantwortungsvollen Quellen
Paper from responsible sources
FSC® C105338

If you have any concerns about our products,
you can contact us on
ProductSafety@springernature.com

In case Publisher is established outside the EU,
the EU authorized representative is:
Springer Nature Customer Service Center GmbH
Europaplatz 3, 69115 Heidelberg, Germany

Printed by Libri Plureos GmbH
in Hamburg, Germany